SECOND EDITION

Visualizing HEALTH CARE STATISTICS

A Data-Mining Approach

Zada T. Wicker, MBA, RHIT, CCS, CCS-P
Professor Health Information Technology
Sullivan University

Dr. J. Burton Browning, EdD
Retired Professor
Author and Entrepreneur

World Headquarters
Jones & Bartlett Learning
5 Wall Street
Burlington, MA 01803
978-443-5000
info@jblearning.com
www.jblearning.com

Jones & Bartlett Learning books and products are available through most bookstores and online booksellers. To contact Jones & Bartlett Learning directly, call 800-832-0034, fax 978-443-8000, or visit our website, www.jblearning.com.

Substantial discounts on bulk quantities of Jones & Bartlett Learning publications are available to corporations, professional associations, and other qualified organizations. For details and specific discount information, contact the special sales department at Jones & Bartlett Learning via the above contact information or send an email to specialsales@jblearning.com.

21529-8

Production Credits
VP, Product Management: Amanda Martin
Director of Product Management: Laura Pagluica
Product Manager: Sophie Fleck Teague
Content Strategist: Tess Sackmann
Manager, Project Management: Lori Mortimer
Project Specialist: Kathryn Leeber
Digital Project Specialist: Angela Dooley
Senior Marketing Manager: Susanne Walker
Production Services Manager: Colleen Lamy
VP, Manufacturing and Inventory Control: Therese Connell
Composition: Exela Technologies
Cover Design: Michael O'Donnell
Text Design: Kristin E. Parker
Senior Media Development Editor: Troy Liston
Rights Specialist: Maria Leon Maimone
Cover Image (Title Page, Chapter Opener):
© Mad Dog/Shutterstock
Printing and Binding: McNaughton & Gunn

Library of Congress Cataloging-in-Publication Data
Names: Browning, J. Burton, author. | Wicker, Zada T., author.
Title: Visualizing health care statistics : a data-mining approach / J. Burton Browning, Zada T. Wicker.
Description: Second edition. | Burlington, MA : Jones & Bartlett Learning, [2021] | Includes bibliographical references and index.
Identifiers: LCCN 2020007462 | ISBN 9781284197525 (paperback)
Subjects: MESH: Data Mining | Data Interpretation, Statistical | Data Collection | Health Information Management
Classification: LCC R859.7.D35 | NLM W 26.55.I4 | DDC 610.285–dc23
LC record available at https://lccn.loc.gov/2020007462

6048

Printed in the United States of America
24 23 22 21 20 10 9 8 7 6 5 4 3 2 1

We wish to thank our parents: Coleen Grower, who passed in April 2019, and Betty Browning. You both taught us to give back to the world and leave things better than we found them. As such, we knew there was a need for an engaging and real-world healthcare statistics textbook to help our students, and this work is the result of that multi-year effort. The effort involved took time away from you both, but we tried to always be available when needed. Your enduring support and encouragement have given us the strength to not only finish the original edition, but revise a second edition.

–J. Burton Browning

As a sincere thought from a daughter-in-law: Betty, thank you for accepting me into the family and being there when I needed you. As a final word to my mother Coleen, thank you for showing me the world and helping me to become the person I am today. As a single mother, you are still my inspiration and have given me the strength and ability I now am trying to impart to others.

–Zada T. Wicker

Brief Contents

Foreword ix
Preface x
Reviewers xi
Acknowledgments xiii

CHAPTER 1 **Introduction to Healthcare Statistics** 1
CHAPTER 2 **Central Tendency, Variance, and Variability** 14
CHAPTER 3 **Patient Data** 29
CHAPTER 4 **Occupancy and Utilization Data** 55
CHAPTER 5 **Morbidity and Mortality Data** 79
CHAPTER 6 **Autopsy Data** 108
CHAPTER 7 **Infection, Consultation, and Other Data** 130
CHAPTER 8 **Health Information Management Statistics** 152
CHAPTER 9 **Research Methodology and Ethics** 170
CHAPTER 10 **Data Collection and Reporting Methods** 189
CHAPTER 11 **The Future of Healthcare Statistics** 210
APPENDIX A **Common Statistical Abbreviations Used in This Text** 229
APPENDIX B **Resources for Further Information** 230
APPENDIX C **Historical Abuse of Human Research Subjects** 231
APPENDIX D **Formulas for Healthcare Statistics** 234
APPENDIX E **CAHIIM Competency Exercises** 237

Glossary 240
Index 249

Contents

Foreword ix
Preface x
Reviewers xi
Acknowledgments xiii

CHAPTER 1 **Introduction to Healthcare Statistics** 1

Introduction 2
History and Rationale of Healthcare Statistics 2
Definition of Statistics 3
The Use of Statistics in Health Care 4
Key Producers and Users of Healthcare Statistics 4
Data Mining 5
Definition 5
History 6
Current Applications 6
Basic Statistical Concepts 7
Dataset 7
Variables 8
Data Distribution 8
Types of Data 8
Types of Data Mining Models 9
Predictive Models 9
Descriptive Models 9
Decision Models 9
Obtaining Data 10
Global Perspective 11
Chapter Summary 11
Apply Your Knowledge 12
References 12
Web Links 12

CHAPTER 2 **Central Tendency, Variance, and Variability** 14

Introduction 15
Measures of Central Tendency 15
Mean 15
Median 17
Mode 17
Frequency Distribution 18
Variance and Measures of Dispersion or Variability 19
Min and Max 19
Range 20
Outlier Data 21
Interquartile Range 21
Standard Deviation 22
Variance 22
Data Harvesting 25
Global Perspective 26
Chapter Summary 26
Apply Your Knowledge 26
References 28
Web Links 28

CHAPTER 3 **Patient Data** 29

Introduction 30
Census Data and Their Importance 31
Calculation and Reporting of Patient Census Data 33
Inpatient Service Days 33
Average Daily Census 34
Data Visualization 35
Visually Examine Data With Sparklines and Microcharts 38
Newborn Services 41
Open-Source Software 41
Freeware and Shareware 42
Types of Databases 44
Flat-File Database 44
Relational Database 45
Data Formats 45
R-Project 46
Data Stored in R 48
Global Perspective 52

Chapter Summary 53
Apply Your Knowledge 53
References 54
Web Links 54

CHAPTER 4 Occupancy and Utilization Data 55

Introduction 56
Bed Count Computation 56
Definition of Inpatient Bed Count 56
Importance of Inpatient Bed Count 57
Bed Occupancy Ratios 57
Certificate of Need 58
Calculating Newborn Bassinet Occupancy Ratio 58
Bed Turnover Rate 59
Two Formulas for Bed Turnover 59
Length of Stay 59
Discharge Days and How to Calculate Them 60
Total Length of Stay 61
Average Length of Stay 61
Median and Standard Deviation for Length of Stay 66
Median Length of Stay 66
Why Might You Use Median Instead of Mean? 66
Standard Deviation of Length of Stay 67
Visually Representing Data 68
PowerPoint Presentation 69
Data Mining: Association Rules With R-Project 70
Global Perspective 74
Chapter Summary 77
Apply Your Knowledge 77
References 78
Web Links 78

CHAPTER 5 Morbidity and Mortality Data 79

Introduction 80
Morbidity Rates 81
Incidence 81
Prevalence 82
Mortality Rates 83
Gross Mortality Rate 83
Fatality Rate 84
Net Mortality Rate 86
Postoperative Mortality Rate 87
Maternal Mortality Rate 89
Maternal Mortality Rate as the Number of Deaths per 100,000 or 10,000 Births 90
Anesthesia Mortality Rate 91
Newborn Mortality Rate 92
Fetal Mortality Rate 92
Mortality Rates for Cancer 94
Mortality-Adjusted Rates 94
Interpreting Mortality Rate Results 95
Conducting Formal Research 95
Research Design 95
The Hypothesis and Null Hypothesis Statements 96
Statistical Measures 96
P Values and Significance 96
Type I Errors 97
Type II Errors 97
Either End of the Curve: Tails 97
The Normal Distribution of Data 98
Parametric and Nonparametric Tests 98
z Score 98
t Test 100
Infant Mortality by Race and County 104
Global Perspective 104
Chapter Summary 104
Apply Your Knowledge 104
References 107
Web Links 107

CHAPTER 6 Autopsy Data 108

Introduction 109
What Is an Autopsy and Why Is It Important? 109
Centers for Disease Control and Prevention Data 112
Types of Autopsy Data 112
Autopsy Rate 112
Net Autopsy Rates 113
Inpatient Hospital Autopsies 114
Adjusted Hospital Autopsy Rate 114
Autopsy Rate for Newborns 115
Fetal Autopsy Rate 116
Statistical Measures 116
F Test: Comparison of Two Variances 116
Analysis of Variance 119
One-Way Analysis of Variance 119
Two-way Analysis of Variance 119

Global Perspective 126
Chapter Summary 126
Apply Your Knowledge 126
References 129
Web Links 129

CHAPTER 7 Infection, Consultation, and Other Data . . . 130

Introduction 131
Infection Rates 131
Infection Control Committee 131
Nosocomial Infection Rate 132
Specific Infection Rate 133
Postoperative Infection Rate 134
Complication Rates 136
Cesarean Section Rate 137
Consultation Rates 138
General Occurrence Rates 139
Statistical Measures and Tools 140
R Commander Graphical User Interface 140
Post-Hoc Analysis of Variance 144
Tukey Honest Significant Difference 144
HAI Reporting and Tracking 145
Text Mining and Visualization Using R-Project and Wordle 146
Global Perspective 149
Chapter Summary 149
Apply Your Knowledge 149
References 150
Web Links 151

CHAPTER 8 Health Information Management Statistics 152

Introduction 153
Functions of Health Information Management 153
Requirements to Work in Health Information Management 153
Labor Cost and Compensation 154
Transcription Cost and New Technology 155
Other Costs Associated With Health Information Management 157
Productivity 159
Healthcare Facility Staffing 159
Measuring Utilization 161
Types of Financial Reports 161
Readmission Rate Reports 161
Discharge Reports 162
Two Types of Budgets: Operational and Capital 162
Operational Budget 162
Capital Budget 163
Data Mining With Naive Bayes and R-Project 166
Global Perspective 168
Chapter Summary 168
Apply Your Knowledge 168
References 169
Web Links 169

CHAPTER 9 Research Methodology and Ethics 170

Introduction 171
Types of Research 172
Research Process 172
Step 1: Identify the Problem 172
Step 2: Research the Problem 172
Step 3: Develop Research Questions 172
Step 4: Determine the Type of Data Needed, Sample Size, and Methods of Analysis 172
Step 5: Collect Data 173
Step 6: Analyze the Data 173
Step 7: Draw a Conclusion 173
Step 8: Report the Results and Implications for Further Research 173
Research Ethics and the Abuse of Human Subjects 173
Late 1700s: Edward Jenner and the First Smallpox Vaccine 173
1850s: J. Marion Sims, the Father of Gynecology 174
1900: Walter Reed and Yellow Fever Transmission 174
1932 to 1972: Tuskegee Study of Untreated Syphilis 175
1964: The Declaration of Helsinki 175
1979: The Belmont Report 175
The Institutional Review Board 176
Review Process 176
Exemption and Types of Review 176
Informed Consent 177
Membership 177
Data Dictionary 177

Statistical Measures and Tools 178
Common Nonparametric Statistics 178
Regression Analysis: Simple Linear Regression . . . 181
Global Perspective 186
Chapter Summary 186
Apply Your Knowledge 187
References 187
Web Links 188

CHAPTER 10 Data Collection and Reporting Methods 189

Introduction 190
Descriptive Research and Information Collection 191
Sampling. 191
Nonprobability Sampling 191
Probability Sampling 192
Random Numbers and Random Sampling. 193
Quality Question Design for Data Collection . . . 196
Bloom's Taxonomy and Question Design 196
Guidelines for Question Writing 196
Types of Questions 197
Types of Studies 199
Longitudinal Study 199
Case Study 200
Documentary Study. 200
Data Collection Methods 200
Survey 200
Interview. 201
Questionnaire 201
Observation. 201
Appraisal. 201
Survey Tools 201
Pivot Tables. 202
Multiple Regression Analysis 205
Global Perspective 208
Chapter Summary 208
Apply Your Knowledge 208
References 209
Web Links 209

CHAPTER 11 The Future of Healthcare Statistics 210

Introduction 211
The Future of Healthcare Statistics 211
Radio Frequency Identification 211
Automatic Medication Dispenser. 212
Health Information Exchange 212
Efforts to Decrease Healthcare Costs 213
Time Series Analysis of Data 213
Forecasting Future Data 215
Project Management General Concepts 217
Analysis of Covariance. 220
Coronavirus—The Next Pandemic 224
Ebola. 225
Global Perspective 225
Chapter Summary 225
Apply Your Knowledge 226
References 227
Web Links 228

APPENDIX A Common Statistical Abbreviations Used in This Text. 229

APPENDIX B Resources for Further Information 230

APPENDIX C Historical Abuse of Human Research Subjects. . . . 231

APPENDIX D Formulas for Healthcare Statistics 234

APPENDIX E CAHIIM Competency Exercises. 237

Glossary. 240
Index 249

Foreword

As a career public health specialist working in a community college setting in a non–public health position, I am often asked to speak to students about public health–related topics. Moreover, as someone who specialized in the translation of data at a state public health level, I am in a position to understand the value of data and how it should be taught in an introductory course.

With this textbook, *Visualizing Health Care Statistics: A Data Mining Approach*, authors J. Burton Browning and Zada Wicker have captured complex topics in a way that makes understanding them much easier and more accessible to the student. Examples include their explanations of biostatistics data points and their relevance to the workplace and in general society across a range of issues impacting health data such as morbidity and mortality. Hopefully, students will see the benefit of these statistical concepts and basic calculations to the health field in general, but will also understand the relevance of this kind of information to themselves personally. Each of us represents a part to a whole in a data point somewhere.

For some, this textbook can help build foundational skills to launch a career pathway in the future working with health statistics in a meaningful way. Browning has nurtured this project with Ms. Wicker as a coauthor to capitalize on her many years of experience in health information technology. Their combined 35-plus years of experience translate into a project that will help students cross the bridge from learning to understanding.

Pamela Federline, MPH
Director, Office of Planning, Research, and Effectiveness
College of the Albemarle Elizabeth City, NC

Preface

Too often, texts on statistics are filled with examples that are designed to teach concepts in a vacuum and not related to real data. More often than that, the critical step of applying statistical knowledge to real-world situations or global situations is missing. Use of massive amounts of data in a data mining context is so new that most texts ignore big data altogether. It is in that light that the authors have tried to create an introductory text that is real-world, helps students develop data-gathering skills, and all the while prepares them for a national Registered Health Information Technician (RHIT) exam, which has required healthcare statistical knowledge.

The authors have designed this textbook as an introductory healthcare statistics text for students in healthcare–related curricula, most specifically the health information technology (HIT) field. It is with the RHIT exam in mind that this text has been written. Yet it is of great value to any student in healthcare informatics or related fields. As such it is most appropriate for second-year students, although certainly other levels could benefit.

A key feature of this text is that it covers basic statistical methods to give students a solid grounding in the theoretical framework of statistical measures. Moreover, it provides all key statistical measures required on the RHIT exam and introduces students to the very important topic of data mining within a framework of harvesting real-world healthcare data via a global perspective. Each chapter reinforces the concept of data visualization, a key aspect of reporting data to end users in a meaningful way.

The chapters are sequenced such that each chapter builds on knowledge from the previous chapter. Thus, it is important to present chapters as sequenced in the text. In most instances, either commonly available commercial software (such as Microsoft Excel or a web browser) or open-source free software is used to minimize costs to both students and schools and to provide students with cutting-edge software.

Key pedagogical features of the text include the following:

- Chapter-opening quote
- Chapter outline
- Learning outcomes
- Chapter key terms list and end-of-book glossary
- Chapter introduction
- Data mining basic concepts
- Hands-on Statistics: A step-by-step numbered list of procedures for performing a statistics-related calculation or task either by hand or using a common computer application, such as Excel
- How Does Your Hospital Rate?: A progressive case study that starts near the beginning of each chapter, resumes in the middle of the chapter, and concludes at the end of the chapter. Each box includes critical-thinking questions
- Global Perspective: An exercise that allows the student to practice data mining from international data sources
- Apply Your Knowledge: End-of-chapter statistics calculation and research exercises
- Software information to highlight integrated software
- Calculation by hand exercises to highlight manual calculation opportunities
- Did You Know?: Sidebar boxes containing interesting information related to statistics and other topics
- Chapter summary
- Screenshots of integrated computer applications

To aid both instructors and students, the authors have provided an Instructor's Guide with ideas on how to present each chapter. PowerPoint presentations are available for each chapter as well. These ancillary resources are available via the Jones & Bartlett Learning website.

Reviewers

Karen Bakuzonis, PhD, MSHA, RHIA
Program Chair
Ashford University
San Diego, California

Lynda Carlson, PhD, MS, MPH
Professor and Program Director
Borough of Manhattan Community College
New York, New York

Cheryl Christopher, MPA, RHIA
Assistant Professor
Borough of Manhattan Community College
New York, New York

Charlotte E. Creason, RHIA
Adjunct Faculty
Tyler Junior College
Tyler, Texas

Kelly Davis, EdD, MSN, RN, CNE
RN to BSN Program Coordinator
Delaware Technical Community College
Georgetown, Delaware

Barbara Dimanlig, RHIA, CHPS
Health Information Technology
City College of San Francisco
San Francisco, California

Sandra Hertkorn
Health Information
Bryan College
Gold River, California

Deb Honstad, MA, RHIA
Program Director and Associate Professor
San Juan College
Farmington, New Mexico

Margo Imel, MBA/TM, RHIT
Instructor
Araphoe Community College
Littleton, Colorado

Mariann Davidson Jeffrey, RHIT
Health Careers
Central Arizona College
Apache Junction, Arizona

Christine Kowalski, EdD, RHIA, CP-EHR
Course Mentor
Western Governors University
Salt Lake City, Utah

Meg Kurek, MSIS, PMP
Workforce Development
Community College of Allegheny County
Monroeville, Pennsylvania

Maribeth S. Lane, RHIA
Program Director
Northwest Iowa Community College
Sheldon, Iowa

Cynthia Lundquist, BHA
Kaplan College
Stockton, California

Barbara Marchelletta, CMA (AAMA), RHIT, CPC, CPT
Program Director
Beal College
Bangor, Maine

Vince Ochotorena, MBA, MAM, BS
Instructor
Anthem College
Phoenix, Arizona

Terri L. Randolph, MBA/HCM
Stratford University
Falls Church, Virginia

Patricia L. Shaw, EdD, RHIA, FAHIMA
Chair and Program Director
Weber State University
Ogden, Utah

Jody Smith, PhD, RHIA, FAHIMA
Professor Emeritus
St. Louis University
St. Louis, Missouri

Karen J. Smith, PhD, RHIA, FAHIMA
Health Informatics and Information Management
Saint Louis University
St. Louis, Missouri

Jasper Xu, PhD, MHA, CHFP
Assistant Professor
University of North Florida
Jacksonville, Florida

Heather Watson, RHIA
Program Director
Davidson County Community College
Thomasville, North Carolina

Mary Worsley, MH, RHIA, CCS
Associate Professor
Miami Dade College
Miami, Florida

Acknowledgments

It is our pleasure to give a special thank you to the following individuals for their input on the second edition: Donna Estes, RHIT, HIMT Program Director at Georgia Northwestern Technical College in Rock Spring, Georgia, CPHQ-Retired, MPM, and Katrina Putman, RHIT.

Thank you for the kind comments and notes from the *First Edition* reviewers and the *Second Edition* reviewers. With reviewers' comments, we were able to make positive changes to an already popular text. Also, a special thank you to Tess Sackmann from Jones & Bartlett Learning, who helped orchestrate this text.

CHAPTER 1

Introduction to Healthcare Statistics

Far and away the best prize that life has to offer is the chance to work hard at work worth doing.

—Theodore Roosevelt

CHAPTER OUTLINE

Introduction
History and Rationale of Healthcare Statistics
Definition of Statistics
The Use of Statistics in Health Care
Key Producers and Users of Healthcare Statistics
Data Mining
  Definition
  History
  Current Applications
Basic Statistical Concepts
  Dataset
  Variables
  Data Distribution
Types of Data
Types of Data Mining Models
  Predictive Models
  Descriptive Models
  Decision Models
Obtaining Data
Global Perspective
Chapter Summary
Apply Your Knowledge
References
Web Links

LEARNING OUTCOMES

After completing this chapter, you should be able to do the following:

1. Define and describe the history of statistics.
2. Describe how statistics are used in health care.
3. Identify common users of healthcare statistics.
4. Define data mining.
5. Describe the history of data mining and how it is used in health care.
6. Explain what a dataset is and how it is used.
7. Identify the four types of data.
8. Discuss five basic elements of data mining.

KEY TERMS

Aspect
Data
Data mining
Data-driven
Dataset
Decision model
Dependent variable
Descriptive model
Evidence-based medicine (EBM)
Flowchart
Independent variable
Interval data
Machine learning
Nominal data
Ordinal data
Parameters
Population
Predictive model
Primary data
Ratio data
Sample
Secondary data
Statistics
Telehealth
Variable
Viewpoint

How Does Your Hospital Rate?

Paul, a 54-year-old man who has been diagnosed with congestive heart failure, is facing heart valve replacement surgery. He and his family would like the best surgeon and hospital available in his area for this major surgery. They are considering three different hospitals. One site for statistical information that Paul and his family might review would be the Health section of the *US News and World Report* website (https://health.usnews.com), specifically the information relating to cardiology. This website ranks the specialist, survival, patient safety, patient volume, and nursing staff.

Consider the following:

1. What statistics should Paul be considering as he makes this decision?
2. How would the hospitals in your area rate?
3. Which one would you pick, and why?
4. Why is it important for a hospital to keep healthcare statistics?

Introduction

Healthcare statistics allow a hospital to assess, improve, and communicate its quality of patient care, develop better policies and procedures for infection control, and achieve many other goals we will be discussing throughout the following chapters. In this chapter, we will introduce to you the history and definition of healthcare statistics and their importance to a hospital. We will also discuss organizations that keep statistics, not just here in the United States, but around the globe, such as the Centers for Disease Control and Prevention (CDC), the World Health Organization (WHO), and other groups. Keeping statistical data allows us to watch for trends in health care and make proactive rather than reactive choices.

Because the word data will be used so often throughout this text and can be used so many different ways in everyday life, it is worth taking a moment to define this term for our purposes. **Data** (singular, *datum*) are units of information, such as measurements, that can be collected and interpreted. They are the commodity in which we deal as healthcare statisticians.

Data mining is another important concept that we will discuss. We will explain its use in the healthcare arena and show you how you can use data mining and benchmarks with your local hospital. We will also cover how to use Excel and R-Project, both powerful data analysis tools, to examine statistical methods.

Let's begin our exploration of healthcare statistics in the United States and around the globe.

History and Rationale of Healthcare Statistics

The history and study of statistics is as much an examination of historical events as it is a study of the probability, logic, and mathematics behind statistical analysis. As the famous mathematics educator Freudenthal and others have noted, learning the history of a topic often aids in the overall understanding of the focus of a student's study (Leen, 1994). It is in this light that this text will discuss statistics in general and statistical analysis in health care.

There are differing views as to the first uses of statistical analysis, and in fact even the word *statistics*, but certainly most would agree that a large majority of the first uses were descriptive in nature, before the term statistics was formally used.

An early example can be found in Mackenzie's book of ancient stories from India. He notes that Rituparna estimated the number of leaves and berries on two branches of a fruit tree and estimated probabilities of dice rolls. With regard to the fruit tree, he estimated the number of leaves and berries on the basis of a twig, which he multiplied by the estimated number of twigs on the two branches. After a night of counting, he found that his estimate was very close to the real number. Most likely the use of two branches provided a way to take the count from each and determine an average, to be used in estimation for the entire tree (Mackenzie, 1913).

In 431 BCE, the author of *The History of the Peloponnesian War*, Thucydides, notes that the origins of probability can be found in the Athenians' evaluation of the height of the wall of Platae. This estimation was done by determining an average size for a brick, counting the number of bricks in an area, and multiplying by the area they were trying to attack to determine the height they would need for ladders to scale the walls. To determine the height, they multiplied the mode (most frequently occurring value of several sampled bricks) by the count of the number of bricks (Thucydides, n.d). As you will later learn, *mode* is the most frequently occurring value in a set of values. Since bricks were not necessarily uniform in size, determining the mode of a standard brick would be important if you were estimating the size of a wall of bricks you needed to scale.

Early accounts of the use of the undefined term *statistics* vary but include the ninth century work "A Manuscript on Deciphering Cryptographic Messages" written by Al-Kindi. It included a thorough account of how to use statistics and frequency analysis to decipher secret encrypted messages. As a formal term to describe the subject of this text, a version of the word first appears as *statistik* by German author Gottfried Achenwall in 1749 in describing data about the state or arithmetic of the state; however, earlier civilizations, such as the Romans, collected state demographic information and other related data earlier than 1749, though not necessarily using the term *statistic* (Johnson & Kotz, 2012). Unrelated to state data, other early recordings of statistical data concerned sailing, temperature, astrology, and related data that were used for predictions. If you are familiar with the *Farmer's Almanac* or similar texts, then you are aware of the direct impact that predictive statistics can have on people and the state.

Today, statistics are used to express everything from temperature and demographic data averages to mortality rates for cardiac surgery in health care and pattern analysis in massive datasets. In a data-driven society, almost every aspect of life in some way has statistical factors associated with it, from your interest rate at the bank to the target heart rate you are trying to meet for a fitness plan. In short, we use statistics to make a great many important decisions, including those relating to finances, work, and, most specifically for the purposes of this text, health care.

DID YOU KNOW? Florence Nightingale was a member of the Royal Statistical Society. As one of the first women to collect statistics on health policy, she led the way for other female statisticians to work in the field. She was also credited with using graphs to present her findings to Queen Victoria in an effort to reform the sanitary conditions in military hospitals, so she was also an early proponent of data visualization methods (Lancaster, 2013).

Now you know!

Definition of Statistics

Before we discuss in detail how statistics are used in health care, identify some of the users of statistics, and examine some of the statistical methods, we should formally define the term statistics as it will apply to our uses. According to Batten (1986), "Statistics is a series of methods to collect, analyze, and interpret masses of numerical data." However, statistics are not typically just an end unto themselves but rather are used to solve real-world problems. Thus, a practical definition of **statistics** might be the collection of data for the purpose of making predictions (inferences) or considerations to answer a question. Along with this definition are many strategies, rules, and procedures that define and formalize our collection and use of data and resultant findings. We will focus on healthcare statistics, but the underlying methods and processes are applicable to other areas, such as research statistics, which is integral to the field of health care.

The Use of Statistics in Health Care

Statistics are frequently used by many healthcare organizations, including hospitals and insurance companies. Such organizations use statistics to aid in making beneficial business decisions based on data they have collected over time. Hospitals collect and summarize data to improve quality of care, analyze cost of patient care, measure utilization of services provided to patients, examine target marketing decisions, and improve potential offerings of services to patients.

For example, a healthcare professional might examine average length of stay at several area hospitals. On finding that one hospital had a considerably longer average length of stay for patients, the hospital administrators might look for some underlying cause. By finding the issue or issues and resolving them, the hospital's quality of service (QOS) would be enhanced.

Hospitals who are accredited are required to retain data from certain areas to maintain their accreditation standards. The main hospital accrediting body is The Joint Commission. As an example of what is required, a hospital might have to keep fetal monitor strips or complete patient records for 7 years on-site, with older data being held off-site in some archived form for long-term storage (perhaps converted to a different medium, such as microfiche, tape, or optical storage).

The federal government is pushing legislation to move forward with the electronic health record. The *Federal Register* contains all the rules and regulations regarding the implementation of the electronic health record. Included in this legislation are antikickback rules for physicians who refer a patient to a lab or other type of facility in which they have ownership. Rules and regulations are also set forth by the Department of Health and Human Services (DHHS) and the Centers for Medicare and Medicaid Services (CMS).

Third-party payers may require hospitals to collect and maintain performance data. Data can be collected and abstracted for these purposes using many different tools and methods. Of critical importance is how the data are analyzed and used. There are some very specific statistical methods that are used to analyze these data for reporting to external and internal consumers of the data, such as third-party payers and marketing data consumers, respectively.

Key Producers and Users of Healthcare Statistics

The Bureau of Labor Statistics, the National Center for Health Statistics (NCHS), the CDC, and CMS are just some of the agencies who produce and maintain healthcare statistics. Vital statistics are also kept in each state and are another source of statistical information, along with census information. These organizations provide and use statistics to improve health care, summarize findings, and examine trends in the United States and around the globe.

The Bureau of Labor Statistics is a unit of the US Department of Labor and serves as the principal agent for the US Federal Statistical System. The primary mission of the Bureau of Labor Statistics is to collect and analyze essential statistical data for use by the public and the US Congress. The most common statistics that are kept by the Bureau of Labor Statistics relate to prices, employment, unemployment, compensation for injuries, and work injuries. For example, according to the US Department of Labor agency, Occupational Safety and Health Administration (OSHA), in 2018 there were 1008 fatalities to workers who performed construction.

The NCHS is the principal agent that delivers statistical information and directs policies and actions that will improve the health of the public. The NCHS is housed within the CDC. The CDC and the NCHS compile statistics for all types of disease for the United States and worldwide. For example, in the United States, rates of the sexually transmitted disease gonorrhea increased by 8% from the year 2010 to 2011, totaling 321,849. By 2017, the number of new gonorrhea cases had climbed all the way to 555,608 (CDC, 2018).

The primary responsibility of CMS is to administer Medicare and Medicaid, the Children's Health Insurance Program, the Health Insurance Portability and Accountability Act (HIPAA), and Clinical Laboratory Improvement Amendments. CMS keeps statistics for each state as to how many people are on Part A or Part B or both for Medicare. As of July 2016, 56.5 million were on only Part A of Medicare, and 52.1 million were on Part B (National Committee to Preserve Social Security and Medicare, 2020). Keeping this type of statistic is vital to health care because people are living longer and will eventually require more healthcare services.

In addition to the organizations already discussed, users of statistical data include the following:

- Federal government agencies gather information that references public health issues such as HIV/AIDS, cancer, births, and deaths.

- Accreditation agencies use statistics to show the most common diagnoses and procedures and the amount of resources used to treat those patients.
- Managed care organizations use statistics to review costs for the level of care that is being provided to their patients.
- Healthcare researchers use the data from health law and regulations, physician practices, and **telehealth** (services that use electronic information and telecommunications technologies to support long-distance clinical health care). Technologies can include videoconferencing, the internet, store-and-forward imaging, streaming media, terrestrial communication, and wireless communication. There are many other types of information used for research.
- Mental health facilities and drug and alcohol facilities use this information to measure the success of the services being provided and success rates of patients.

Note that these parties are external to the healthcare provider (e.g., the hospital). In fact, the hospital would be a primary consumer of this valuable data. Activities such as QOS, as mentioned earlier, are but one reason for this information. Another is governmental oversight via the certificate of need (CON), which is a statement issued by a government agency for projected construction or modification of a healthcare facility. The facility must meet the requirements statistically to meet the CON criteria. It ensures that the new facility will be needed at the time of completion for those additional services. Basically, you might consider that a CON is an assurance that facilities are not built or expanded beyond the requirements of the community.

Hospitals and health information management organizations use statistical information as well. Both of these organizations typically use similar types of information in their statistical analysis. Let's examine the departments that would perform and use the various data and findings.

- Healthcare administrators use health statistics to make data-driven decisions. Think of **data-driven** as making decisions based on statistical information instead of by guessing. For example, if a hospital had data for 3 years on nursing needs of the emergency department on New Year's, you might be able to expand staffing on that day, based on previous years' data. Or if data showed that the number of patients in labor and delivery increased by 10% each year over a span of 4 years, the facility might consider adding more beds and staff to that area.
- Healthcare department managers use statistics to set goals for the department, such as annual budgets.
- Cancer registries use statistical information regarding the different types of cancer, stages, and treatment. They also maintain survival rates of cancer patients. Cancer registries can receive accreditation by the American College of Surgeons (ACS). The cancer registry must meet the standards set by the ACS to be accredited.
- Nursing facilities use statistical information to review the different types of payers of insurance their patients have.
- Home healthcare organizations keep statistical data to track patients and their outcomes. The information includes the following: the number of visits, dressing changes, oxygen machine use, and many other services that home health organizations make. Further data include how many patients are taking their medication as directed, how many are improving, and how many are being readmitted.
- Hospice provides services either in the home or in a healthcare facility. The services they provide are linked to the patient's diagnosis.

Data Mining

Now that we have learned about some consumers of statistical data, we should examine sources for these data and how they are collected.

Definition

The process of extracting information from a large set of data is known as **data mining**. Tan, Steinbach, and Kumar (2003) note that data mining is a "confluence of many disciplines" and show an overlap of statistics, data mining, and artificial intelligence, machine learning, and pattern recognition. As a process, data mining involves the steps of defining, finding, and extracting data or knowledge that is buried in large sets of raw data. In this process, "data is retrieved, consolidated, managed and prepared for analysis" (Valova & Noirhomme, 2009). Following data mining, the resultant data are analyzed and then organized into a usable format.

History

To better understand this concept, let's cover some of the history of data mining while setting some boundaries for how we will obtain data, review some processes and strategies to make sense of mined data, and integrate these data and methods into software analysis tools.

Data mining to make healthcare medical treatment decisions is not new, although the use of formal computerized data mining tools is. In fact, data mining and **evidence-based medicine (EBM)**—medical practice that is based on the best available current methods of diagnosis and treatment, as revealed in research—have existed since the time of Hippocrates (460 BCE) in ancient Greece. Other notable "data miners" in history include Aulus Cornelius Celsus of ancient Rome, and John Snow, the father of modern epidemiology. Celsus wrote that wound cleansing and hygiene were important in health care, though these practices did not take hold until the late 1800s. John Snow tracked via maps the source of cholera in 1854. Thus, we see that statistics has a role in not only medical administration, but treatment as well.

In the 1960s, when the computer age was just starting to become commercialized and heavily adopted by the business community, data were collected on magnetic tapes, punch cards, and disks—media that offered considerably less storage capacity than today's options. Data mining actually took a major step in revolution in the 1980s when relational databases and structured query languages (SQLs) were developed. By using the data sorting and retrieval power afforded by structured query language, hospitals were better able to analyze and make sense of large sets of data.

Data warehousing, the centralized storage of large sets of data, was introduced in the 1990s. It supported online analytic processing and multidimensional databases, which helped it to grow rapidly. In today's world of big business, hospitals and other large corporations use collected information to make large-scale assessments, such as predicted growth over the next 5 years or total revenue over the last 3 years.

Three different areas have provided the growth to make data mining what it is today: statistics, artificial intelligence, and machine learning. Statistics has enabled organizations like the CDC to provide better health care and services to patients, enabling its top 10 achievements in the 20th century: improved vaccinations, improved motor vehicle safety, safer and healthier foods, better control of infectious diseases, improved workplace safety, reduced deaths from heart disease and stroke, better family planning, increased awareness of tobacco as a health hazard, fluoridation in drinking water, and healthier mothers and babies. Statistics play a vital role in keeping the general population healthy. With the advent of computers and the internet, we can now harvest and refine medical decisions to an even finer degree, to the benefit of patients and medical establishments.

Current Applications

Today, data mining or predictive analysis in health care is a growing field. In fact, it is being used more and more to not only predict trends and analyze findings, but as an important tool for medical establishments to improve patient care, improve service offerings, and decrease losses, not the least of which is lost revenue in patient billing. Data mining has long been used in other fields but is still an emerging area within the healthcare industry. Canlas notes that data mining tools are being used not only for e-business and marketing but also by healthcare providers for "analysis of health care centers for better health policy-making, detection of disease outbreaks and preventable hospital deaths, and detection of fraudulent insurance claims" (Canlas, 2009). Moreover, the recent changes in health care, billing, and administration in the United States offer great opportunities for data mining.

Another current trend in data mining is **machine learning**. In addition to predictive analysis for the future, this process can be used to analyze past historical data. Therefore, data mining is both a process and an analysis method. It is a process, as there are procedures for collecting and preparing the raw data. The appropriate analysis method is chosen based on the type of data needed and the desired results. You will examine data mining in more detail in this chapter as a formal (and modern) method. Lastly you will examine some statistical terms and processes and examine how to handle some common statistical formulas using computer applications.

How Does Your Hospital Rate?

Now that Paul has found a website, he can compare facilities. The hospital Paul should review is Hospital A because it has a score of 100 out of 100, but what if that hospital is located in Ohio and Paul lives in Maryland, which is quite a distance to travel? In that case, Paul should look at Hospital B, which is nearby and has a score of 76.6 out of 100.

After choosing to review Hospital B, he finds that the reputation with the specialist is only 23.3%, survival rate is better than expected, the safety rating for patients is moderate, there is a high volume of cardiac patients, it has the highest rating for magnet nursing recognition, and, most important, it is rated seven out of seven for advanced technologies and key patient services, including an advanced trauma center and intensivist staff, meaning they have a staff physician in the intensive care unit at all times. This information has hopefully answered some of Paul's questions about the physicians and quality of care for cardiac patients.

Consider the following:

1. What other information should Paul consider before making his decision?
2. Which facility would you choose if you were Paul?

Basic Statistical Concepts

Although a detailed discussion of statistics goes beyond the scope of this text, the next sections will introduce some of the major statistical concepts relevant to healthcare professionals. Let's start with some basic items applicable to most statistical measures.

Dataset

A **dataset** is the data collected on a subject under examination. When your dataset includes information on every member of the group being investigated, you are examining a **population**. Or you may have a **sample** of data, which is a subset of data that statistically represents the entire population, ideally. A capital letter N is used to represent a population size and a lowercase n refers to a sample size. These concepts are important and ones we will refer to throughout the text, so it is important to note them.

As an example of the previous definitions, consider the Nielsen media rating group. Suppose they survey a sample of 1 out of 1,000 households, asking them which television shows they like or dislike. Considering how many households there are in the United States, even using this seemingly small sample size will yield thousands of survey results. Similarly, in the healthcare setting, collecting data from 50 randomly selected hospitals from across the country might be a fair and manageable representation of the greater population, which consists of every hospital in the United States. Imagine trying to survey all the hospitals in the United States! Obviously, it would be far easier to sort through and make inferences or summaries about the data when working with a sample rather than an entire population.

Typically, data mining involves very large samples or sets of data, or "big data," as you may hear it referred to. There are significant qualitative differences between the conclusions you can draw when examining a population versus a sample. When examining an entire population, you are not dealing with estimates or probability of an outcome but actual facts about an outcome. For example, consider a case in which out of 100 people surveyed, only 10 had a terminal disease. You know in this case that 10 of the 100 have a terminal disease—not just 10% but 10 real people. In this case, you are dealing with a population and not statistical or inference data. You have all the data and know all of the variables, so there is no need to estimate or infer. On the other hand, if the 100 people surveyed are only a random sample of a larger population rather than the entire population itself, you could generalize that the 10% who reported having a terminal disease represent the entire population. This, however, would be an inference only, not the statement of a fact.

Moving along, **parameters** are descriptive measures of a population and are sometimes referred to as fixed references (Batten, 1986). The use of parameters is different from that of statistical data, in which you are making assumptions about a larger population based on a random sampling of the population. An example of a parameter could be that we have 10,000 people under study in a population. We know we are examining 10,000 people. For example, imagine that you have 10,000 people to make a generalization about. In contrast, using statistical data methods (strictly speaking), you would randomly gather information from, say, 500 of the 10,000 and assume that the other 9,500 would answer the same way to the questions asked. As you can infer, having all the data is typically better than relying only on a sample of data.

However, in statistics, you typically do not have all the data, so you must collect data from many similar sources, which hopefully can lead to a valid finding for the unique situation at hand. In other words, we have data on the basis of which we can generalize findings that *should* be applicable to other groups.

For example, of 10,000 people surveyed, we found that 99% believed that hospital stays following major surgery, as covered by third-party payers, were not long enough. That being determined, we could infer that most others in the population would be closely aligned to the 99% mark, given that we could survey them as well. This of course would still be an assumption as to the outcome, but because the percentage of people who responded in this manner is so high (99%), it is probably true. For example, the data from five hospitals in our state of North Carolina seem to show that if patients are given information on a certain lifestyle change (such as hand washing) on their first visit, their rate of secondary infection is lowered by over 50%.

Variables

A **variable** is a characteristic or property of something that may take on different values. For example, the number of patients admitted to a hospital between the hours of 12 am (midnight) and 7 am might be tabulated and examined over a week's time frame. Each day's data would be a variable. An **independent variable** (also known as an experimental or predictor variable) is a factor we can measure, manipulate, or control for to produce a change in another variable, which is known as the **dependent variable** or outcome variable. For example, imagine that you would like to determine whether using twice the amount of a certain drug for a disease will help patients recover more quickly. In this example, the amount of the drug administered is the independent variable, and the recovery rate of the patients is the dependent variable.

Data Distribution

Data distribution refers to the characteristic pattern that data assume when represented in graphical form. A normal distribution of data is one that is characterized by data that average around a central value with no real tendency to skew left or right. In this case, 50% of data falls to the left of the peak of the curve, and 50% falls to the right, forming a symmetric curve that possesses a single peak when graphed. Because the curve is shaped like a bell, you will hear it referred to as a "bell curve" (**Figure 1.1**). This data distribution pattern is one of the most important statistical concepts to understand.

However, if the distribution is not normal, data could be spread out to the left or to the right, or it could be randomly distributed.

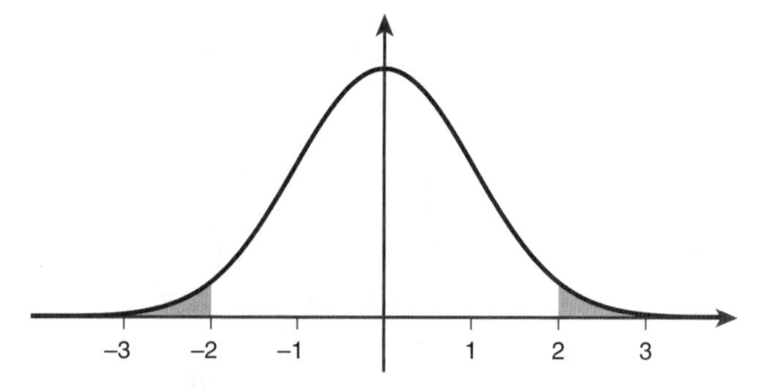

Figure 1.1 Bell curve.

Types of Data

The four types of data you might encounter are nominal (categorical), ordinal, interval, and ratio.

1. **Nominal data**: This type of data indicates categories and cannot be ordered. Examples include the following: techniques (technique A, technique B), gender (male or female), or occupation (e.g., students, professional programmers).
2. **Ordinal data**: This type of data can be ordered (e.g., in terms of size), but the difference between any of the two values may not always be equal. Examples include responses to a Likert-scale question, such as the following: *use it every day*, *use it once a week*, *use it once a month*, *use it once a year*, and *have never used it*. Likert-scale questions will be discussed in more detail in a later chapter.
3. **Interval data**: This type of data can be ordered, and the difference between any two consecutive values is the same, but there is no absolute zero, which allows you to have meaningful negative values. The most famous example of this type of data is temperature in degrees Celsius (°C) or Fahrenheit (°F). The zero values (i.e., 0°C and 0°F) are artificially defined, and negative temperature values are possible. But the difference between any two consecutive values at any point on the scale (i.e., between 0°C and 1°C and between 100°C and 101°C) is the same.

   Note that when working with data from Likert-scale questions, if you can assume that the differences between any two options are equal, you can treat them as interval data. For instance, if your options are *strongly agree, agree, neutral, disagree*, and *strongly disagree*, you may be able to treat them as interval data, which is handy for computer tabulation. You simply assign a numeric value for each selection (e.g., strongly agree = 1, agree = 2) then count the occurrences of each from the dataset, such as "out of 50 people sampled, 5 chose that they strongly agreed with the first question, 15 reported that they only agreed," and so on.

4. **Ratio data**: This type of data can be ordered, the distance between any two consecutive values is the same (interval data), and there is an absolute zero. This means that a meaningful negative value of interval data does not exist (in statistics). Consider as examples the weight of a llama, the height of a tree, the length of something, or a recorded time or speed. A count could be considered as a ratio, as well.

Different statistical methods are designed only for certain types of data. As you work through this text, you should pay particular attention to this fact, as well as the size of the sample you are working with. Both choice of method and sample size are critical to choosing an appropriate statistical measure.

Types of Data Mining Models

Because this text offers the opportunity to mine real data for use in statistics, it will be helpful to discuss some key data mining models in detail here, including predictive, descriptive, and decision models. Subsequent chapters will give you hands-on experience with data mining methods.

Predictive Models

A **predictive model** explicitly predicts future behavior based on past trends. In each case, a model is created or chosen to try to best predict the probability of an outcome. When a predictive model is used in data mining, its main purpose is to forecast probabilities and trends in the future. Data are collected for pertinent predictors. The model is formulated, and finally it is validated and revised as new data become available.

For example, predictive models can be used by insurance companies and government programs such as Medicaid to assist in the prediction of future medical needs. A predictive model can also identify those who may be at high risk for developing a chronic disease or having poor health outcomes. However, initial medical information (data) on these patients is needed to make such predictions. Without their personal health information, identification of their risks would be significantly delayed. This could potentially lead to unfavorable outcomes or a delayed improvement in the patient services offered.

Descriptive Models

A **descriptive model** describes patterns in existing data, including the main features or a summary of the data under examination by the researcher. It provides a hypothetical basis for the system under study. Descriptive models can recognize relationships while being used with other models to make predictions. In this approach, maps, charts, and graphics portray and promote understanding of real-world, complex, and sometimes redundant systems or services. Think of it as a way to describe visually the data in question. There are two general dimensions that are used for this description of mined data, viewpoints and aspects.

A **viewpoint** is a specific context or approach that you intentionally adopt when examining the data that allows you to focus on relevant details in the study and ignore irrelevant data.

An **aspect** is a specific category of data that is used in conjunction with your viewpoint. You might have one or more aspects associated with each viewpoint. Think of a viewpoint as a book chapter, with aspects being headings within the chapter. For example, a viewpoint might be data on patients being treated for lung cancer, such as survival rates and length of survival times. After patients who are cured are removed from consideration, aspects would include data on patients who survived 1 to 6 months, 7 to 12 months, and 1 to 2 years.

Decision Models

A **decision model**, also known as a business rule or business logic, is a logic system to determine desired actions for the business based on thresholds, conditions, or events. When using decision models, all elements in a relationship are examined to forecast or predict results. Units are arranged into groups according to the relationships between the units. All information would be organized into a logic **flowchart**, which is a graphical way to depict a series of actions, given a certain series of events. For example, a "Get up and go to school" flowchart would include events in the order in which they should occur, such as "wake up and turn off the alarm," "take a shower," "dry off," "eat breakfast," etc. Note the importance of the sequence of events in a flowchart. If you took a shower after you dried off, you might be a bit wet at the breakfast table!

As another example, consider diagramming a process to detail patient admittance to a medical facility. Each step would be graphically presented, with details and substeps, as a part of the flowchart. The first step might be to determine whether an arriving patient is an emergency case. This would be a logical true/false decision, with one of two paths being taken, depending on the case. The flowchart then would present the alternate steps for true and for false responses in a graphical fashion.

With decision models, the relationships among all the known data of a situation and the result of the logical decision process can be used to forecast future patterns and thus better prepare the organization with regard to planning. Thus, this model can help optimize and streamline the business and help the business to provide better end service and maximize cost savings. In later chapters we will examine some specific formulas and tools used for flowcharting and decision logic, such as for forecasting. Although there are many ways to examine data, this text will generally be limited to models most appropriate for the healthcare facilities.

Obtaining Data

If you are conducting an exhaustive search for information from many sources, use of data mining and statistical inference might be the best way to proceed. Regarding how to refer to your data, if a dataset was collected by other people, it is considered a **secondary data** source; if you collected the dataset yourself, it is a **primary data** source.

Related to obtaining data, there are five basic steps in data mining:

1. Extract and transform the data and load the data into the data warehouse system.
2. Store and manage data using a relational database system.
3. Provide data access to users.
4. Use application software to analyze the data.
5. Present the gathered information in a meaningful manner.

Keep in mind that if you are examining all of the data available, say from a local source such as the hospital where you work, then you are examining a population (capital *N*) and are using local data, otherwise known as a primary data source. Regardless of where you get your data, you should cite your sources, along with what measures and tools you used. Your results might be suspect if you do not validate where your information came from, so give as many details as you can.

Some may assume that secondary data are inferior to primary data, but make no mistake, using mined secondary data in health care will improve patient care and increase benefits in the future. As paper medical records are replaced with electronic ones, mined secondary data will increase significantly, allowing for quality improvement in patient care and increases in cost savings, patient satisfaction, and revenue. Thus, a leading goal for healthcare facilities at this time is increased use of mined secondary data internally in their facilities, especially on identified areas that are in need of quality improvement.

How Does Your Hospital Rate?

Paul and all Americans want high-quality health care by the best professionals. However, we certainly cannot afford to travel to other states to receive treatment from the top-rated hospitals unless we are rich. When shopping for a medical care facility, consider that you might want the following: timely service, a safe environment, current medical technologies and treatments, and reliable patient-centered care. As Paul found out, the Hospital B has a high score for the use of technology. However, their moderate safety score concerns him; no matter how great the care may be in other areas, receiving a staph infection while in the hospital would hinder his journey back to full health.

To facilitate the sort of rating process that Paul undertook, the US federal government, via Medicare.gov, has created Hospital Compare, a website that allows people to compare area healthcare facilities based on zip code or by facility name.

Consider the following:

1. Use the following website: https://www.medicare.gov/hospitalcompare/search.html. How does your hospital rate for cardiac procedures?
2. Would you have a cardiac procedure performed in your area facility, or would you travel to another facility?
3. What do you think Paul should do?

Global Perspective

Taking our healthcare statistics globally allows us to compare diseases of other countries to those in the United States. In this Global Perspective section, we will examine rheumatic heart disease. This disease is caused by rheumatic fever, which affects the mitral valve in the heart. If a child has strep throat or scarlet fever that is not treated properly with antibiotics, he or she will have a higher risk of acquiring rheumatic heart disease. The mitral valve between the left atrium and left ventricle is affected by this condition. This condition affects children mainly between the ages of 5 and 15 years. Surgery to repair the mitral valve is one way to treat this condition.

The data below were retrieved from WHO. **Table 1.1** shows data for rheumatic heart disease in the United States and India, comparing males and females between the ages of 0 and 14 years in 2008. The population of males followed in India was 195,436, and the population of females was 179,323. The population of males in the United States was 32,646, and the population of females was 31,056.

On an interesting note, further research could be done to understand why India has a 2.3% rate of rheumatic heart disease in males and 3.0% rate in females. Questions that could be asked might include the following:

- Why is the female rate of rheumatic heart disease higher than that of males?
- Do children in India receive the antibiotic at all?
- Are incorrect dosages of the medication given?
- Are children not able to get the surgery necessary to repair the valve?
- Why do children in India acquire strep throat or rheumatic fever?

Table 1.1 Incidence of Rheumatic Heart Disease in India and the United States in 2008

Country	Sex	Population with Rheumatic Heart Disease
India	Male	2.3%
	Female	3.0%
United States	Male	0%
	Female	0%

Many other questions could be asked, and further research could be done with this information to follow up with types of treatment given to children in India compared with those in the United States.

This comparison of a disease's prevalence in the United States versus India is just a small sample of some of the interesting global trends and statistics that we will cover throughout subsequent chapters.

Hands-on Statistics 1.1: Examine Other Data from WHO

1. Using the same data source (WHO) as referenced in the Global Perspective section, compare India with a different country and discuss the results of your comparison. See http://www.who.int/healthinfo/global_burden_disease/estimates_country/en/index.html.
2. What are some strengths and weaknesses of this data source?

Chapter Summary

This chapter established a foundation for learning more about healthcare statistics and examining sources of data. It also examined some parts of the process involved in data analysis. The study of statistics is complex, involving extensive mathematical algorithms and research considerations, but rest assured that subsequent chapters will further explain and provide real-world examples of how to use the statistical methods involved in healthcare statistics. When you finish this text, you will be confident and well versed in the subject and will be able to put your knowledge to immediate practical use.

Also covered in this chapter is an introduction to data analysis and presentation of your findings. Hopefully, this chapter has whet your appetite for these subjects. In coming chapters, you will learn more topics related to statistics, data harvesting, data analysis, and presentation of your findings, all emphasizing real-world data. This chapter is only a small step toward an exciting journey to find out how powerful statistics can be.

Apply Your Knowledge

1. Consider users of healthcare statistics other than those mentioned in the chapter. List at least three users of healthcare statistics, and describe the statistics they keep.
2. Other than Florence Nightingale, who were some of the first users of statistics, and what statistics did they keep?
3. Name at least two countries that use telehealth.
4. Write a pro and con for some health issue, such as "smoking increases your risk of lung disease" or "running on pavement is bad for your knees." Be creative!
5. Discussion: When researching *data mining*, you may come across the terms *data dredging* or *data snooping*. What do they mean? What implications should a researcher know about these terms? How might they be avoided?
6. Go to the following website: http://www.who.int/healthinfo/global_burden_disease/estimates_country/en/index.html. Review data for males and females in India and in the United States, comparing the percentages between the different age groups of those with rheumatic heart disease: 15 to 50 years and 60+ years. Write a brief explanation of your findings.
7. Go to http://www.cdc.gov/DataStatistics/. Click on a topic of your choice and compare health statistics for your state with those of another state. Did your state have a higher statistic than your comparison?

References

Batten, J. (1986). *Research in education* (rev. ed.). Greenville, NC: Morgan Printers.

Canlas, R., Jr. (2009). *Data mining in healthcare: Current applications and issues* (Master's thesis). Adelaide, South Australia: Carnegie Mellon University in Australia.

Centers for Disease Control and Prevention. (2018). Table 10. Selected nationally notifiable disease rates and number of new cases: United States, selected years 1950–2017. Retrieved from https://www.cdc.gov/nchs/data/hus/2018/010.pdf

Johnson, N., & Kotz, S. (2012). *Leading personalities in statistical sciences: From the seventeenth century to the present*. Wiley Online Library. Retrieved from http://onlinelibrary.wiley.com/doi/10.1002/9781118150719.ch2/summary.

Lancaster, L. (2013). Celebrating statisticians: Florence Nightingale. JMP Blog. Retrieved from https://community.jmp.com/t5/JMP-Blog/Celebrating-statisticians-Florence-Nightingale/ba-p/30247

Leen, S. (1994). The legacy of Hans Freudenthal. *Educational Studies in Mathematics, 25*(1-2), 164.

Mackenzie, D. (1913). Indian myth and legend. Sacred Texts. Retrieved from http://www.sacred-texts.com/hin/iml/index.htm

National Committee to Preserve Social Security and Medicare. (n.d.). Medicare. Retrieved February 21, 2020, from https://www.ncpssm.org/our-issues/medicare/medicare-fast-facts/

Tan, P.-N., Steinbach, M., & Kumar, V. (2006). *Introduction to data mining*. Boston, MA: Pearson Education.

Thucydides. (n.d.). *The history of the Peloponnesian War* (R. Crawley, Trans.). Project Gutenberg. Retrieved from http://www.gutenberg.org/files/7142/7142-h/7142-h.htm

Valova, I., & Noirhomme, M. (2009). Comparative analysis of advanced technologies for processing of large data sets. *Information Technologies and Control*, 1-13.

Web Links

Using Excel to do Basic Statistical Analysis: https://www.statisticshowto.datasciencecentral.com/mode/#excel

Excel Tutorials for Statistical Data Analysis: http://www.stattutorials.com/EXCEL/EXCEL_TTEST2.html

Introductory Statistics: Concepts, Models, and Applications: http://www.psychstat.missouristate.edu/introbook/sbk25m.htm

MS Excel: How to Use the QUARTILE Function (WS): http://www.techonthenet.com/excel/formulas/quartile.php

Excel and Quartiles: http://www.meadinkent.co.uk/excel-quartiles.htm

Drawing a Normal Curve: http://www.tushar-mehta.com/excel/charts/normal_distribution/

Sexually Transmissible Infections: http://www.abs.gov.au/AUSSTATS/abs@.nsf/Lookup/4102.0Main+Features10Jun+2012

Sexually Transmitted Diseases (STDs): Data and Statistics: http://www.cdc.gov/std/stats11/trends-2011.pdf

Medicare Enrollment—Aged Beneficiaries: As of July 2010: http://www.cms.gov/Research-Statistics-Data-and-Systems/Statistics-Trends-and-Reports/MedicareEnrpts/Downloads/10Aged.pdf

Has Statistics Made Us Healthier? The Role of Statistics in Public Health: http://www.statisticsviews.com/details/feature/5025891/Has-statistics-made-us-healthier-The-role-of-statistics-in-public-health.html

Descriptive Statistics: http://www.businessdictionary.com/definition/descriptive-statistics.html

Discrete/Continuous: http://www.chegg.com/homework-help/definitions/discrete-continuous-31

Hospital Compare: https://www.medicare.gov/hospitalcompare/search.html

© Mad Dog/Shutterstock

CHAPTER 2

Central Tendency, Variance, and Variability

Facts are stubborn, but statistics are more pliable.

—Mark Twain

CHAPTER OUTLINE

Introduction
Measures of Central Tendency
- Mean
- Median
- Mode

Frequency Distribution
Variance and Measures of Dispersion or Variability
- Min and Max
- Range
- Outlier Data
- Interquartile Range
- Standard Deviation
- Variance

Data Harvesting
Global Perspective
Chapter Summary
Apply Your Knowledge
References
Web Links

LEARNING OUTCOMES

After completing this chapter, you should be able to do the following:

1. Discuss the uses of mean, median, and mode in a hospital setting.
2. Discuss the uses of measures of central tendency.
3. Demonstrate how to find mean, median, mode, and standard deviation.
4. Describe the difference between continuous data and discrete data.
5. Describe the term *skewed*.

KEY TERMS

Continuous data
Descriptive statistics
Discrete data
Frequency distribution
Interquartile range
Max
Mean
Median
Min
Mode
Outlier data
Range
Skewed
Standard deviation
Variance

How Does Your Hospital Rate?

Dr. Barker, chief of the medical staff at his hospital, is reviewing cases of heart failure and the length of stay, readmission rates, and average age of patients with this condition. The purpose of this review is to improve patient care, reduce readmission rates, and potentially reduce length of stay in the hospital. Heart failure is the heart's inability to pump enough blood and oxygen to support the other organs in the body. Heart failure expenditures average about $32 billion each year in the United States. This cost includes healthcare services, medication, and missed days of work. Dr. Barker went to the website of the Centers for Disease Control and Prevention (CDC) to find data on his county (CDC Interactive Heart Disease site: https://nccd.cdc.gov/DHDSPAtlas/).

Consider the following:

1. If you were in Dr. Barker's position, what specific goals might you have to improve patient care in this situation?
2. What might a high readmission rate indicate, and what are measures that could be taken to reduce this number?
3. Why would Dr. Barker want to reduce length of stay among heart failure patients? How does length of stay relate to the quality of patient care?

Introduction

The Mark Twain quote that introduces this chapter is quite telling in that although the data you have may be reliable and accurate, the interpretation of the data can be questionable, as this and other chapters will reveal. In this chapter, you will learn more about key statistical methods and provide simple and real-world examples of each. These examples will involve computing your results with a Microsoft Excel spreadsheet or other software tools. First you will learn about measures of central tendency, including mode, mean, median, variables, and frequency distribution. These measures are part of **descriptive statistics** and are concerned with summarizing and interpreting some of the properties of sets of data, but they do not suggest necessarily the properties of the entire population from which the sample was taken (which would require determining whether your collected data were in fact representative of the entire population).

Following the discussion of measures of central tendency, you will learn about measures of difference among the numbers in a dataset. Standard deviation, standard error, range, and variance are covered in this section and are known as measures of dispersion or variability.

Measures of Central Tendency

Statistical methods used to determine the shape of a distribution of data are mean, median, and mode. These methods are known as measures of central tendency, or measures of central location and summary statistics. Each is a valid measure of central tendency but is used under different conditions. You will recall from the first chapter how mode was used to estimate the height of a brick wall, which is only one example of this measure.

If all the data are perfectly normal, the mean, median, and mode will be identical in reliability and will represent summary values accurately in a given dataset. Follow along and try the examples to see how these methods work.

Mean

Mean is a measure of central tendency that can be determined by mathematically calculating the average of observations (e.g., data elements) in a frequency distribution. Mean is the most common measure of central tendency. The mean can be used with discrete data (e.g., "choose 1, 2, or 3") but is most commonly used with continuous data (e.g., weight of a patient). In this case, the mean is just a model of the dataset.

One of the most important factors relating to the mean is that it minimizes errors in predicting any one value in the dataset. In fact, it produces the fewest errors compared with median and mode. Another important characteristic of this measure is that it includes all values in a dataset. Remember that a population is all of your data, whereas a sample is just a subset, preferably a random but representative one.

However, using the mean has one disadvantage: it is susceptible to the influence of outlier data, or data elements that are very different or far away from most of the other data elements. Outlier data, discussed later in the chapter, may also be referred to as *flier*, *maverick*, *aberrant*, or *straggler*, although *outlier* is most commonly used. Outlier data may be detected by statistical tests (such as a Dixon or Grubbs test) or by a graphical display of the data.

Moreover, the more **skewed** (or varied) the data become, the less effective the mean is in locating a central tendency and typical value. Skewed simply means that the dataset contains both very small and very large numerical values. Mean is best used with data that are not skewed, such as in the set 1, 3, 2, 1, 3.

There are actually several types of calculation for mean. We have already covered the standard mean, also known as the arithmetic mean. Another is the sample mean, also known as the average. By using the sample mean, outlier data can be used. Skewed data will make no difference in the sample mean. Sample mean is an estimate of the population, and the dataset may be quite varied, such as 8, 34, 56, 25, 41, 2, 17, 25.

Other means are the harmonic mean, for rate or speed measures, and the geometric mean, for when the ranges you are comparing are different and you need to equalize them. Next we will compute a sample mean.

Hands-on Statistics 2.1: Use Excel to Find Mean

To compute the mean, add all items together and divide by the number of elements in the dataset. To perform this calculation in Excel, examine **Figure 2.1** and follow these directions:

1. Open a new, blank spreadsheet in Excel.
2. Key in the following data about inpatient hospital days, entering one numeral per cell in consecutive cells of the same row, beginning in row 2, column C (C2), and ending in cell G2:

   1, 3, 2, 1, 3

3. Use the average function to find the mean of the dataset. To do so, click in cell H2 and then select the average function from the Formulas tab, or type the following formula in cell H2:

   =AVERAGE(C2:G2)

4. Click out of cell H2, and the mean should appear in this cell: 2.

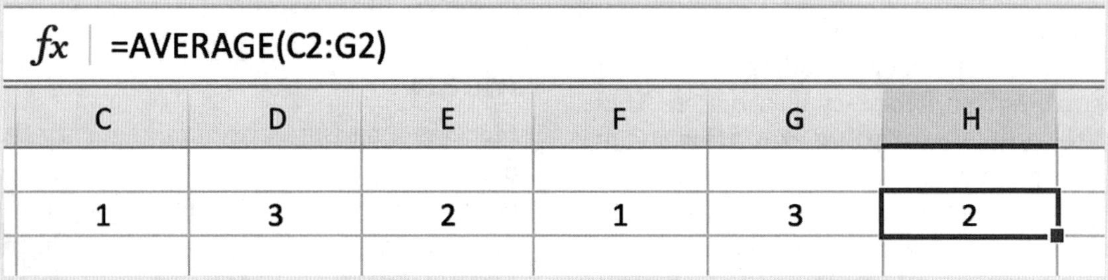

fx =AVERAGE(C2:G2)

C	D	E	F	G	H
1	3	2	1	3	2

Figure 2.1 Using the average function in Excel.

DID YOU KNOW? One in 33 babies are born with some type of birth defect. Birth defects are one of the leading causes of infant deaths and account for more than 20% of infant deaths. Statistics show that 1 in 691 infants born will have Down syndrome, meaning there are 6,037 infants diagnosed with Down syndrome each year. Anencephaly (congenital absence of all or a major part of the brain) occurs in 1 in 4,859 newborns, and 859 cases are reported per year (CDC, 2018b). What can a pregnant woman do to aid in the prevention of these types of birth defects? She should take folic acid every day, avoid alcohol and smoking, prevent infections, discuss current medications with her physician, maintain a healthy weight, and maintain regular office visits.

Now you know!

Median

Median is a measure of central tendency that reflects the midpoint of a frequency distribution when observations are arranged in order from the lowest to the highest. Median is used more often when data are skewed because it can retain its position and is less affected by skewed values. As stated by Agresti and Finlay (1997), "It is a measure of central tendency that better describes a typical value when the sample distribution of measurements is highly skewed." The more skewed the data, the larger the difference between the median and the mean, which is the rationale for using the median in such cases. Median is best used with ordinal and interval/ratio (skewed) types of variables.

For example, consider the following dataset: 72, 2, 60, 44, 38, 51, 83, 25, 16, 69, 45. To calculate the median of this group of numbers, first put the data in order from the lowest to the highest, as follows: 2, 16, 25, 38, 44, 45, 51, 60, 69, 72, 83. In an odd set of numbers, the median will be the number in the very center of the series when arranged in order; that is, there will be exactly as many numbers before the median as after it. Our sample is an odd set of numbers, meaning our median is the number in the center, 45.

For an even set, you find the middle two numbers, add them together, and divide by 2. Your result will be the median value, which could well be a decimal value. You can round the resultant value depending on the type of number you need to obtain (i.e., whole number versus a fractional value).

Mode

Mode is a measure of central tendency that consists of the most frequent observations in a frequency distribution. Although mode is also regularly used in dataset analysis, Batten notes that mode "is considered to be less important than either the mean or the median" (Batten, 1986). The mode need not be unique, as it is possible to have two or more values that are distributed equally. You may have two numbers, 3 and 45, for example, that occur several times in a dataset along with other values. However, mode does not work well with continuous data and should not be used in that case, unless you are working with ratings that are assigned a numeric value, such as with Likert response variables assigned a numeric value (e.g., strongly disagree = 1, disagree = 2). Continuous data are basically any type of data that have infinite values with connected data points, which can result in an unlimited selection of data. For example, a measure of height could be recorded as a round 72 inches, or it could have infinite values—72.2, 72.24, and so on. It is generally accepted in the reporting of this example that you would say, "The average height was . . ." instead of reporting the most frequently occurring height; though both would work, the average better describes height data. Also, mode does not offer an effective measure of central tendency if the most common data are located far away from the rest of the data. Mode is best used when using a nominal type of variable.

Hands-on Statistics 2.2: Use Excel to Find Median

To find the median using Excel, see **Figure 2.2** and follow these directions:

1. Begin with a blank spreadsheet open in Excel.
2. Key in the following data, entering one numeral per cell in consecutive cells of the same row, beginning in cell A2 and ending in cell K2:

   2, 3, 4, 5, 5, 6, 9, 77, 5, 4, 3

3. Use the average function to find the median of the dataset. To do so, click in cell L2 and then select the median function from the Formulas tab, or type the following formula in cell L2:

   =MEDIAN(A2:K2)

4. Click out of cell L2, and the median should appear in this cell: 5.

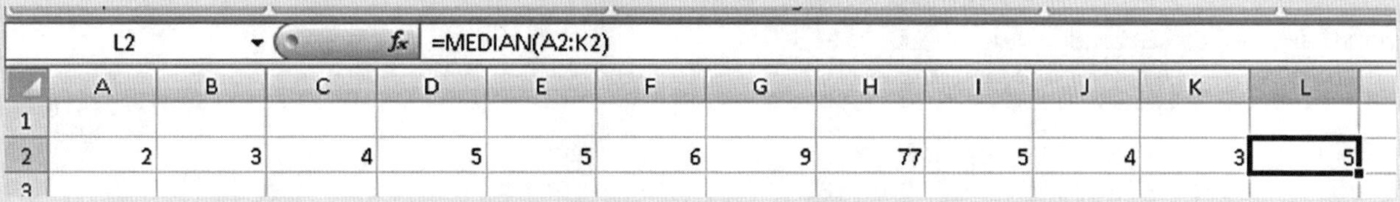

L2 | fx =MEDIAN(A2:K2)

	A	B	C	D	E	F	G	H	I	J	K	L
1												
2	2	3	4	5	5	6	9	77	5	4	3	5
3												

Figure 2.2 Median using Excel.

Hands-on Statistics 2.3: Use Excel to Find Mode

To find the mode using Excel, see **Figure 2.3** and follow these directions:

1. Begin with a blank spreadsheet open in Excel.
2. Key in the following data, entering one numeral per cell in consecutive cells of the same row, beginning in cell G2 and ending in cell M2:

   4, 4, 5, 6, 4, 2, 3

3. Use the mode function to find the median of the dataset. To do so, click in cell F2 and then select the median function from the Formulas tab, or type the following formula in cell F2:

   =MODE(G2:M2)

4. Click out of cell F2, and the median should appear in this cell: 4.

This is a simple example, and you might be thinking, "I could have figured that out in my head!" and you would be correct. With larger datasets, however, it becomes much more difficult to figure out the answer without using a spreadsheet. Consider that even with a dataset as small as 35 items, it becomes very difficult to "eyeball" the results to find the mode.

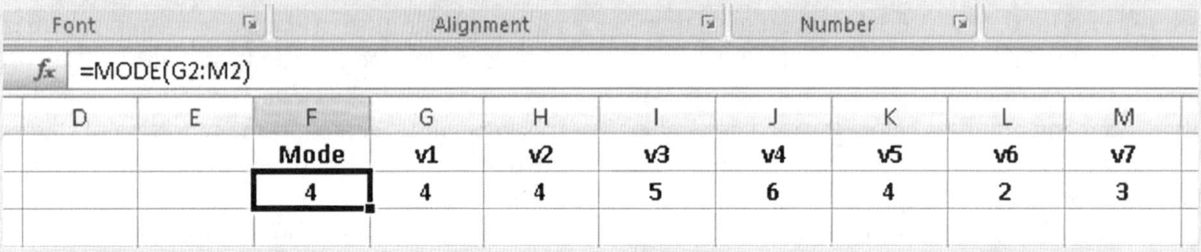

Font | Alignment | Number

fx =MODE(G2:M2)

D	E	F	G	H	I	J	K	L	M
		Mode	v1	v2	v3	v4	v5	v6	v7
		4	4	4	5	6	4	2	3

Figure 2.3 Mode using Excel.

At this point, two main types of data should be considered, **discrete data** and **continuous data**. Discrete data are defined and finite. For example, a survey that asks you to choose 1, 2, or 3 is discrete, as there is no 1.2 or 2.5 data point. Use of continuous data is well illustrated in a measurement of height. There certainly could be fractional points of measurement, and not only a set number defined values. Then you will use Excel to find the mode of a dataset.

Frequency Distribution

Frequency distribution describes how often different values are found in a set. Simply put, this function counts the frequency of values and tallies them for you, even in a chart or histogram, should you choose it.

Hands-on Statistics 2.4: Use Excel to Find Frequency Distribution

For this exercise and many that follow, note that you generally will need to install the free data analysis toolpack into Excel, unless it has been installed for you.

To find the frequency distribution using Excel, follow these directions:

1. Open the spreadsheet titled "CH02_Frequency_Distribution.xlsx" located in Chapter 2 of the eBook.
2. Highlight the two columns of data.
3. Select the *Data* menu, then *Data Analysis*.
4. Select *Histogram* and click *OK*.
5. To the right of *Input Range*, click the red arrow, then highlight the dataset named "data."
6. To the right of *Bin Range*, click the red arrow, then highlight the possible values in the set (the heading in the spreadsheet is named "bin values").
7. Click the down red arrow to maximize the control box again.
8. Check *Chart Output* if you want a chart, then click *OK*.
9. A new sheet will be inserted in your workbook, with the Bin, or value, and the frequency of occurrence to the right.

How Does Your Hospital Rate?

Dr. Barker continues to review the CDC website (http://nccd.cdc.gov/DHDSPAtlas/), which provides him with in-depth information on the statistics of heart failure. His results span the years 2008 through 2010. In his comparison, he looks at race, demographics, and age to see if there is a correlation between them. Dr. Barker finds that in Brunswick County, North Carolina, 78.4% of black male Medicare beneficiaries hospitalized for heart failure were discharged home following treatment, compared with 71.1% of white males in the same circumstances. He also finds that heart disease occurred in 1528 per 100,000 black males in Brunswick County, compared with a national average of 1695.6 per 100,000 among this demographic; in contrast, it occurred in 1413.6 per 100,000 white males in Brunswick County, compared with a national average of 1469.3 per 100,000 in this demographic. These statistics for Brunswick County are very close to the national average.

Consider the following:

1. Visit the website http://nccd.cdc.gov/DHDSPAtlas/. Look up and compare the percentages of black and white male Medicare recipients discharged home after a hospitalization for heart failure. How do these rates compare?
2. How do these rates compare with the same demographic groups for the state as a whole? For the nation?
3. What might we learn about discharges following hospitalization for heart failure among whites and blacks? What additional information should we gather to shed light on this issue?

Variance and Measures of Dispersion or Variability

These measures include minimum (min) and maximum (max) values, range, outlier data, interquartile range, standard deviation, and variance. We will first examine min and max.

Min and Max

Min is simply the minimum or smallest value in a dataset, and **max** is the maximum or largest value in a dataset. Excel can return these values to you automatically, which can be very useful for large datasets.

We will use both of these functions in combination to work with range, or the distance between two possible values. This combination will be very handy for many situations when dealing with large datasets.

Hands-on Statistics 2.5: Use Excel to Find Min and Max

For a given set of values, the formula =MAX(*start:end*), where *start* and *end* are the starting and ending cells, gives you the maximum value from a set of values. The min function, =MIN(*start:end*), gives you the minimum, using the same logic. To find the min and max using Excel, see **Figure 2.4** and follow these directions:

1. Begin with a blank spreadsheet open in Excel.
2. Key in the following data, entering one numeral per cell in consecutive cells of the same row, beginning in cell B1 and ending in cell K1:

3, 1, 66, 4, 7, 9, 1, 2, 8, 7

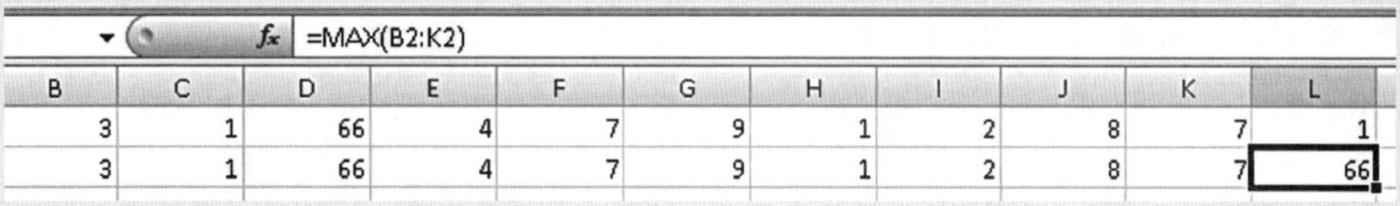

fx =MAX(B2:K2)

B	C	D	E	F	G	H	I	J	K	L
3	1	66	4	7	9	1	2	8	7	1
3	1	66	4	7	9	1	2	8	7	66

Figure 2.4 Finding the min and max using Excel.

(*continues*)

3. Either highlight the cells indicated in the previous step and drag them down to the row below to autofill it or reenter these numbers manually in cells B2 to K2.
4. Use the min function to find the min value of the dataset. Type the following formula in cell L1:

=MIN(B1:K1)

5. Click out of cell L1, and the min should appear in this cell: 1.
6. Use the max function to find the max value of the dataset. Type the following formula in cell L2:

=MAX(B2:K2)

7. Click out of cell L2, and the max should appear in this cell: 66.

Range

Range is the difference between the maximum value in a dataset and the minimum value. When you compute the range, most of the data are ignored because you are using only the largest and smallest extremes. Your range statistic provides information on the statistical dispersion of a data sample, or the start and end points.

To see how range works, consider the following dataset: 53, 64, 78, 98, 58, 61, 83, 89. Noting that the highest value is 98 and the lowest value is 53, you can find the range by subtracting the minimum number from the maximum number: 98 – 53 = 45.

Hands-on Statistics 2.6: Use Excel to Find Range

To find the range using Excel, you must first find the max and min values for the dataset and then use the subtract function to calculate the difference between them. See **Figure 2.5** and follow these directions:

1. Begin with a blank spreadsheet open in Excel.
2. Key in the following data, entering one numeral per cell in consecutive cells of the same row, beginning in cell A2 and ending in cell K2:

2, 3, 4, 5, 5, 6, 9, 77, 5, 4, 3

3. Use the max function to find the max value of the dataset. Type the following formula in cell L2:

=MAX(A2:K2)

4. Click out of cell L2, and the max should appear in this cell: 77.
5. Use the min function to find the min value of the dataset. Type the following formula in cell L3:

=MIN(A2:K2)

6. Click out of cell L3, and the min should appear in this cell: 2.
7. Use the subtract function to find the range of the dataset. Type the following formula in cell L5:

=L2-L3

8. Click out of cell L5, and the range should appear in this cell: 75.

L5 fx =L2-L3

	A	B	C	D	E	F	G	H	I	J	K	L
1												
2	2	3	4	5	5	6	9	77	5	4	3	77
3												2
4												
5											range =	75

Figure 2.5 Calculation of range in Excel by subtracting min from max.

Outlier Data

Outlier data are elements of a dataset that lie an abnormal distance from other values in the dataset or sample. You, the researcher, must decide what is and is not an outlier. In the preceding example, the 77 would be considered outlier data. If you have decided to eliminate outlier data, you would remove them from the dataset but might make note of them.

Interquartile Range

The **interquartile range** is a measure of the dispersion within a dataset. It is the difference between the third quartile and the first quartile. Quartiles are the three points that divide a dataset into four equal groups, each group making up a quarter of the data. The interquartile range, therefore, is the breadth of the interval that encompasses 50% of the sample. Usually it is smaller than the range and is less affected by outlier data. The interquartile range also tells you the size of the box in a box plot chart, which we will address in later chapters. The quartile function in Excel can be used to find the interquartile range, which gives a measure of the spread of the distribution, ignoring outlier data (**Figure 2.6**).

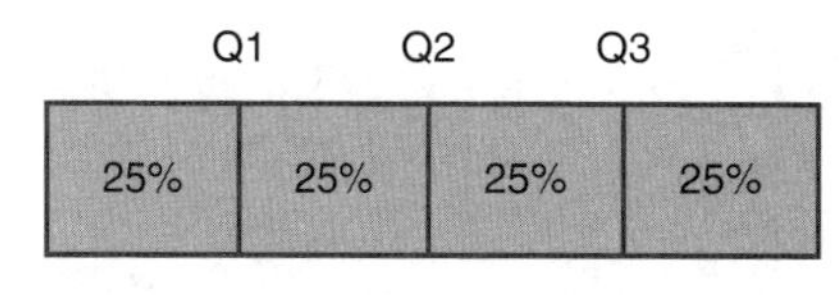

Figure 2.6 Quartiles.

Next let's examine how to summarize just how different sets of numbers are.

Hands-on Statistics 2.7: Use Excel to Find Interquartile Range

The function in Excel for quartile is as follows:

=QUARTILE({array data}, ***n*th quartile**)

where *array data* is a sequence of like data (e.g., 1, 4, 5, 9) and the **n***th quartile* is the quartile value you wish to return. The quartiles are represented as follows: 0 = the smallest value in the dataset; 1 = the first quartile (Q1), or 25%; 2 = the second quartile (Q2), or 50%; 3 = the third quartile (Q3), or 75%; and 4 = the largest value.

For example, based on the formula of Q3 – Q1, we can get the interquartile range, or one-half the total range. See **Figures 2.7** and **2.8**, and follow these directions:

1. Begin with a blank spreadsheet open in Excel.
2. Use the quartile function to find the Q3 value of the dataset. Type the following formula in cell B1:

=QUARTILE({1,2,3,4,5,6,7,8,9,10},3)

ard | Font | Alignment

fx =QUARTILE({1,2,3,4,5,6,7,8,9,10},3)

B	C	D	E	F	G	H
7.75		3.25		4.5		

Figure 2.7 Quartile function in Excel for Q3 in cell B1, with interquartile range in cell F1.

fx =QUARTILE({1,2,3,4,5,6,7,8,9,10},1)

B	C	D	E	F	G	H
7.75		3.25		4.5		

Figure 2.8 Quartile function in Excel for Q1 in cell D1, with interquartile range in cell F1.

(continues)

3. Click out of cell B1, and the Q3 value should appear in this cell: 7.75.
4. Use the quartile function to find the Q1 value of the dataset. Type the following formula in cell D1:

   =QUARTILE({1,2,3,4,5,6,7,8,9,10},1)

5. Click out of cell D1, and the Q1 value should appear in this cell: 3.25.
6. Use the subtract function to find the interquartile range of the dataset. Type the following formula in cell F1:

   =B1−D1

7. Click out of cell F1, and the interquartile range should appear in this cell: 4.5.

Thus, in this example we learn that the difference between the 75th and the 25th percentiles is 4.5, which represents the size of the spread of the middle 50% of the data in the distribution.

Standard Deviation

Standard deviation is a measure of variability that describes the deviation from the mean of a frequency distribution of data. The standard deviation is symbolized by *sd* or *s*. Note that the more the data are spread, the greater the standard deviation. For example, if you have test scores that range from 48% to 98%, the standard deviation will be higher than test scores that range from 95% to 98% because in the latter case the data are not as varied as in the first range.

Hands-on Statistics 2.8: Use Excel to Find Standard Deviation

The formula in Excel for finding the standard deviation is =STDEV(*start:end*), where *start* and *end* are the starting and ending cells of the range. To find the standard deviation using Excel, see **Figure 2.9** and follow these directions:

1. Begin with a blank spreadsheet open in Excel.
2. Type "Scores" in cell B2 and then key in the following data, entering one numeral per cell in consecutive cells of the same column, beginning in cell B3 and ending in cell B12:

   1, 2, 3, 1, 2, 4, 1, 1, 3, 2

3. Use the standard deviation function to find the standard deviation of the dataset. Type the following formula in cell B13:

   =STDEV(B3:B12)

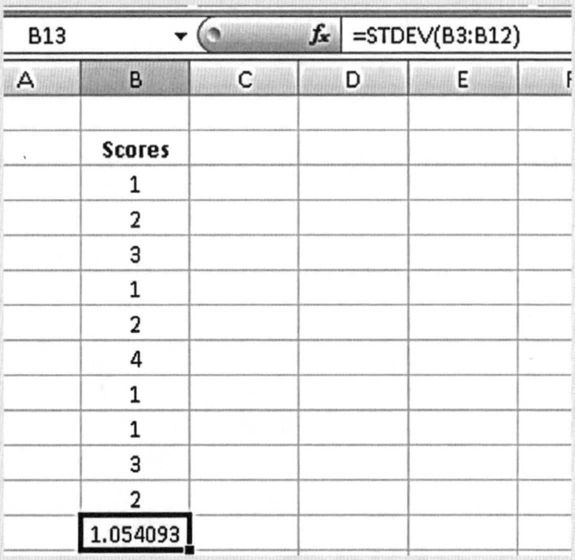

B13 | *fx* =STDEV(B3:B12)

A	B	C	D	E	F
	Scores				
	1				
	2				
	3				
	1				
	2				
	4				
	1				
	1				
	3				
	2				
	1.054093				

Figure 2.9 Standard deviation function in Excel.

4. Click out of cell B13, and the standard deviation value should appear in this cell: 1.054093.

Variance

Variance is a measure of the spread of observations in a distribution of data. It is equal to the square of the standard deviation, which we just examined. When stating the variance of a dataset, you are giving a measure of how closely aligned your expected value is to the distribution. It is basically a spread of the distribution and its average value.

When you have a large variance, individual values of random variables will tend to be farther from the mean. The smaller the variance, the closer the individual values and random variables tend to be to the

Hands-on Statistics 2.9: Use Excel to Find Variance

There are several variance functions in Excel, and we will use the simplest right now, the VAR function. However, note that there are others, and you will see them via the autocomplete feature of Excel when you go to insert a function.

To find the variance using Excel, see **Figure 2.10** and follow these directions:

1. Begin with a blank spreadsheet open in Excel.
2. Key in the following data, entering one numeral per cell in consecutive cells of the same row, beginning in cell B1 and ending in cell J1:

   3, 4, 3, 2, 6, 8, 9, 1, 3

3. Use the VAR function to find the variance of the dataset. Type the following formula in cell K1:

   =VAR(B1:J1)

4. Click out of cell K1, and the variance value should appear in this cell: 7.5.

fx =VAR(B1:J1)

B	C	D	E	F	G	H	I	J	K	L
3	4	3	2	6	8	9	1	3	7.5	

Figure 2.10 Variance function in Excel.

mean. You should also note that variance and standard deviation of random variables are always non-negative.

A sample variance is the sum of the squared deviations from their average divided by one less than the number of observations in the given sample, written as follows:

$$\text{Variance} = ((a - \text{mean})^2 + (b - \text{mean})^2 + (c - \text{mean})^2) / \text{total number of values} - 1$$

In this formula, a, b, and c are all values in the dataset, and mean is the average of all the values in the dataset.

Let's apply this formula to the dataset 10, 20, 30. The average of this dataset is 10 + 20 + 30 = 60 / 3 = 20. The equation for variance is:

$$((10 - 20)^2 + (20 - 20)^2 + (30 - 20)^2) / 3 - 1$$
$$= (-10^2 + 0^2 + 10^2) / 2$$
$$= (100 + 0 + 100) / 2$$
$$= 200/2 = 100$$

The final computation of variance for this dataset is 100.

This type of statistical information has many purposes and can be used in clinical trials of new prescription drugs or patient treatments, for example.

DID YOU KNOW? According to the *Washington Post*, the cost of an emergency department visit can vary in cost depending on what kind of medical insurance you have and where you seek treatment (Kliff, 2013). A trip to the emergency department can cost you 40% more than the average cost of rent. For example, for sprains and strains, the median cost was $1,051.00 (982.00–1,110), the mean was $1498.00 (1,304.00–1,692.00), and the interquartile range was $1,018. Charges for sprains and strains ranged from $4.00 to $24,110. Kidney stones ranked the highest in interquartile range of the top 10 diagnoses treated in the emergency department, with a median of $3,437.00 (2,917.00–3,877.00), mean of $4,247.00 (3,642.00–4,852.00), and interquartile range of $3,742.00. The cost of kidney stones ranged from $128.00 to $39,408. How do the charges for these conditions rank at your local hospital?

Now you know!

Hands-on Statistics 2.10: Examine Typical Hospital Data

Now we will use the knowledge just covered to examine a 6-hour time frame from a hypothetical hospital emergency department. Using Excel, see **Figure 2.11** and follow these directions:

1. Open the file "CH02_Hospital_Time_Frame.xls" located in Chapter 2 of the eBook. The data in that file should appear as in Figure 2.11.
2. From left to right, observe the data in the following columns. The *Medical Record Number* is just a number representing a patient. *Sex* is noted as 1 for male and 2 for female. *Age* is as you would assume. *Diagnosis Code* is an alphanumeric value for what the patient is diagnosed as having—for example, a cold. *Procedure Code* is a code for a specific procedure (such as an X-ray scan) that was performed on the patient. Lastly, *Ins. Code* (Insurance) is a numeric value, with 1 being Blue Cross, 2 being VA, 3 being Kaiser, etc. Now, considering the aforementioned items, let's examine some different functions and insert them into the spreadsheet to make sense of the raw data.
3. Use the MODE function to find the most frequently occurring sex and insurance payment type. Type the following in cell F51:

   =MODE(F4:F50)

   Then type the following in cell B51:

   =MODE(B4:B50)

   As you click out of these cells, the mode values will appear in them.
4. Use the standard deviation function on the age. Type the following formula in cell C51:

   =STDEV(C4:C50)

   As you click out of this cell, the standard deviation value should appear in it.

Next, find the range value for age by following these steps:

1. Use the max function to find the max value of the dataset. Type the following formula in cell C52:

   =MAX(C4:C50)

   Click out of cell C52, and the max should appear in this cell.
2. Use the min function to find the min value of the dataset. Type the following formula in cell C5:

   =MIN(C4:C50)

   Click out of cell C53, and the min should appear in this cell.
3. Use the subtract function to find the range of the dataset. Type the following formula in cell C54:

   =C52–C53

   Click out of cell C54, and the range should appear in this cell.

	A	B	C	D	E	F	G	H
1		Patient admittance data for 6 hr time frame for Cape of Good Hope Hospital						
2								
3	Medical Record Number	Sex	Age	Diagnosis Code	Procedure Code	Ins. Code		
4	112211	1	22	359.21		2		
5	112201	1	63	346.91	3.31	3		
6	112233	2	47	873.51	86.59	3		
7	112275	1	58	305.01		5		
8	112281	2	30	305.01		4		
9	112287	1	21	873.53	86.59	2		
10	112292	2	34	303.01		2		
11	112298	2	59	780.31		1		

Figure 2.11 Sample data from Cape of Good Hope Hospital.

For some items, such as Medical Record Number, there is no analysis to be performed, perhaps other than a count of the number of patients, performed as follows:

4. Use the Count function to find the number of medical records listed. Type the following formula in cell A51:

   =COUNT(A4:A50)

   Click out of cell A51, and the number of medical records should appear in this cell.

Data Harvesting

In the previous example, our data were already in an Excel spreadsheet. However, that will not always be the case. Raw data often come in different file types, such as a Word document, Excel spreadsheet, or image file such as JPEG. Raw data formats that are common include comma-separated values (CSV), in which each cell or data item is separated in the file with a comma, and tab-delimited values, in which

Hands-on Statistics 2.11: Harvest Real-World Data

In this example, we will examine hospital data using the Washington State Department of Health's Comprehensive Hospital Abstract Reporting System (CHARS). Follow these steps:

1. Visit the CHARS primary webpage at http://www.doh.wa.gov/DataandStatisticalReports/HealthcareinWashington/HospitalandPatientData/HospitalDischargeDataCHARS.aspx. Alternatively, you can enter the following phrase in a search engine: "Comprehensive Hospital Abstract Reporting System (CHARS) from the Washington State Department of Health."

   Scroll down the page and click on *Chars Reports*. You will have the option of downloading reports as either PDF (Adobe Acrobat) or Excel files.
2. Scroll down to the bottom of the page to the section titled *2015 Full Year Standard Reports Discharges: Inpatient / Observation*. Or use a newer year if desired.
3. In the subsection titled *Excel files*, click on the link *Hospital Census and Charges*. Save this file to an appropriate folder on your computer and then open it.
4. Find the min and max for discharges (total) for all hospitals. To do so, you will have to write a function that examines only certain cells in the spreadsheet. To get started for min, type "=MIN(c9,c12.....)" in an open cell at the bottom of the document, adding the cell numbers for each bold total for each hospital. Use the same approach to find max.

How Does Your Hospital Rate?

Dr. Barker also considers how Brunswick County compares with the rest of the counties in the state by using the County Health Rankings website. Dr. Barker finds that Brunswick County ranked 54th out of 100 in mortality, 23rd in morbidity, 47th in health behaviors, and 42th in social and economic factors. He finds that preventable hospital stays decreased from the year 2003 to 2010. He finally concludes that social and economic factors and health behaviors had the most influence as well as genetics for those with heart failure. Some of the questions that could be asked are: Are the patients taking their medications correctly, or at all? How are their eating habits?

Consider the following:

1. Visit the County Health Rankings website (http://www.countyhealthrankings.org/app/), and look up your own county. How does your county rate compare with the state average in mortality, morbidity, health behaviors, and social and economic factors?
2. In which area does your county receive its highest ranking? In which area does it receive its lowest ranking?
3. Which health behaviors in your county appear to be contributing to poor health?

each cell is separated by a uniform amount of space, such as a tab. There are other formats, including some that are specifically formatted for an Apple Mac or MS-DOS. Just choose the type that is most compatible with the system you are using.

That was a simple example of harvesting data, but taking a quick look at the spreadsheet, you will probably notice many other items worth examining. In later chapters we will dig deeper into the data. For now, though, let's step back and examine what the research process looks like and a few key considerations.

Global Perspective

Consider the following US healthcare statistics:

- The average life expectancy in the United States is 78.6 years according to the CDC for 2017 (CDC, 2018a).
- The infant mortality rate is 5.8 deaths per 1,000 live births (CDC, 2018a).
- In 2007, there were 96 preventable deaths per 100,000 people.
- In 2007, there were 2.4 physicians per 1,000 consumers.
- In 2017, 18% of the gross domestic product was related to health expenditures; this is more than double the average spent among other developed countries (Committee for a Responsible Federal Budget, 2018).

So how do we as a nation compare with other countries? The United States, as of 2020, spends more than any country on health care—in fact, disproportionately more when viewed graphically. An internet search for "How does health spending in the United States compare to other countries?" reveals very interesting data. When measured as a percentage of gross domestic product, the United States spends more than any other nation; yet US citizens have comparably poor health outcomes. In comparison, Japan spends much less on health care but has the world's lowest infant mortality rate (2.17 deaths per 1,000 live births) and a much higher life expectancy (82 years). Even though healthcare systems vary substantially from one country to another, the goal of these comparisons is to improve health care across the globe. That the United States has the most expensive healthcare system in the world there is no doubt.

Chapter Summary

This chapter covered the basic statistical measures of central tendency—mean, median, and mode—and explained their importance in calculating hospital statistics. It also discussed variability in data, or ways to examine how spread out items in a dataset are from the mean. In doing so, it applied two statistical standards: standard deviation and variance. Standard deviation, the square root of the variance, returned a measure of how far each value was from the computed mean of the dataset. Variance returned a similar value but also returned a value of zero if all values in the set were the same, a low positive value if they were similar, and a high value if they were dispersed and quite separated from the mean. Discrete data, such as a number of patients, were contrasted with continuous data, such as patient weight. The chapter also compared nominal (named) data, such as types of cars or gender, to ordinal (ordered) data, such as a scale of 1, 2, 3, etc. Next, among other key hospital statistics, the concept of outlier or skewed data—that is, values quite different from the majority of other values in the dataset—was examined. Last, but not least, the chapter introduced descriptive statistics, descriptive models, data harvesting, and many other meaningful tools used in healthcare statistics.

Apply Your Knowledge

1. Discussion: Describe in your own words how using the different types of statistical measures we have covered in this chapter would be of benefit to a physician's office or hospital to improve patient care and reimbursement methodologies.
2. Fifty children were diagnosed with leukemia during the past year. The weight of each child was recorded (in pounds) at the time of diagnosis. Listed are the weights from low (lightest) to high (heaviest): 20, 22, 23, 26, 27, 28, 28, 29, 30, 31, 31, 32, 33, 34, 35, 37, 38, 39, 40, 41, 42, 42, 43, 43, 44, 45, 47, 48, 48, 49, 49, 51, 52, 52, 52, 54, 55, 56, 58, 58, 58, 60, 61, 62, 63, 64, 65, 67, 68, 70. Use an Excel spreadsheet to complete the following:
   a. Calculate the mean, median, and mode.

   b. Calculate the variance and standard deviation computed from a frequency distribution with a class interval of 1.
   c. Calculate the weights that are one standard deviation above and one standard deviation below the mean.
   d. Calculate the weights that are two standard deviations above and two standard deviations below the mean.
3. Compute the mean for the following values *by hand*. Do not use Excel.

   5, 6, 7, 1, 2, 5, 6, 7, 1, 9, 33, 2, 9

4. Find the variance (by hand, not using Excel) for the following values. Show your work.

   300, 400, 150, 510, 430, 611

5. Use the Excel STDEV function to find the standard deviation for the following heights (cm) of 12 students in a class:

   170, 160, 165, 161, 163, 164, 170, 164, 163, 170, 166, 165

6. Use data about the ages of 80 nursing home residents, provided in **Table 2.1**, to complete the following:
   a. Calculate the mean, median, mode, variance, and standard deviation for the ages listed in the table.
   b. Prepare a frequency table in Excel that includes the ages listed in the table.
7. The lengths of hospital stay for patients with diverticulitis who had a partial bowel resection performed are recorded as follows. Calculate the mean, median, and mode using Excel.

   5, 7, 9, 4, 6, 5, 7, 3, 6, 6, 4, 5, 5, 7, 7, 3

Table 2.1

65	68	71	73	75	76	78	80
65	68	71	74	75	76	78	80
65	69	72	74	75	76	78	82
66	69	72	74	75	77	80	82
66	69	72	75	77	77	79	80
67	70	72	75	77	78	80	83
67	70	73	75	77	79	80	83
68	70	73	75	79	80	80	84
68	71	73	76	79	80	80	84
69	71	73	76	79	80	80	84

8. For this question, use data below in **Table 2.2**, in a spreadsheet. A given medical clinic takes in the specified amounts of money per day in the form of cash co-payments. Find the range and mode for this one day. Use the SUM function to find the day's total cash intake.

Table 2.2

Patient Number	Payments by Patient
1	25
2	10
3	250
4	75
5	35
6	25
7	150
8	100
9	25
10	0
11	25
12	25
13	25
14	250
15	100
16	150
17	75
18	50
19	35
20	50
21	75
22	15
23	35
24	25
25	50
26	150
27	250
28	25
29	25
30	25
31	25

References

Agresti, A., & Finlay, B. (1997). *Statistical methods for the social sciences*. (3rd ed.). Upper Saddle, NJ: Prentice-Hall.

Batten, J. (1986). *Research in education* (rev. ed.). Greenville, NC: Morgan Printers.

Centers for Disease Control and Prevention. (2018a). Mortality in the United States, 2017. Retrieved June 2018, from https://www.cdc.gov/nchs/products/databriefs/db328.htm

Centers for Disease Control and Prevention. (2018b). Updated national birth prevalence estimates for selected birth defects in the United States, 2004–2006. Retrieved June 2018, from https://www.cdc.gov/ncbddd/birthdefects/features/birthdefects-keyfindings.html

Committee for a Responsible Federal Budget. (2018, May 6). American health care: Health spending and the federal budget. Retrieved August 2019, from https://www.crfb.org/papers/american-health-care-health-spending-and-federal-budget

Kliff, S. (2013, March 2). An average ER visit costs more than an average month's rent. *Washington Post*. Retrieved May 2014, from http://www.washingtonpost.com/blogs/wonkblog/wp/2013/03/02/an-average-er-visit-costs-more-than-an-average-months-rent/

Web Links

Discrete/Continuous: http://www.chegg.com/homework-help/definitions/discrete-continuous-31

Descriptive Statistics: http://www.businessdictionary.com/definition/descriptive-statistics.html

MS Excel: How to Use the QUARTILE Function (WS): http://www.techonthenet.com/excel/formulas/quartile.php

Excel and Quartiles: http://www.meadinkent.co.uk/excel-quartiles.htm

Drawing a Normal Curve: http://www.tushar-mehta.com/excel/charts/normal_distribution/

Interactive Atlas of Heart Disease and Stroke: http://nccd.cdc.gov/DHDSPAtlas/reports.aspx?state=NC&themeId=13#report

Data and Statistics on Birth Defects: http://www.cdc.gov/ncbddd/birthdefects/data.html

An Average ER Visit Costs More Than an Average Month's Rent: http://www.washingtonpost.com/blogs/wonkblog/wp/2013/03/02/an-average-er-visit-costs-more-than-an-average-months-rent/

Life Expectancy: http://www.cdc.gov/nchs/fastats/life-expectancy.htm#

Relative to the Size of Its Wealth, the US Spends a Disproportionate Amount of Health Care: https://www.healthsystemtracker.org/chart-collection/health-spending-u-s-compare-countries/#item-start

CHAPTER 3

Patient Data

He who builds on the people builds on mud.

—Niccolò Machiavelli

CHAPTER OUTLINE

Introduction
Census Data and Their Importance
Calculation and Reporting of Patient Census Data
- Inpatient Service Days
- Average Daily Census
- Data Visualization
- Visually Examine Data With Sparklines and Microcharts
- Newborn Services

Open-Source Software
Freeware and Shareware
Types of Databases
- Flat-File Database
- Relational Database

Data Formats
R-Project
Data Stored in R
Global Perspective
Chapter Summary
Apply Your Knowledge
References
Web Links

LEARNING OUTCOMES

After completing this chapter, you should be able to do the following:

1. Describe census data and their use in health care.
2. Describe the admissions, discharge, and transfer report and explain how to calculate it.
3. Define inpatient service days, describe this statistic's importance, and explain how to calculate it.
4. Define average daily census and explain how to calculate it.
5. Create sparklines and microcharts to present data.
6. Create a line chart to present data.
7. Define and explain the differences between open-source software and freeware/shareware.
8. Describe types of databases.
9. Import raw data from healthcare websites into Excel.
10. Describe data formats.
11. Use R-Project for basic statistical analysis.

KEY TERMS

Admissions, discharge, and transfer (ADT)
Average daily census
Binary code
Census data
Code
Comma-separated values (CSV)
Commercial code
Flat-file database
General Public License (GPL)
GNU's not Unix (GNU)
Infographics
Inpatient service days
Intensive care unit (ICU)
Language
Matrix
Microchart
Open-source software
Patient care unit (PCU)
Recapitulation algorithm
Relational database
Scalar
Script-mode interface
Source code
Sparklines
Sunset
Tab-delimited values
Vector

How Does Your Hospital Rate?

Veronica is a nurse at Hospital A, a 25-bed licensed facility located on the southeast coastal region in North Carolina (**Figure 3.1**). She has been assigned the task of calculating the hospital's rates of admissions, discharges, and transfers.

Consider the following:

1. What uses might hospital administrators have for this information?
2. What do these measures indicate about the operation of this hospital?
3. Where might Veronica find this information?

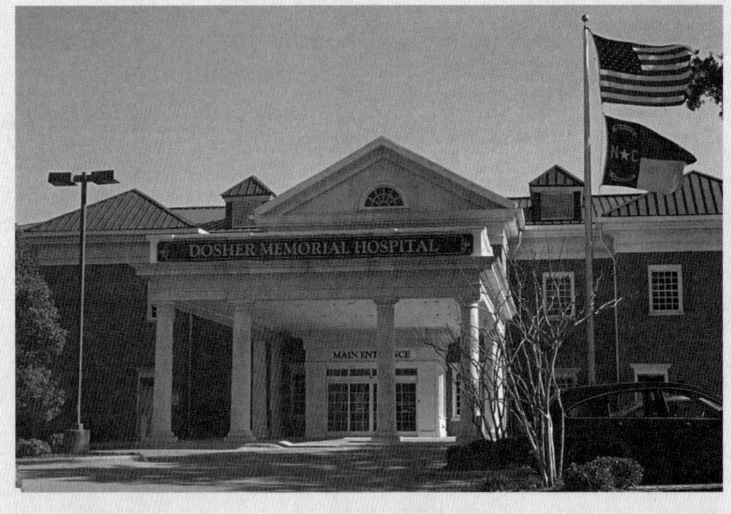

Figure 3.1 Dosher Memorial Hospital, Southport, North Carolina.
Photo by J. Burton Browning.

Introduction

The **admissions, discharges, and transfer (ADT)** list aids hospitals in calculating how many patients were admitted, discharged, and transferred each month and provides yearly totals for each patient care unit in the facility. This information is useful for forecasting staffing issues, implementing new services, and reviewing the age and sex of the population. Average length of stay is another important statistic that may be viewed. The average length of stay can reveal trends of patients with certain diagnoses and procedures. For example, imagine that eight patients come in with pneumonia; three of these patients go home in 3 days, and the other five stay 6 days. The medical record could be reviewed to determine why those five patients had a longer length of stay. The transfer part of the ADT indicates how many patients were transferred out and why. Based on this information, facility administrators should consider questions such as the following: Do we provide the services that patients needed? Were we lacking specialists for certain procedures?

Accumulating, analyzing, and reporting facility data not only help in many ways with daily operation of the hospital but also provide benchmarking and forecasting information important to hospital administration. In this chapter you will learn how to calculate these data, ways to report them, and other information related to patient data. Specifically, you will review census data, ADT reports, inpatient service days, and calculations for these vital hospital data. Next you will learn how to present statistical data using tools such as sparklines and microcharts. Data formats and data mining will be examined, as well, building on topics from the previous chapters. You will also gather statistical data from the World Health Organization (WHO), compare data on diseases, and present the data visually.

Census Data and Their Importance

A census is a regular, ongoing count of the people who make up a population. Besides the US national census that is conducted every 10 years, many smaller censuses are routinely conducted by organizations throughout the country. In the hospital setting, **census data** are data that describe patients who are currently in the hospital during a certain time frame. Each **patient care unit (PCU)** reports the number of patients admitted that day, discharged, or transferred in or out, perhaps to another hospital or unit. Note that these data include only *inpatients*, who are the patients who have been admitted to the healthcare facility for an overnight or longer stay.

The statistical data provided from the hospital census is important, as it affects patient care and operational efficiency of the hospital and helps determine financial performance. These data are used by hospitals to make decisions regarding staffing, budgeting, and planning for the future. Census data can also be used to keep track of patients' age, ethnicity, and management of chronic conditions. All of these data are invaluable in longitudinal studies the facility conducts. For example, if over a 10-year time frame you determined that the patient rate for retirees was increasing at a 5% rate every year, you might determine that specialized services this patient group requires should be expanded, so you would plan for future growth expenditures in this area. As another example, if you determined that there was no steady increase in child care services, then you would not want to direct financial expansion monies to increasing pediatric services. Most importantly to consumers, however, is that this information can be used to improve patient care.

Haley and Bregman in 1981 noted that staffing decisions, based on census data, had an effect on the spread of infectious diseases among neonates. When the infant-to-nurse ratio was lower, there was a lower rate of infection among patients. This observation, made over 30 years ago, holds true today. Georgetown University School of Nursing and Health Studies in 2008 provided research results on lower staffing levels and the increased risk to patients of contracting nosocomial infections—that is, infections originating in the healthcare facility (Cronin, Leo, & McCleary, 2008). More recently, a culmination of findings from many research studies has reaffirmed this finding (Mitchell, Gardner, Stone, Hall, & Pogorzelska-Maziarz, 2018). The conclusion was that staffing shortages increased the risk of HAIs.

According to the Centers for Disease Control and Prevention (CDC), 90,000 of the 1.7 million people who acquired a nosocomial infection in 2002 would die (Klevens et al., 2007). It is probably no surprise that nosocomial infections not only affect the patient, but the whole healthcare system. Nosocomial infections increase medical costs of patients each year by $4.5 billion to $5.7 billion. The Institute of Medicine reported that 98,000 deaths occurred unnecessarily, in part related to nursing shortages (Kohn, Corrigan, & Donaldson, 2000). The study noted that when the nurse-to-patient ratio declined, the amount of time spent with the patient decreased, leading to an increase in nosocomial infections among patients. Without accurate census data, these important findings may never have been appreciated.

Regarding the census of patients, to conduct a census in the hospital, each inpatient care area counts the number of patients who are in that area or unit each day. Examples of inpatient care areas include oncology, pediatrics, and surgical services, with the number and types of units in a given hospital depending on the size of the facility. Note that the timing of when a census is taken is important. Most hospitals will take the census count at midnight because most patients are asleep then. If it is done at any another time, patients could be undergoing surgery, having x-ray scans taken, or otherwise occupied. The time the census is taken needs to always be consistent, though, regardless of when it is done.

Before computer use became commonplace, census reporting and analysis were done by hand. However, as hospital size and admissions increased, computer use became a necessity. As a result, it is now much quicker and easier to view report data and find errors. With the computer it is easy to see when patients have been transferred to another location, the date of transfer, the date of admission, and the date of discharge, allowing the census taker to efficiently follow the patient during his or her admission to the facility. It may seem strange that computers have not always been used for such tasks, but in some facilities, their integration into healthcare management did not begin until the 1980s. Moreover, other emerging technologies, such as radio frequency identification (RFID), use of barcodes, wireless technologies, and integrated systems, are also helping improve patient care in many ways.

Ashar and Ferriter noted in 2007 that "two general categories are being established for using RFID technology in health care settings." One category is based on retail use of RFID for inventory control of drugs and devices. The second category relating to RFID use involves the capture of streaming data related to patient point of care. Likewise, using barcodes to track patients reduces errors and ensures that correct patient information has been entered in the system. This system can also be used to allocate beds and bed transfers in real time, making it easier to move patients within the hospital. Wireless technologies are increasing the use of mobile devices in the delivery of health care. Mobile apps are now available that will check various vital signs, including electrocardiographic signals and blood glucose levels. Some other advancements in wireless technologies allow the devices to be extremely compact and portable, reaching millions of people in third-world countries as well as many rural parts of the United States, affording them healthcare benefits associated with RFID technology. Such devices include a stethoscope that transmits heart sounds to a physician allowing him or her to make either a diagnosis or order tests without being present. With the growing use of this technology, in the future we may not have to visit a doctor's office in person as often as before such technologies were invented.

Many integrated systems assist in the administration and effective delivery of patient care. Current systems such as those from Novant, Athena Health, and NextGen, and versions of the Veterans Information Systems and Technology Architecture (VistA) used by the US Department of Veterans Affairs, are integrated throughout the facility and can keep track of all needed information from the time of patient admission to discharge in a database known as the master patient index. This index contains data on bed tracking, transfers, discharges, and many other features. When all of the information has been put into the computer system, an ADT list is generated, showing which patients are still categorized as inpatients in the facility and which have been either discharged or transferred (**Figure 3.2**). Certainly all of these technologies, including the use of security controls, will only grow in use by medical facilities.

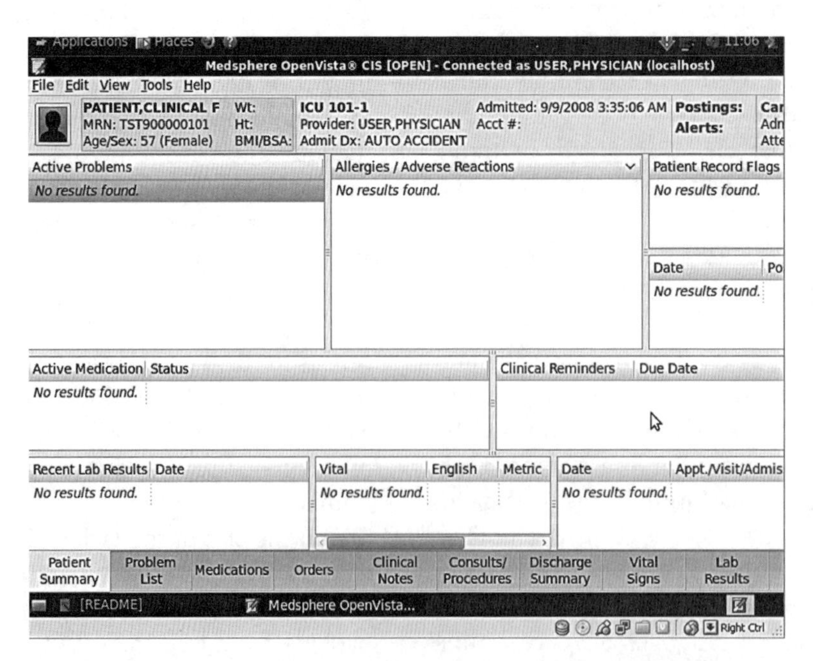

Figure 3.2 The Veterans Information Systems and Technology Architecture (VistA) used by the US Department of Veterans Affairs.

Although these integrated software systems, which are commonly in use at all medical facilities, have effectively automated the ADT list, it is still important for a professional in this field to be able to compute it by hand. In fact, it is critical to understand the statistical processes behind what the automated systems produce because it is the only way we can check behind the system, perhaps detect gross anomalies, and really understand what is taking place.

Therefore, in this chapter and throughout the text, you will learn how to calculate healthcare statistics by hand using formulas and software tools to assist in visualizing the results of your findings.

Calculation and Reporting of Patient Census Data

Two types of patient data that you will need to learn to calculate and report are inpatient service days and the average daily census. Note that although outpatient data are typically not gathered during a census, this information must still be captured so that it can be included in the census report. For instance, consider that you have a patient who was admitted at 9:30 am on June 5 and discharged at 8:30 pm on June 5, the only outpatient in your unit on this day. This patient would not be included in a census taken at 12 midnight on June 5, as he would already be gone. However, he would still need to be counted in the census report. Thus, if you count 45 patients in your care unit during the census, then you would just add the one outpatient to this number, making the total number 46.

Inpatient Service Days

Inpatient service days are a way to measure services that a patient receives within a 24-hour time frame. Other terms for inpatient service day are *patient day*, *census*, *occupancy day*, or *inpatient day*. The *total* inpatient service days refers to all of the inpatient service days for a given period. Keep in mind that physicians and the hospital must ensure that the patient meets severity of illness and medical necessity criteria before admitting the patient. As mentioned, data regarding any outpatients must be added to the inpatient service days and counted on the ADT report. Keep the following points in mind when calculating inpatient service days:

- The reporting period for inpatient service days begins at 12:01 am and ends at midnight.
- A leave of absence occurs when a patient who has been admitted but not yet discharged is not at the facility at the census-taking hour. Any absence of less than 1 day does not constitute a leave of absence. Normally, leaves of absence are not an everyday occurrence due to shorter length of stays in the healthcare facility. Leaves of absence are more common in long-term care and rehabilitation clinics and for those who are developmentally disabled. Note that patients are now allowed to leave without a physician's order stating the patient has permission to leave or return. If the administrators of a facility do not recognize a leave of absence, they may choose to discharge the patient and then readmit the patient after a few days.
- The day of discharge is not counted for inpatient service days, but the admission date is counted.
- Service days are not reported as fractions or divided for a unit of service.
- If a patient is admitted at 3:00 pm and dies at 7:00 pm, that patient will not be counted in the inpatient service days.
- Administrators of facilities will choose how they want to count their inpatient service days, which could be monthly, quarterly, semiannually, or annually.
- Some facilities may choose to exclude obstetrical and intensive care units due to the moderation of services administered.

One important use of inpatient service days is to measure how well the facility is performing year-to-date and do a comparison with the previous year's performance (**Figure 3.3**). This measure also assists the hospitals and physicians in determining the quality of medical care they are providing each patient and also delivering information about the overall health of patients.

	12:01 a.m census		ADM		trf		Total	DIS	DIS	trf	11:59 p.m. census				serv days
Day	A/C	Nb	A/C	b	in	A/C	Nb	A/C	Nb	out	A/C	Nb	a/d	A/C	Nb
7-Dec	50	5	4	2	2	56	7	3	1	2	51	6	3	54	6

Figure 3.3 Example of hospital census. *Abbreviations*: A/C, adults and children; ADM, admissions; DIS, discharges; Nb, newborns; trf, transfers.

Hands-on Statistics 3.1: Calculating and Reporting Inpatient Service

Now we will begin our calculations for inpatient census:

1. The starting point, which was the last census, is 50 adults and children (A/C) and five 5 newborns (Nb).
2. Next, the current census is as follows: 50 A/C + 4 admissions + 2 transfers in = 56 A/C.
3. For Nb, the current census would be: 5 Nb + 2 births = 7 Nb.
4. Next, subtract from the census of A/C discharges and transfers out: 56 – 3 discharges (A/C) – 2 transfers out = 51 A/C.
5. Lastly, subtract from the census of Nb discharges: 7 – 1 discharge (Nb) = 6 Nb.

Let's review some topics for clarification:

- The terms *transfer in* and *transfer out* refer to changes within the hospital only. For example, a patient may be admitted to the surgical floor but the next day transferred to the **intensive care unit (ICU)**, the area of the hospital reserved for patients with severe illnesses or injuries that require constant monitoring. This simply means that the patient was transferred off of the surgical floor and into the ICU. These transfers will be included in the ADT report. Alternatively, the physician may say the patient was "transferred," meaning the patient went to a rehabilitation facility. In this case, the patient is actually discharged from the facility but would still be counted in the ADT report. If the transfers are not counted, the data will be out of balance and a particular care unit will fail to report the transfers in, transfers out, or discharges correctly. Transfers in and out of the PCU will not always be equal, but they will be equal with the overall recapitulation. A **recapitulation algorithm** is one that verifies the data either monthly or yearly, which means it provides a summary of the data collected for a given period.
- For census purposes, newborns are counted separately. However, this is an administrative decision by medical staff or others who would be using the statistical data. Any birth in the facility is considered a newborn admission.
- Another question that might be asked regarding the inpatient census is, how many are in the house? The remainder of patients that are still admitted in the hospital after midnight or as of 11:59 pm are considered "in house."

Average Daily Census

Average daily census is the average number of patients in a facility on a given day. It provides the administration with information on a specific unit: Are additional beds or services needed? Should new services be added for patient care? It informs even large-scale questions like, is construction of a particular care unit required? Staffing and budgeting of supplies and equipment are affected by average daily census results.

For each PCU in the healthcare facility, the following formula is used to calculate the average daily census:

$$\frac{\text{Total inpatient service days for a period (excluding newborns)}}{\text{Total number of days in the period}}$$

To calculate the census for a month, you need to know the number of days in each of the 12 months (and don't forget the leap year). Remember that adults and children are counted separately from newborns unless your facility directs you otherwise. For example, consider the following: Ocean View Hospital had a total of 5,321 inpatient service days for the month of January. Divide 5,321 by 31 days in January: 5,321/31 = 171.6. Then round the answer up: 172.

Hands-on Statistics 3.2: Oceanside Hospital Case Study of Average Daily Census

Try your hand at calculating the average daily census in the following exercise:

1. Oceanside Hospital has a 20-bed ICU and has 635 inpatient service days for the month of October.
2. Divide 635 inpatient service days by 31 days in the month of October to get 20.4, which you will round to 20.
3. The census shows that the ICU is filled to capacity each day of the month since you only have 20 beds.
4. Administration will use this information to determine whether to add additional beds in that unit.

That was fun to work out by hand, but almost all computations are now done on a computer. So, now we will work out an average daily census with a spreadsheet application.

Hands-on Statistics 3.3: Southport Hospital Case Study of Average Daily Census Using Excel

To calculate the average daily census for Southport Hospital using Excel, examine **Figure 3.4** and follow these directions:

1. Open a new, blank spreadsheet in Excel.
2. In a cell in row 2 cell, type "**Average Daily Census Southport Hospital**" as a title in bold and 12-point font.
3. Type "Inpatient service days" in cell A4, "Month" in cell B4, "Days in Month" in cell C4, and "Max number of beds = 20" in cell D4, as in Figure 3.4.
4. Type "635" under *Inpatient service days* in cell A5, "October" under *Month* in cell B5, "31" under *Days in Month* in cell C5, and the formula "=A5/C5" under *Max number of beds = 20* in cell D5.
5. Right click on D5, select *Format cells*, then *Number*, and then change the decimal places to zero (0), as you will not have a fractional bed available for use.
6. When you click out of cell D5, the average daily census should appear: 20.

Save this document, as you will add to these data in the next exercise.

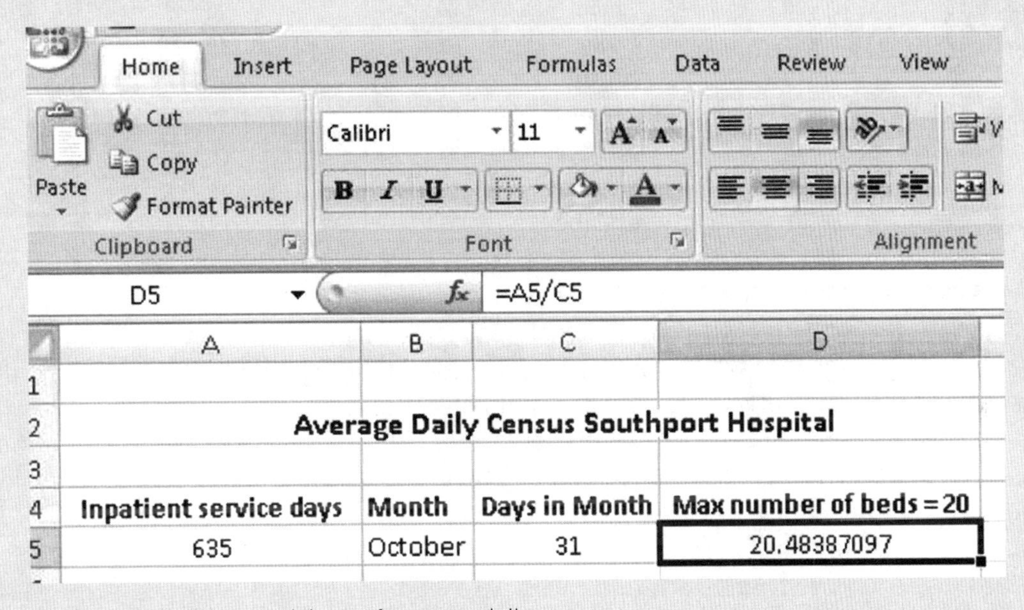

Figure 3.4 Excel spreadsheet of average daily census.

Data Visualization

Data visualization lets us make better sense visually of complex data. We live in a society in which images are of paramount importance, so it stands to reason we should try to present our data graphically where it is appropriate to do so. David McCandless gives an eloquent description of this in his TED Talks video *The Beauty of Data Visualization*. The old proverb "A picture is worth a thousand words" certainly applies to data visualization; however, in a clinical or scholarly setting, where data are being summarized and reported, both words and graphics work in harmony to represent culminated data accurately to consumers.

To properly choose the right data visualization tool, you must first understand the audience you will be presenting to, as real- world data can be complex and overwhelming for the average person. There are three steps for providing great data visualization: First, know your target audience. Second, make a clear framework for the information you are displaying. Third, make sure it tells a story. Some of the more common forms of data visualization are line graphs, flowcharts, diagrams, pie charts, bar graphs, infographics, maps, cluster charts, and word clouds. These types of data visualization are used in large corporations and all other fields of business, including health care.

Hands-on Statistics 3.4: Chart Average Daily Census Data Using Excel

In this example, we will change a previous example of Southport Hospital to have 12 months of data. As each month has a differing number of days, we will adjust for that. Examine **Figure 3.5** and follow these directions:

	A	B	C	D
1				
2	**Average Daily Census Southport Hospital**			
3				
4	**Inpatient service days**	**Month**	**Days in Month**	**Max number of beds=20**
5				
6	600	January	31	19
7	615	February	28	22
8	635	March	31	20
9	611	April	30	20
10	590	May	31	19
11	600	June	30	20
12	601	July	31	19
13	612	August	31	20
14	613	September	30	20
15	635	October	31	20
16	623	November	30	21
17	601	December	31	19

Figure 3.5 Average daily census: Step 1.

1. With the Excel spreadsheet you created in the previous exercise open, edit it and add the data shown in Figure 3.5 so that you have 12 full months of data.
2. Next, highlight cells B6 through B17, then hold the CTRL key down and also highlight cells D6 through D17. Select Insert, Bar Chart, and 2D Bar. You should now have a chart showing the average daily census for each month. Note **Figure 3.6**.

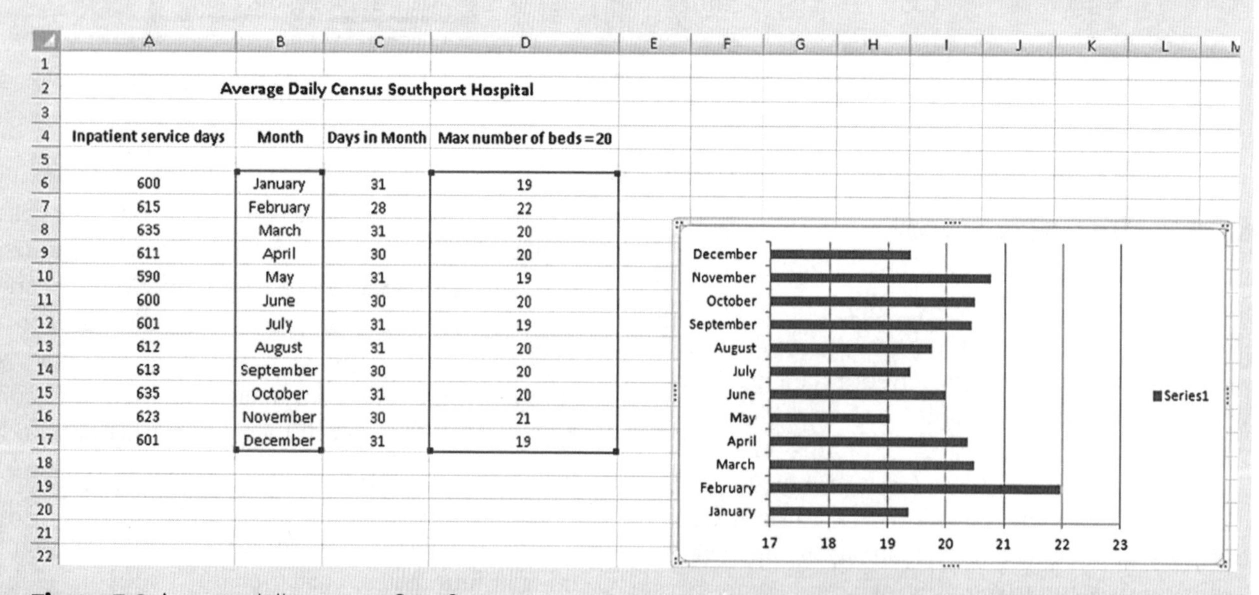

	A	B	C	D
1				
2	**Average Daily Census Southport Hospital**			
3				
4	**Inpatient service days**	**Month**	**Days in Month**	**Max number of beds=20**
5				
6	600	January	31	19
7	615	February	28	22
8	635	March	31	20
9	611	April	30	20
10	590	May	31	19
11	600	June	30	20
12	601	July	31	19
13	612	August	31	20
14	613	September	30	20
15	635	October	31	20
16	623	November	30	21
17	601	December	31	19

Figure 3.6 Average daily census: Step 2.

? DID YOU KNOW? Since 45 BCE, and Julius Caesar's Julian Calendar, the number of days in each month has stayed the same. Many school children learn the days of each month with the following rhyme: 30 days hath September, April, June, and November, all the rest have 31, excepting February alone, and that has 28 days clear, with 29 in each leap year.

Now you know!

You now can visualize the usage by month for a calendar year. It might be handy to be able to see whether there are any trends that can be noticed over several years' time frame. In the next Hands-on Statistics exercise, let's assume you have computed the census by month for 3 years. You will learn how to use a line chart to examine whether there are any patterns that can be derived from the data that might not be as noticeable by just looking at the raw numbers.

Hands-on Statistics 3.5: Three-Year Census Data Represented by a Line Chart Using Excel

1. Create in Excel a spreadsheet like the one shown in **Figure 3.7**.

Inpatient Service Days			
Oceanside Hospital	**85 Bed Hospital**		
	Year 2010	**Year 2011**	**Year 2012**
Jan.	1823	1871	1856
Feb.	1856	1858	1839
March	1844	1851	1801
April	1855	1822	1818
May	1821	1830	1825
June	1833	1834	1844
July	1849	1829	1836
August	1837	1811	1827
September	1858	1826	1843
October	1844	1802	1826
November	1816	1799	1800
December	1841	1812	1822

Figure 3.7 Creating a line chart in Excel: Oceanside Hospital raw data.

2. Next, highlight all data, as shown in **Figure 3.8**.
3. Lastly, select *Insert*, *Line*, and then *Line with Markers*, as shown in **Figure 3.9**.
4. Note that there are definite recurring patterns over 3 years in September through December, as shown in **Figure 3.10**.

(continues)

Inpatient Service Days			
Oceanside Hospital	85 Bed Hospital		
	Year 2010	Year 2011	Year 2012
Jan.	1823	1871	1856
Feb.	1856	1858	1839
March	1844	1851	1801
April	1855	1822	1818
May	1821	1830	1825
June	1833	1834	1844
July	1849	1829	1836
August	1837	1811	1827
September	1858	1826	1843
October	1844	1802	1826
November	1816	1799	1800
December	1841	1812	1822

Figure 3.8 Creating a line chart in Excel: Highlighting data.

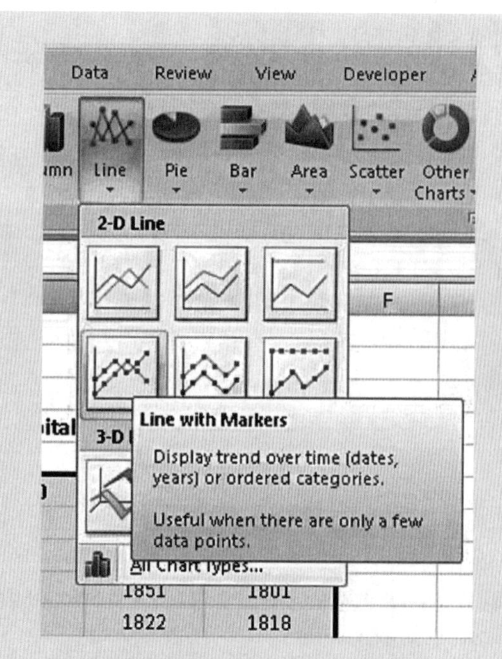

Figure 3.9 Creating a line chart in Excel: Choosing line with markers.

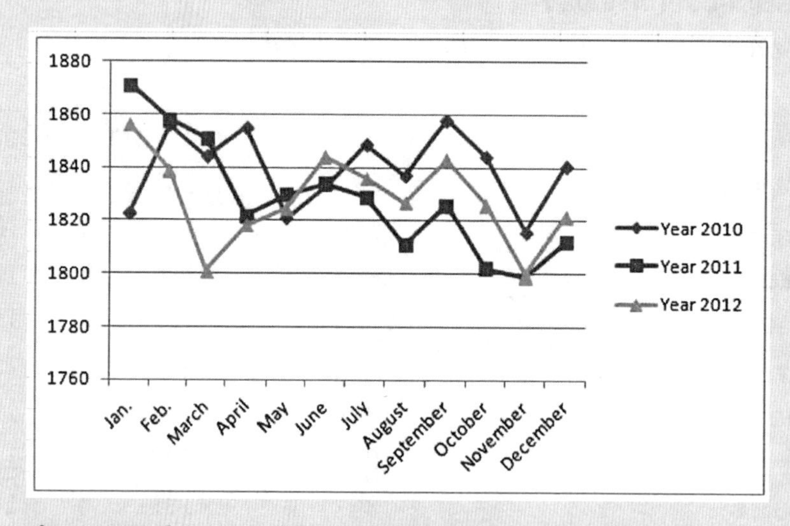

Figure 3.10 Creating a line chart in Excel: Line chart with markers.

Visually Examine Data With Sparklines and Microcharts

We have examined census data for each month as a total for each month. However, what if you wanted to examine census data for every day total in the month? Perhaps you would notice some trends, such as usage increasing typically on Saturdays or Mondays? That might be useful information to justify adding another staff member on those days. Remember, administration as a general rule likes decisions that are driven by data, and this would be one way you could make data-driven staffing decisions.

Sparklines and microcharts are examples of **infographics**, or graphical representations or charts of data or knowledge, in this case of cells with numeric data; think of them as an electroencephalographic readout of your data. Edward Tufte invented **sparklines** as a way to display data in small graphics. He referred to them as "intense, simple, word-sized graphics," which is a great way to think about them.

If you have examined any popular stock market site, you might also notice the use of this tiny but powerful infographics tool.

Microcharts, which are known by several different terms, and which are the subject of a patent dispute involving Microsoft over ownership, are similar graphic representations of data. Other terms you may hear for this tool include *in-cell micro charts* and *microcharting*. Try the next Hands-on exercise to see how easy they both are to use.

Hands-on Statistics 3.6: Average Daily Census for Each Day in a Month With Sparklines and Microcharts Using Excel

Practice creating microcharts and a sparkline for a month's worth of average daily census data. For microcharting, we use the rept function. The formula for rept is as follows:

=REPT("character",# times to repeat)

In the following example, a repeating character will represent the count for the census for each day, noted in a cell next to the numeric value visually represented by the character.

Before beginning the exercise, download and install the proper version (based on your version of Office) of Sparklines from the following website: http://sparklines-excel.blogspot.com/. The software is free. Enable macros when asked. Examine **Figures 3.11** through **3.13** and follow these directions:

1. Open a new, blank spreadsheet in Excel.
2. Type in the title "January 2010 Daily Census Report" in cell A4, "Day" in cell B5, and "Census" in cell C5, as in Figure 3.11.

Clipboard | Font | Alignment | Number

D6 =REPT("*",C6)

	A	B	C	D
1				
2				
3				
4		January 2010 Daily Census Report		
5		Day	Census	
6		1	55	***
7		2	62	**
8		3	60	**
9		4	66	**
10		5	60	**
11		6	56	**
12		7	51	***
13		8	56	**
14		9	63	***
15		10	55	***
16		11	61	***
17		12	56	**
18		13	55	***
19		14	63	***
20		15	57	***
21		16	53	***
22		17	62	**
23		18	61	***
24		19	64	**
25		20	62	**
26		21	57	***
27		22	60	**
28		23	56	**
29		24	61	***
30		25	62	**
31		26	60	**
32		27	58	**
33		28	62	**
34		29	59	***
35		30	57	***
36		31	53	***

Figure 3.11 Microcharts.

(continues)

3. Type in the dates for January 2010 under *"Day"* in column B and the census number for each date in column C, as indicated in Figure 3.11.
4. Type the formula "=REPT("*", C6)" in cell D6.
5. To duplicate this formula for the remaining days, highlight cell D6 and drag its contents down to cell D36.

Now that you have a spreadsheet like what you see in Figure 3.11, add a Sparkline for the month.

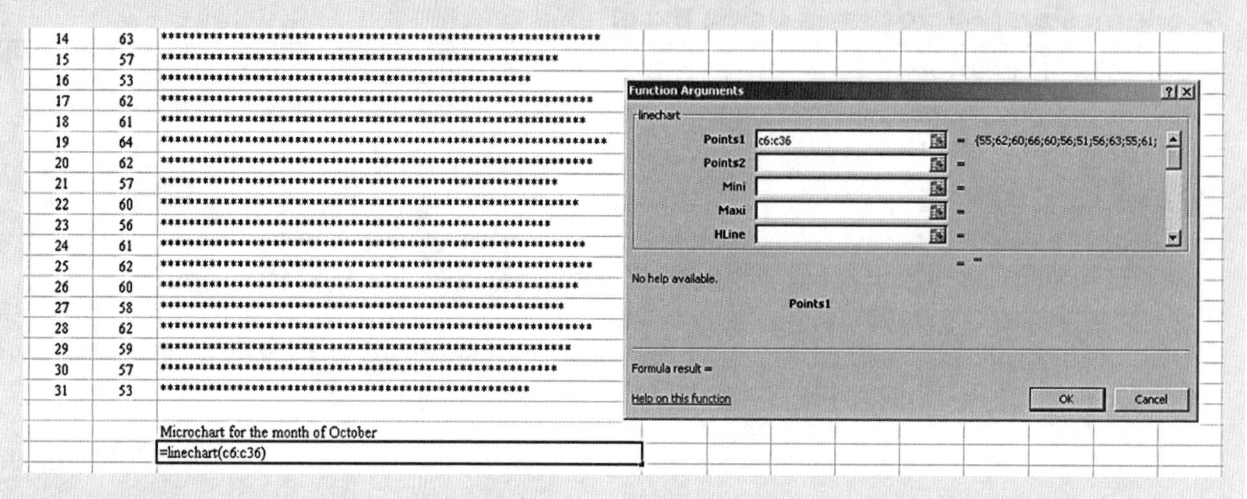

Figure 3.12 Sparkline: Selecting range.

6. Type "Sparkline for the month of January" in cell D38.
7. Click in cell D39, and then go to the *Add-Ins* menu; select *Sparklines* and then *Line chart*. Note that Microsoft often moves items around in different versions of their software, so depending on which version of Excel you have, you may have a different location for certain features.
8. Key in "c6:c36" in the *Points1* field in the pop-up box that appears, as that is your data range to chart, as shown in Figure 3.12.
9. Click *OK* for this output and you are done! (See Figure 3.13.)

27	58	**
28	62	**
29	59	***
30	57	***
31	53	***
		Microchart for the month of October

Figure 3.13 Sparkline: Output.

Now you have a microchart for each census day count, as well as a sparkline for the whole month of January.

Newborn Services

Newborn services should be counted separately from other services due to the unique differences in what is provided. Average daily newborn census is calculated with the same general process as that used for the average daily census for adults and children. The formula is as follows:

$$\frac{\text{Total newborn inpatient service days for a period}}{\text{Total number of days in the period}}$$

Hands-on Statistics 3.7: Oceanview Hospital Case Study of Average Daily Newborn Census

Practice calculating an average daily newborn census by following these directions:

1. Oceanview Hospital has 85 bassinets and 925 newborn inpatient service days in the month of October.
2. Divide the total number of inpatient service days by the number of days in the month (31 days): 925/31 = 30 (round up because you cannot have a fractional baby!).

Note, to perform this calculation using Excel, just insert a division formula similar to the one you used in the earlier example in this chapter: =cell#/cell#. For example, use "=A5/C5" (assuming you had data in each cell). Also, remember to right click on the formula cell when done, select Format and Number, and select no decimal places, as you do not want fractional babies!

Open-Source Software

Now we will examine some options for commercial software, data tools, and data formats applicable to healthcare statistics studies. **Open-source software** is software for which both the source code and binary code are given away freely. **Code**, in this context, refers not to a password code or some form of identification but to computer programming code, or a computer **language**, such as C++, BASIC, or Java. **Source code**, or precompiled code, is the form of a program as it was written by a programmer in a computer language that is readable by humans. **Binary code**—also known as the executable program, machine language file, or compiled code—is the form of the program that users download, install, and run. Usually, open-source software is covered under a **General Public License (GPL)** such that users are free to modify the original source code (if they are computer programmers) and give it away as long as they give it away for free and reference the original author's work. The global movement for open-source software is very strong, as indicated by the popularity of such organizations as the Free Software Foundation, Creative Commons, and others.

The advantage of open-source software is that, if you are a programmer, you can change or improve the source code of the software for your own purposes, in addition to just using it as-is. Thus, there are new releases of better code fairly often. Regardless of how you use it, however, you typically would not have to pay a fee to use the software, as you do for commercial applications such as those from Microsoft. In fact, when you install such software, it will most likely have a requirement that you will share the software you download and not convert it to closed-source or **commercial code**, which is code that is owned by a company and that is not free to share with others. An example of such a license is from the Free Software Foundation titled "GNU AFFERO GENERAL PUBLIC LICENSE." GNU stands for **GNU's not Unix** and is an open-source version of the Unix operating system. However, it has expanded over the years to relate to applications and not just operating systems.

In contrast to open-source software, commercial software, such as most office suite software, is not free to share. Office suite software is a set of bundled software applications that a typical business or office user would likely need, usually including a word processor, a spreadsheet, a presentation package, an image editing software, and a database application. Besides the familiar Microsoft Office, this category of software includes OpenOffice and LibreOffice, to name just a couple. Most are compatible with each other and can be used to open, edit, and save files from other office suite applications. So, just because you created something in MS Word 2007 does not mean you cannot open it in another software package that supports it.

So, why would one choose open-source software instead of commercial software, such as Microsoft Office? Price is certainly a major reason. There is a trend with bigger software companies to force users to upgrade to newer versions (for a fee) or lose technical support and features. Often these upgrades can cost tens of thousands of dollars or more. Planned obsolescence is another factor. How many times have you just purchased a version of Microsoft Office only to find a new version just came out? Or perhaps the version you have at home is not compatible with what you use at work or that your friends use. As such, when companies **sunset**, or discontinue, a product in this fashion, forcing an upgrade or replacement, users may determine that free and open-source software is a better choice, as they can have greater control of when they upgrade their software.

Consistency is another factor. For instance, OpenOffice has been found to be more consistent in terms of the layout of the interface and interchangeability of files than Microsoft Office. Depending on the version of MS Office you have, if you save in the latest version, older versions of Word cannot open the file without the addition of special import converters. OpenOffice has avoided this issue, keeping the same basic file format. MS Office has also changed the interface (e.g., location of menus) almost every time they have released a new version, which is something OpenOffice has not done. From a user perspective, relearning where the menu is to change line spacing, for example, is not an efficient use of time, so in this respect, OpenOffice has been more consistent and user friendly.

Add to these strengths the fact that open-software packages tend to be more interchangeable than commercial packages, and you have a good reason to consider switching to open-source software. As an example, two very popular noncommercial email clients, Mozilla Thunderbird and Postbox, use the same extensions. An extension is an add-on product that enhances the capabilities of an application, such as to add scheduling capabilities with Lightning or to add Pretty Good Privacy (PGP) encryption capabilities with the Enigmail add-on. This is not to say that commercial packages do not have add-ons, but there is no financial incentive for a revenue-driven firm to offer an add-on that a competing firm could use. Open-source software, however, is user supported however, so the profit incentive would not be present.

One very stable, flexible, and large-scale package hospitals are using for patient data is the previously mentioned VistA system (not to be confused with the outdated Microsoft Vista operating system). This patient management system was developed by the Department of Veterans Affairs (VA) and is now given away at no charge. It has been used for over 20 years by all of the 160-plus VA hospitals in the United States and their outpatient clinics.

A quick Internet search for open-source software or a glance at Sourceforge.net will reveal that there are thousands of open-source projects. Some programs are started as open source as incubators for future versions that might become commercial projects. Others are simply written and released for the general good and used by the public, with no intention of ever creating commercial versions.

Despite the many advantages of open-source programs, there are a few drawbacks. With most open-source programs, there is no single source to contact for support. There is no support contract, and there is no assurance that a new version will not be buggy, although commercial applications are often buggy, as well. All in all, however, open-source software should be strongly considered by the savvy business professional.

Freeware and Shareware

Freeware and shareware software are offered under a different model than open-source software. Freeware and shareware are typically not given away with the source code. Only the compiled code or executable code is given away. In the case of freeware, it is given away with full functionality and not limited in any way by such things as time restrictions on use or the number of files it can save or open. It essentially is given away free for the user to use, but not change, unlike with open source, in which users can recompile and change the original source code.

Shareware is similar to freeware except that it is more likely to be limited in either ability or duration of use. For example, a shareware package may be fully functional for 30 days or 30 uses. After that period or number of uses, certain features may be disabled, such as the ability to save changes. The big advantage of shareware over commercial software is the ability to try out a fully functional software package for free to evaluate whether it is worth buying.

Hands-on Statistics 3.8: Sparklines Example with OpenOffice Calc

To practice using sparklines with OpenOffice Calc, examine **Figure 3.14** and follow these directions:

1. Download and install OpenOffice for your operating system from the following website: http://www.openoffice.org/.
2. Download and install sparklines for OpenOffice from the following website: http://extensions.openoffice.org/en/project/eurooffice-sparkline.
3. Note you will have to restart your computer for this add-on to take effect. Also note that you may have to manually install the extension in some cases.
4. To manually install an extension, first download to a location on your computer where you can access the extension. Note that it will have a zip file name extension, indicating that it is a compressed archive file.
5. Click on the Tools menu then Extension Manager, then use the Add button to browse to where you have downloaded the extension you wish to install. Click on the zip file, and accept the usage agreement. The manager will automatically unzip (uncompress) and install the file. If the program successfully installed, you will now see the extension listed in your Extensions window. Reboot the computer to finish the process.
6. Open the application Open Office by clicking on the shortcut now installed on your computer's desktop.
7. Click on *"Spreadsheet"* to open a new blank spreadsheet.
8. Type the data indicated in Figure 3.14 into the blank spreadsheet (note that this is the same data you entered into Figure 3.11 in Excel). You should note that OpenOffice Calc works about the same as Excel. You can also open the "CH03_January_OpenOffice" document (found in Chapter 3 of the eBook) with the raw data.

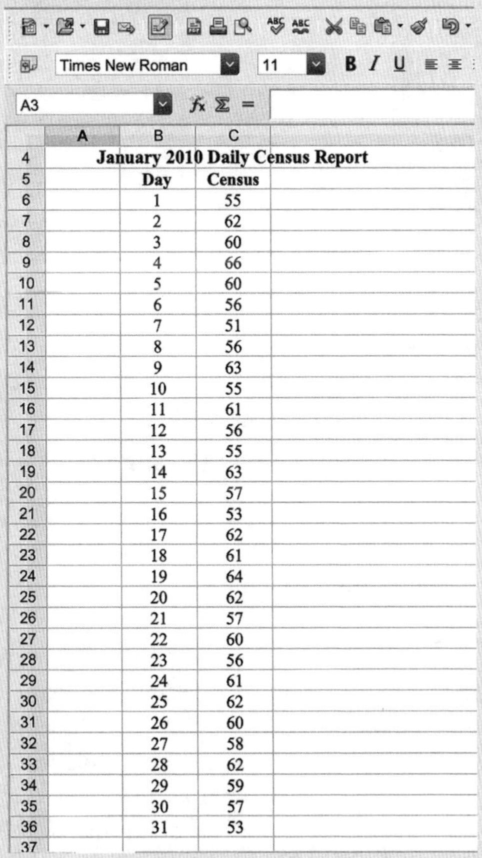

Times New Roman 11 B I U

A3

	A	B	C
4		January 2010 Daily Census Report	
5		**Day**	**Census**
6		1	55
7		2	62
8		3	60
9		4	66
10		5	60
11		6	56
12		7	51
13		8	56
14		9	63
15		10	55
16		11	61
17		12	56
18		13	55
19		14	63
20		15	57
21		16	53
22		17	62
23		18	61
24		19	64
25		20	62
26		21	57
27		22	60
28		23	56
29		24	61
30		25	62
31		26	60
32		27	58
33		28	62
34		29	59
35		30	57
36		31	53
37			

Figure 3.14 OpenOffice raw data.

(continues)

9. Highlight your data range, then select *Insert* and *EuroOffice Sparkline*, then select whether you want Start, High, Low, and End marks, and click *Create*. (See **Figure 3.15**.)

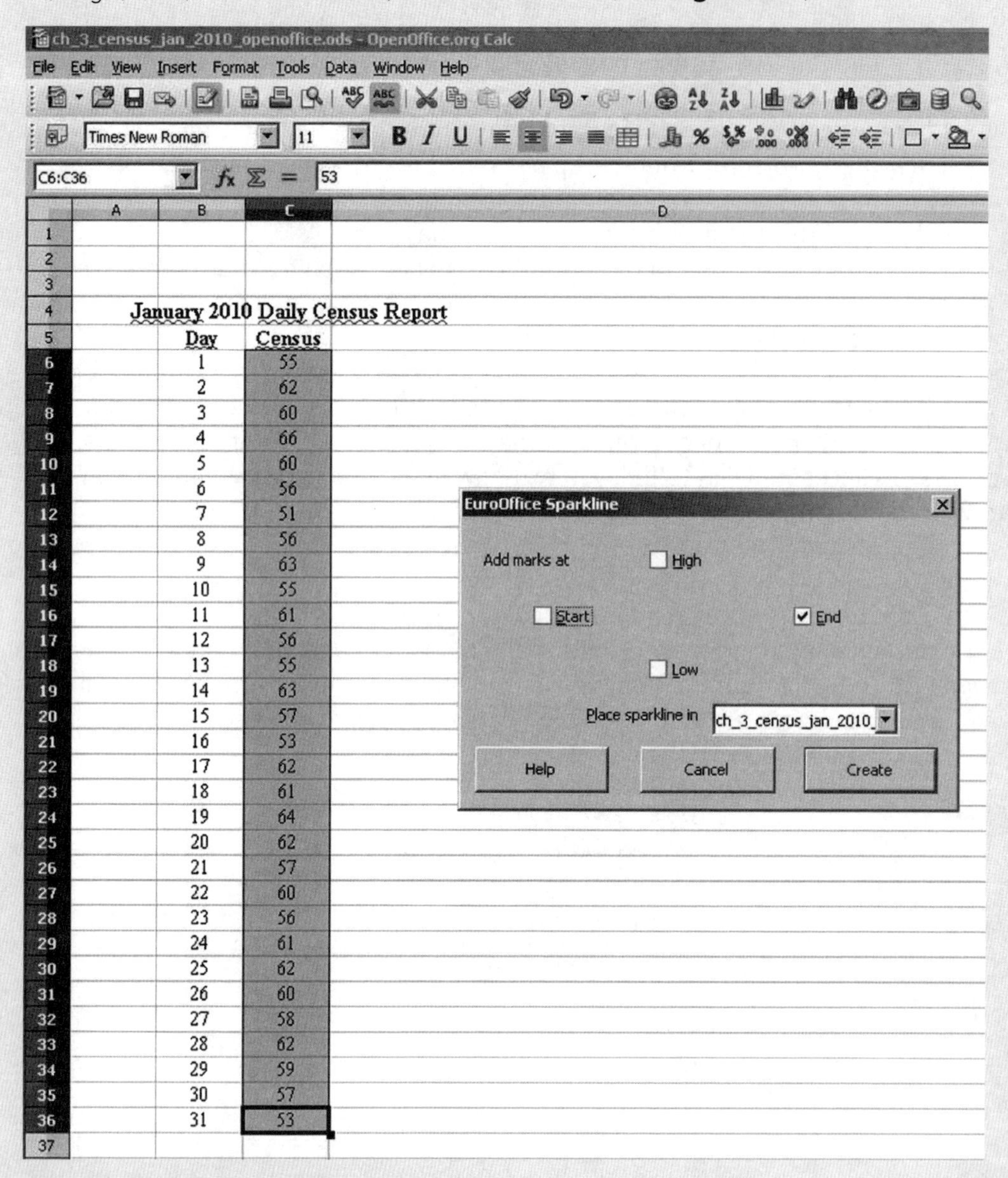

Figure 3.15 Inserting a sparkline with OpenOffice.

10. Move and resize the sparkline to where you want it. Also note that you can right click on the line itself to change properties of the sparkline.

Types of Databases

Next we will discuss two types of database technology relating to structure of operation: flat-file and relational.

Flat-File Database

A **flat-file database** features a table of data or multiple tables that may not be linked together. This type of database is designed for single users with no particularly specialized database system knowledge (e.g., the home user, a public school student) and is the simplest type of database. A modern and full-featured database application such as MS Access, IBM DB2, MySQL, and others can make a flat-file database, but a spreadsheet such as MS Excel could as well since it is only a text file with some form of delimiter, such as a comma or tab, to differentiate fields. This database structure has only one table of records and is not designed for tables of data to be linked together. Certainly the program would allow for many tables to be created, but they are not linked together; they are separate entities. Flat-file databases will not be used often in a medical setting.

As stated, a spreadsheet program can also be used for flat-file database work. For example, an Excel spreadsheet with one sheet in a workbook, with each row acting as a record of database data, would be considered a flat-file database. In fact, it would offer very similar features to most flat-file database applications, such as sorting and formatting.

Relational Database

A **relational database** has one or, more often, many tables that are linked for organizational purposes. Microsoft Access, Oracle, and MySQL are a few examples of relational database programs that support very large datasets and multiple users. Enhanced security is also a hallmark of these very large and comprehensive systems. In the case of Microsoft Access, the size and number of concurrent users and security are somewhat limited when compared with a larger program such as MySQL. Also, larger systems are a combination of server software—to manage access and transactions between clients (users)—and the data on the server, whereas smaller programs, such as Microsoft Access, are often used as a single program and database file all on the same computer. However, the Access database could be maintained in a folder on a file server with client computers, each with Access installed, able to access the database on the server at the same time up to the maximum number of users allowed (255 with Microsoft Access).

In light of the capabilities of relational databases, you may be wondering whether two different computer users could access the same patient's record at the same time. If so, this could be a big problem. Think about it. If several clients opened the same record, all made different changes, and each saved his or her changes, whose changes would take precedence? To prevent this problem, multiuser systems have a record-locking feature such that when the first client enters a record, later clients either cannot open the record or are allowed only to view the record until the first client leaves the record.

Data Formats

Regardless of the database, data are primarily available for download in two different formats: **comma-separated values (CSV)**, in which fields are separated by commas, and **tab-delimited values**, in which fields are separated, or delimited, by tab spaces. When you import a file into a program, such as Excel, the comma or tab, depending on the format, tells the application where the next field or record starts.

Hands-on Statistics 3.9: Import a CSV File Into Excel

To import a CSV file into Excel, see **Figure 3.16** and follow these directions:

1. Start Excel.
2. In the student resources in the eBook, you should see the file named "ch3csv1.txt." Download this file to a folder on your computer.
3. In Excel, select *Get External Data From Text*.
4. Browse to the folder on your computer containing the file you just downloaded.
5. Pick *delimited* from the list and select *Next*.
6. Select the text file and check only comma-separated values, then select *Next*.
7. You should see your data. Select *Finish* and *OK*, and you now have data imported. At this point, you can format it into the shape you want.

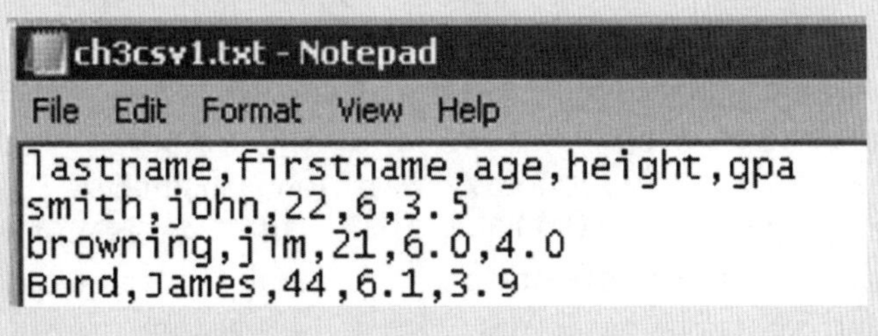

Figure 3.16 Comma-separated raw text file.

(continues)

Repeat this process, but this time, check the *Tab* checkbox instead of the *comma-separated value* checkbox. **Figure 3.17** shows a tab-delimited file, opened in Notepad.

```
sdf.txt - Notepad
File Edit Format View Help
lastname        firstname       age     pay_hr
Smith   John    34      15
Thompson        Pete    55      20
Jackson Fred    32      15
```

Figure 3.17 Tab-separated raw text file.

Remember to press the tab key just one time after each data item. Although pressing tab more than once may allow you to line things up better, an extra tab may hinder your ability to properly import your data into a spreadsheet or other application, as the extra tab is interpreted as a blank field. A comma-separated file does not have this issue.

How Does Your Hospital Rate?

The Assistant Director of the Health Information Management Department at Hospital A during a telephone interview provided the following details regarding the ADT report and its importance to the facility.

- Question: How does Hospital A calculate their ADT report?
  Answer: Hospital A calculates their ADT report electronically. The Chief Nursing Officer is the one responsible for the ADT report. I review the ADT several times weekly to ensure that the report is calculating properly and to know if we have a full house. This report is then used in forecasting higher volumes of admissions and which months have high admission rates. The report enables us to forecast for additional staffing if the need arises.
- Question: How many people on your staff understand how to manually calculate an ADT report?
  Answer: There are two people in the Health Information Management Department that know how to manually calculate the ADT report if the situation would warrant. Remember, Hospital A is only licensed for 25 beds.

Consider the following:

1. Do you believe this situation is satisfactory? Why or why not?
2. Contact a local hospital in your area that is comparable in size to Hospital A and conduct the same interview to learn about this practice.

R-Project

R-Project (R) is both a software application and a programming language for statistical computing and graphics. It has an interface similar to Python, Ruby, and other rapid application development (RAD) languages and offers a line-by-line and **script-mode interface**. It is powerful, cross-platform (runs on many different types of computers), free (open source), and well maintained. Add to these facts that you can add in more features (other functions), and you have a fantastic application to add to your statistical toolbox. The R language is based on an earlier statistical programming language named S, developed by John Chambers.

The R programming language, R-Project, and R-Studio integrated development environment (IDE) software provide a variety of statistical programming and visualization tools. In fact these tools are constantly being improved and added to by the global community of R contributors. Chambers even lists "collaborative" as one of the "six facets" of R (Chambers, 2009). When comparing it to commercial packages such as MINITAB, SPSS, and SAS, users likely will find that R fits their needs without the large price tag of commercial applications. For most of the statistical analysis examples in this text, R-Project and Excel will be used. However, tools that augment these processes, such as R-Studio, will also be highlighted so that the reader is made aware of the many tools available. Note that the R-Studio IDE has some limitations with regard to displaying very large datasets. So, if you are working with large sets, you would be better off working just with R-Project to eliminate this issue.

Hands-on Statistics 3.10: Install R-Project and the Integrated Development Environment

To install R-Project and the IDE software, follow these directions:

1. Go to http://www.r-project.org/.
2. On the left side of the screen, click on the hyperlink "CRAN" to pick a mirror site from which to download the installer. (A mirror website is an Internet location with a copy of the file you are downloading. If a primary site is busy with other people downloading, you can go to, or be automatically redirected to, a less busy mirror so your download time is not as long as it would be on the primary site.)
3. Click on the URL under 0-Cloud to automatically redirect to an available mirror site, then select Download R for Windows.
4. Select *Base*, then Download R for Windows at the top, and save to a location you can access. *This is critical: do not just save!* Make sure you know where you are saving to, as you will need to click the installer in the next step. The save location might be a "downloads" folder or other area.
5. From the save location, right click the file, select *Run as Administrator*, and install with all defaults. If you have a 64-bit version of Windows, select the 64-bit version; otherwise, select the 32-bit version. Follow all of the prompts of the Setup Wizard until installation is complete.
6. Lastly, download R-Studio from the following website: http://www.rstudio.com/ide/download/.
7. Select the *desktop version* and install it as well. Follow all of the prompts of the Setup Wizard until installation is complete. You will now have a nice IDE that can make things easier when using R, although you can always just use R.

Now that we have R installed, let's explore some features of R and examine some simple statistical methods with R. We will use R for some of the methods discussed in other chapters.

R has an interactive prompt—a red greater-than symbol (>)—for line-by-line commands you can use, or you can write a script with multiple commands (a program) that you could save, edit, and re-run later. In this way, R is like other programming languages and IDEs, such as Python, Ruby, and PHP: Hypertext Preprocessor (PHP). For the next Hands-on exercise, you will be in interactive, or line-by-line, mode.

Both R and R-Studio can be used interchangeably, but R-Studio gives you a few more features.

Hands-on Statistics 3.11: Demo of Graphics in R

For a simple introduction to R-Studio and R, follow these directions:

1. Start R-Studio.
2. From the interactive prompt of R (left side pane), type "demo(graphics)" and press Enter; you will see some of the graphics capabilities of R in the lower right hand side window. Just follow the instructions (telling you to press Enter again) to view. You will see several samples of what R can do. Close R-Studio.
3. Next, start R. Type "demo()"; a pop-up window will display showing various options. Close R.

Hands-on Statistics 3.12: Simple Analysis of Data With R

To practice a simple analysis of data using R, see **Figure 3.18** and follow these directions:

1. Start R.
2. Make a data vector by typing "age=c(12,13,9,14,12,11,10)" next to the prompt, where *age* is just a name for your dataset and the numbers are ages of students. Press Enter. Vectors keep the order of data such that each element (every age entered is an element) is separate and individually accessible.
3. To display data from a set, just type the name of the set (in this case "age") and press Enter. All the data you just entered in the previous step should now be displayed, preceded by "[1]," which indicates that the data are a *vector*.
4. Finding the mean and median is just as easy. Simply use the functions and give them the vector of data under examination. To see the results, type "mean(age)" and press Enter, and then type "median(age)" and press Enter. Your results should match those in Figure 3.18.

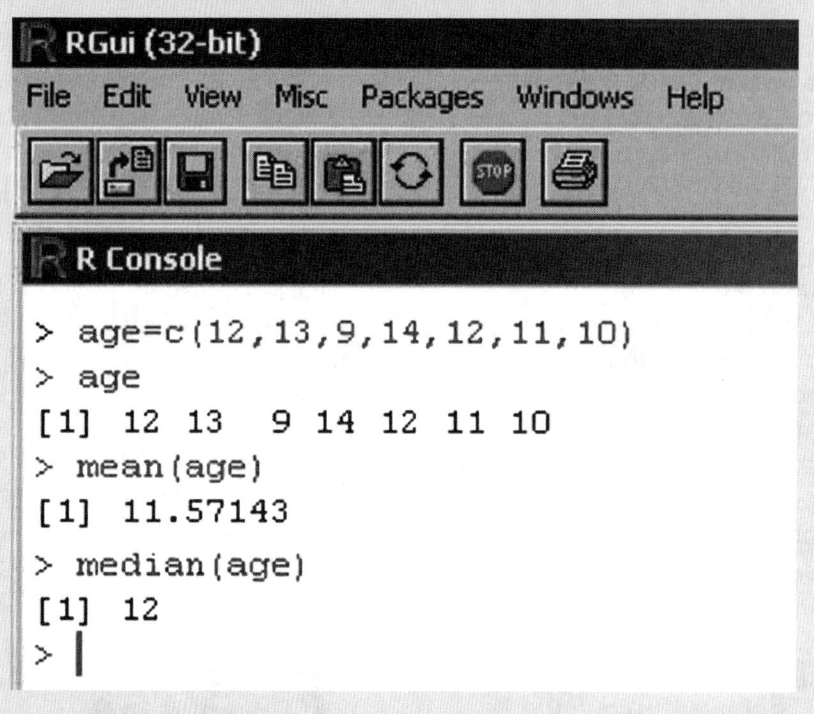

Figure 3.18 Simple analysis with R.

5. Type "var(age)" to see the variance.

In the next example, we will examine some of the data visualization capabilities of R. Nothing adds impact like good graphics!

Data Stored in R

Data, which are groups of elements, may be of the four following types in R:

- Numeric: numbers, either integer or floating point
- Character: alphanumeric data, stored as a char
- Boolean or logical: elements that can be defined as TRUE or FALSE
- List: any type of object, including lists of list

Three types of numeric data are stored in R. A **scalar** is a single number, a **vector** is simply a row of numbers and is one dimensional, and a **matrix** is data stored in a table-type format. A vector is a commonly used basic data object in R.

Hands-on Statistics 3.13: Visualize Data With R—The Line Chart

To create a line chart in R, see **Figures 3.19** and **3.20** and follow these directions:

1. Open R-Studio.
2. From the command prompt in R, let's first see how easy it is to graph a few numbers in R.
   a. Create a vector named "rates" with data by typing:

      rates <-c(7.6,7.9,6.3,5.4)

   b. Next plot the data in red by typing:

      plot(rates, col="red").

   c. Change to a line chart in blue by typing:

      lines(rates, type="o", col="blue").

   At any point, you can press the Print Screen button (PRTSC) to copy the graphic to your clipboard for cleanup in a paint program so that you can use it in a presentation or report.
3. Now we will chart infant mortality rates.
   a. Copy the R Script file "CH03_Mortality.r" found in Chapter 3 of the eBook to a folder on your computer or a removable drive you can easily access. Using R-Studio, select File, then Open File and browse to where you copied the file, select it, and open it.
   b. In the upper left window, you will see the R script with mortality data from the CDC for 2006 for the United States for "all," "black," and "white" ranked by four education groups, with the lowest of 0–8 years of education being 1. See Figure 3.19.

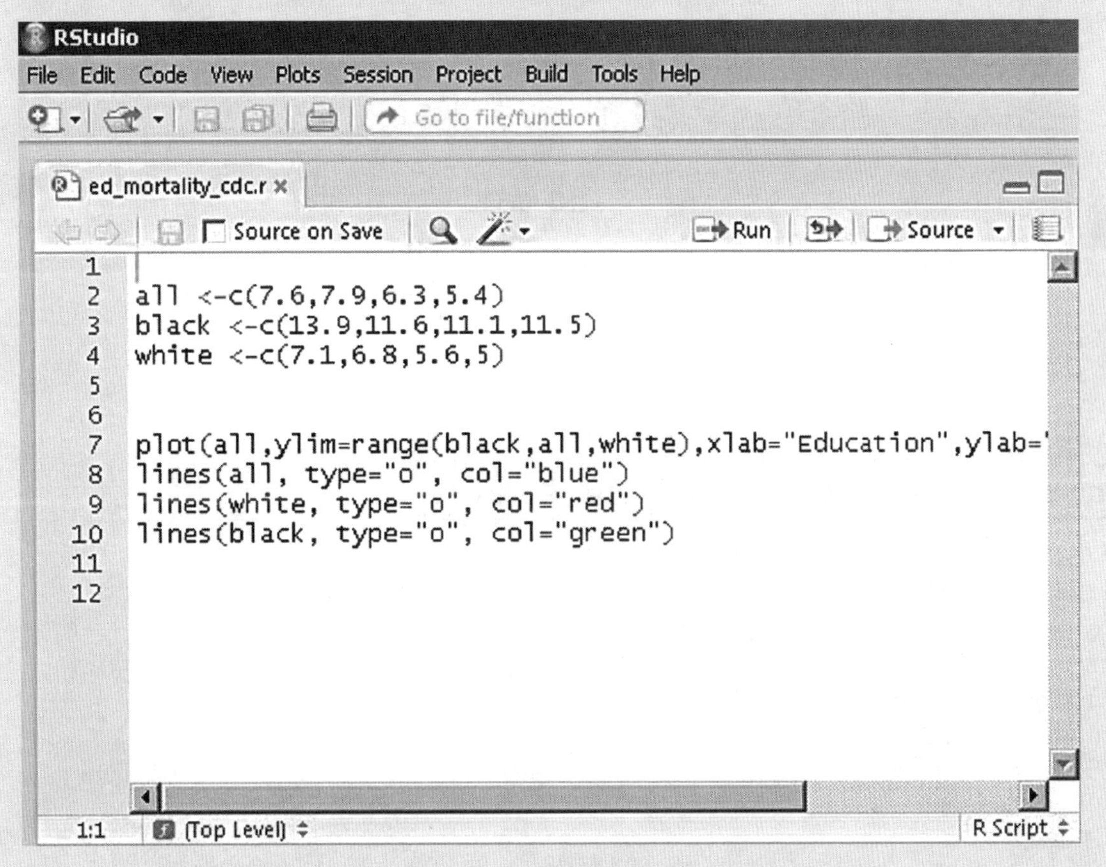

Figure 3.19 Line chart script for R.

(continues)

c. Click the *Run* button in the upper left code window until it executes each line and you see the chart produced in the lower right window. You can run all of your code at once (which is what you would do normally) by selecting *Code*, *Run Region*, or *Run All* or by using the hot keys CTRL + ALT + R. See Figure 3.20.

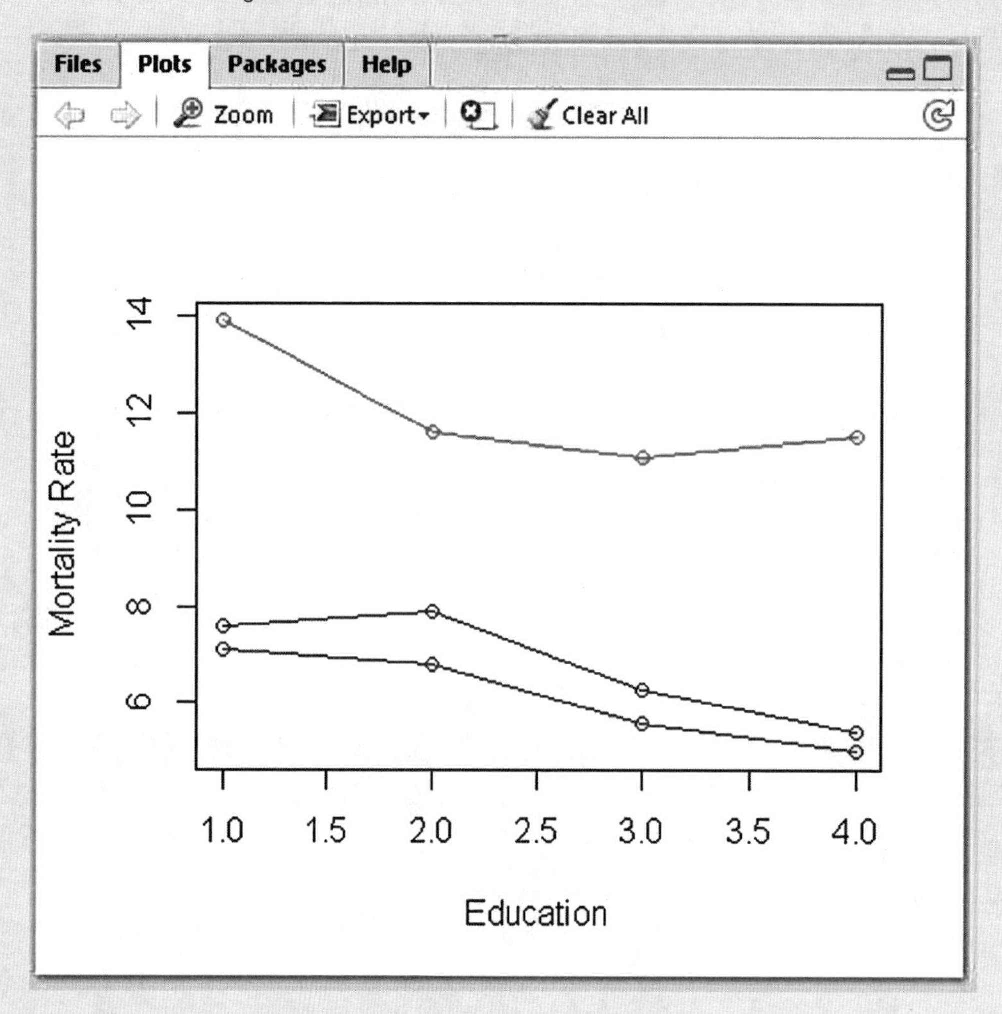

Figure 3.20 Line chart created by R.

Hands-on Statistics 3.14: Import a CSV File Into R

To create a dataset, save it as a CSV file, and import it into R for analysis, follow these directions:

1. First we will create a sample data file using Notepad.
   a. Start Notepad from the Accessories group in Windows, or type "notepad" and press Enter from the Windows search box.
   b. Type the following in the blank Notepad file that appears:
      name,age,salary,sex
      Smith,45,25000,m
      Jones,44,44000,f
      Brown,34,41000,m
      Yades,56,40000,m
2. Save the file as a CSV file by selecting *File*, *Save As*, then *Save as type: All Files*, and name it "sample.csv." Save it to **x**:\, the root of your **x**: removable drive. Note that *x* is a letter designating

your removable pen drive, which may be F:, D:, E:, etc., depending on the number of drives on your unit.

3. Start R-Studio, in the upper left code window, type the following, pressing Enter after each line. Note that *x* is your removable drive letter:

```
data <- read.csv(file="x:/sample.csv",head=TRUE,sep=",")
data
```

You will see your data is displayed in the lower left window. At this point, the data are a vector with your dataset in it. Note that normally in DOS (Windows) the slash (\) leans left and not right, but as R has UNIX roots (a more robust computer operating system than Windows), the directory format of UNIX is used, in which this symbol leans to the right (/) instead.

Because the data in the Notepad file are separated by commas, we used the code `sep=","` to indicate that the separator is a comma. The code `head=TRUE` indicates that the first line has column headers. Lastly, the code `file="c:/sample.csv"` tells the program where to find the file, using a UNIX-style front slash instead of a backslash. Data are being stored into a variable we named "data." There are many other types of files you can import, such as fixed-width format (fwf) and Excel spreadsheets.

Hands-on Statistics 3.15: Math With R

In this example, we will perform some simple math on the file you just used. Using either R-Project or R-Studio, follow these directions:

1. To find the minimum age in column 2 (the age column), key in the following and press Enter:
   ```
   mage<-min(data[,(2)])
   ```
   mage is a variable that stands for "**m**inimum **age**," and *data* is the dataset you imported from the sample.csv file.
2. Show minimum age by typing "mage" and pressing Enter.
3. Max works just as easily. To find the highest salary, type the following, and press Enter after each line:
   ```
   max_sal<-max(data[,3])
   max_sal
   ```
4. To sum your salary data, type the following, and press Enter after each line:
   ```
   sal_sum<-sum(data[,3])
   sal_sum
   ```

We will work more with R throughout the text, but for now you have a sense of what the software can do. Consider it one of many tools you have available.

Hands-on Statistics 3.16: Explore Other R Import Options

There are many other options for importing data into R, and a good way to explore them is to examine *The R Manuals*, available at http://cran.r-project.org/manuals.html.

Additionally, there are many websites that provide information on using R. Visit the following resources, or just do an Internet search for "R-Project tutorials" or similar key terms, and then list four interesting functions or tips you learned from them about using R.

- Quick-R by DataCamp: http://www.statmethods.net/input/importingdata.html
- *An R Tutorial*: http://pages.pomona.edu/~jsh04747/courses/RTutorial.pdf

How Does Your Hospital Rate?

Benchmarking against other healthcare facilities that are of the same size and offer the same types of services can assist in improving processes in your facility. If your facility has only one person who can manually do the census, that is probably not a good idea.

Even if your census is done electronically, you still need to know how to calculate manually and have more than one person who knows how to take the census.

Global Perspective

Using data visualization enables us to compare real-world health trends nationally and internationally. It allows healthcare facilities and other organizations such as the CDC and WHO to track and monitor for any outbreaks of diseases. Databases are used to store information that can be abstracted and published for the public on diseases and deaths for a specific period of time. For example, with the AIDS virus, data visualization allowed the CDC to track and monitor the number of cases and which states had the highest number of patients. This information was also used to observe the declining number of AIDS cases around the world as new medications were introduced to reduce the viral load of the virus. Data visualization has enabled us to view healthcare trends from a different perspective.

Hands-on Statistics 3.17: Download Real-World Data From a National Data Site

To download a dataset from the Centers for Medicare and Medicaid Services (CMS), follow these directions:

1. Visit the CMS homepage at http://www.cms.gov/ and explore the site. List at least two interesting features of the site.
2. From the homepage, click on *Research, Statistics, Data & Systems.*
3. In the search box, enter "Public Use File" and visit the page at https://www.cms.gov/Research-Statistics-Data-and-Systems/Statistics-Trends-and-Reports/Medicare-Geographic-Variation/GV_PUF.
4. Click on the zip file titled "State County Table All Beneficiaries" from the *Downloads* section at the bottom of this page.
5. Save the zipped Excel file to your computer.
6. Open the file, and in the Excel file, examine the *year 2011* tab, as shown in **Figure 3.21**.
7. Using the sort feature of Excel, find the top three states with the highest average age. To do this, highlight the column *Average Age*, then select *Data* and *Sort*. If a *Sort Warning* pop-up appears, leave the option *Expand the selection* checked and click on the *Sort* button. The *Sort* pop-up window will display. In the dropdown menu under *Column*, select *Average Age*. Leave the field under *Sort* set to *Values*. In the dropdown menu under *Order*, select *Z to A*.
8. Using the sort feature again, as in the previous step, find the top five states for "Hospital 30-day death (mortality) rates for pneumonia patients."

	A	B	C	D	E	F	G	H	I
1	State	Count of Beneficiaries	Average Age	Percent Female	Percent Male	Percent Non-Hispanic White	Percent African American	Percent Hispanic	Percent Asian American/Pacific Islander
2	National	25,832,920	76.5	57.9	42.1	84.5	7.3	5.0	2.1
3	AK	43,023	74.5	52.0	48.0	76.5	2.1	1.9	3.3
4	AL	464,378	75.9	58.9	41.1	84.8	14.2	0.4	0.3
5	AR	326,430	76.0	57.9	42.1	91.0	7.5	0.5	0.3
6	AZ	412,646	75.7	53.8	46.2	87.8	1.7	6.2	0.9
7	CA	2,010,957	76.4	56.4	43.6	68.0	4.4	14.7	10.9
8	CO	282,500	76.0	55.6	44.4	88.3	2.4	7.4	1.1
9	CT	361,347	77.6	59.4	40.6	90.6	4.7	2.9	1.0
10	DC	42,868	77.3	61.0	39.0	33.0	61.6	2.8	1.5
11	DE	105,118	75.9	57.0	43.0	85.5	11.4	1.3	1.1
12	FL	1,846,584	76.6	56.6	43.4	84.8	5.5	8.2	0.9
13	GA	726,234	75.6	58.7	41.3	81.4	16.3	1.0	0.8
14	HI	85,685	76.4	55.5	44.5	25.1	0.7	5.4	58.1
15	IA	363,657	77.3	58.7	41.3	97.9	0.8	0.5	0.3
16	ID	123,344	76.1	54.5	45.5	95.9	0.2	2.2	0.6

Figure 3.21 Hospital 30-day death rates in Excel.

Note that the sort feature can be used to put items in ascending or descending order, based on your needs. It can be used to sort either string (text) data or numeric data, such as "highest to lowest."

Chapter Summary

In this chapter you have learned about census reporting from several different angles, for several types of purposes. As the data can be difficult to interpret, you reviewed a few ways to visually show patterns in numeric data via infographics tools such as sparklines and microcharts. As powerful tools are needed for this visual representation, the chapter discussed and used both commercial and open-source/freeware software tools to create and summarize the information. Data formats and tools for raw data also were explained, because most large datasets will be available either in a database or extracted to some universal format such as CSV for you to import into your application suite.

Apply Your Knowledge

1. Locate on the Internet a copy of a GPL license. List one advantage and one disadvantage of the license.
2. Examine an open-source office suite package. Imagine that your firm needs a free alternative to Microsoft Office. Examine another office suite, other than OpenOffice, and list three positive and three negative features.
3. Use the data in **Table 3.1**, collected from St. Pawley's Hospital records, to calculate the average daily inpatient census for each of the three periods listed.

Table 3.1

Time Period	Inpatient Service Days	Bed Count
January through April	43,725	400
May through August	59,218	500
September through December	65,383	550

4. A 250-bed hospital with 20 newborn bassinets records the following inpatient service days for the month of March:
   - Adults and children: 7,380
   - Newborns: 558

   Based on these data, calculate the following:
   - Average daily adults and children census
   - Average daily newborn census
5. A hospital reported the following statistics shown in **Table 3.2** for September. Based on these data, calculate the following:
   - Inpatient census for midnight on September 30
   - Average daily inpatient census for September
   - Average daily newborn census for September

Table 3.2

Counts	Adults and Children (Beds = 150)	Newborns (Bassinets = 15)
Census (midnight August 31)	140	11
Admissions	310	90
Discharges (live)	300	92
Deaths	15	2
Fetal deaths		5
Inpatient service days	4,236	410

6. The information from **Table 3.3** is reported for three clinical units during the month of November.

Table 3.3

Count	Pediatrics	Orthopedics	Psychiatry
Bed count	12	15	10
Beginning census	10	12	6
Admissions	85	122	54
Discharges (live)	86	120	53
Deaths	1	3	0
Inpatient service days	344	433	284

Based on these data, calculate the following:
- Average daily inpatient census for each of the three clinical units

- Ending census (November 30) for each of the three clinical units
- Total inpatient service days for the three clinical units for November

7. Define the term *intrahospital transfer* and give an example.
8. Ocean View Hospital has reported the information in **Table 3.4** for the month of October. Using these data, calculate the month's average daily inpatient census for each unit.
9. What are other terms for *inpatient service days*?
10. When is the best time to take an *inpatient census*? Why?
11. Define the term *"leave of absence,"* and explain when it would be used.
12. Is there a time when the *transfers in* and *transfers out* must be equal? If so, give an example.

Table 3.4

Unit	Inpatient Service Days	Average Daily Inpatient Census
Medicine	5,948	
Obstetrical	432	
Surgery	3,586	
Pediatric	2,470	
Total adult and children	12,436	
Newborn	388	
Neonatal ICU	108	
Total nursery	496	

References

Ashar, B. S., & Ferriter, A. (2007). Radiofrequency identification technology in health care: benefits and potential risks. *The Journal of the American Medical Association, 298*(19), 2305–2307.

Chambers, J. M. (2009). Facets of R. *The R Journal, 1*(1), 5–8.

Cronin, S., Leo, F., & McCleary, M. (2008). Linking nurse staffing to nosocomial infections: a potential patient safety threat. *Georgetown University Journal of Health Sciences, 5*(2). Retrieved from https://blogs.commons.georgetown.edu/journal-of-health-sciences/issues-2/previous-volumes/vol-5-no-2-december-2008/linking-nurse-staffing-to-nosocomial-infections-a-potential-patient-safety-threat/

Haley, R. W., & Bregman, D. A. (1981). The role of understaffing and overcrowding in recurrent outbreaks of staphylococcal infection in a neonatal special-care unit. *The Journal of Infectious Diseases, 145*(6), 875–885.

Klevens, M., Edwards, J., Richards, C., Horan, T., Gaynes, R., Pollock, D., & Cardo, D. (2007). Estimating health care-associated infections and deaths in US hospitals, 2002. *Public Health Reports, 122*(March–April). Retrieved July 2007, from http://www.cdc.gov/hai/pdfs/hai/infections_deaths.pdf

Kohn, L., Corrigan, J., & Donaldson, M. (Eds.). (2000). *To err is human: Building a safer health system*. Committee on Quality of Health Care in America, Institute of Medicine. Washington DC: National Academy Press.

McCandless, D. (2010). The beauty of data visualization. TEDGlobal. Retrieved from http://www.ted.com/talks/david_mccandless_the_beauty_of_data_visualization.html

Mitchell, B. G., Gardner, A., Stone, P. W., Hall, L., & Pogorzelska-Maziarz, M. (2018). Hospital staffing and health care–associated infections: a systematic review of the literature. *The Joint Commission Journal on Quality and Patient Safety, 44*(10), 613–622.

Web Links

Comma-Separated Values: http://en.wikipedia.org/wiki/Comma-separated_values

Open-source version of VistA: http://www.openclinical.org/os_OpenVistA.html

Forecasting Patient Census: Commonalities in Time Series Models: http://www.ncbi.nlm.nih.gov/pmc/articles/PMC1071910/?page=1

10 Statistics Your Hospital Should Track: http://www.beckershospitalreview.com/hospital-management-administration/10-statistics-your-hospital-should-track.html

Microcharting in Excel—7 Alternatives Reviewed: http://chandoo.org/wp/2008/09/05/microcharting-excel-howto/

Sparklines for Microsoft Excel: http://sourceforge.net/projects/sparklinesforxl/

Lightweight Data Exploration in Excel: http://www.juiceanalytics.com/writing/lightweight-data-exploration-in-excel/

Excel Dashboards—Tutorials, Templates, and Examples: http://chandoo.org/wp/excel-dashboards/

Graphing With Excel: http://www.ncsu.edu/labwrite/res/gt/gt-menu.html

Six Ways to Transform Data Into Real Information That Drives Decision Making—Tools, Tips, and Training: https://www.allbusiness.com/transform-data-real-information-drives-decision-making-16096-1.html

© Mad Dog/Shutterstock

CHAPTER 4

Occupancy and Utilization Data

Common sense is not so common.

—Voltaire

CHAPTER OUTLINE

Introduction
Bed Count Computation
- Definition of Inpatient Bed Count
- Importance of Inpatient Bed Count
- Bed Occupancy Ratios

Certificate of Need
Calculating Newborn Bassinet Occupancy Ratio
Bed Turnover Rate
- Two Formulas for Bed Turnover

Length of Stay
- Discharge Days and How to Calculate Them
- Total Length of Stay
- Average Length of Stay

Median and Standard Deviation for Length of Stay
- Median Length of Stay
- Why Might You Use Median Instead of Mean?
- Standard Deviation of Length of Stay

Visually Representing Data
- PowerPoint Presentation

Data Mining: Association Rules With R-Project
Global Perspective
Chapter Summary
Apply Your Knowledge
References
Web Links

LEARNING OUTCOMES

After completing this chapter, you should be able to do the following:

1. Define and discuss the importance of inpatient bed count.
2. Describe and compute bed and bassinet occupancy rates.
3. Define certificate of need.
4. Describe and compute bed turnover rate.
5. Define key features of length of stay.
6. Describe and compute discharge days.
7. Describe and compute total, average, median, and standard deviation for length of stay.
8. Create visual presentations.
9. Create association rules with R-Project.

KEY TERMS

A priori
Algorithm
Bed count
Bed count day
Bed occupancy ratio
Bed turnover rate
Benchmark
Certificate of Need (CON)
Confidence
Empirical
Forecasting and trend analysis
Hybrid
Left-hand side (LHS)
Length of stay (LOS)
Lift
Lift chart
Market basket analysis
Object linking and embedding (OLE)
Percentage
Posteriori
Rate
Ratio
Right-hand side (RHS)
Support

How Does Your Hospital Rate?

Richard, an 80-year-old man who is 5 feet 11 inches tall and weighs 285 pounds, is scheduled to undergo a total hip arthroplasty at a local hospital. When discussing the upcoming surgery with his surgeon, he is surprised to learn the length of the hospital stay following the procedure. He asks the surgeon the reason for the lengthy stay.

Consider the following:

1. What factors can determine a patient's length of stay in the hospital?
2. Contact your local healthcare facility and determine its average length of stay for a total hip arthroplasty, without complications. The national average is 8 days. How does your hospital compare?

Introduction

A major challenge hospitals face is the efficient use of their resources. This is an area where statistics and data mining can significantly help hospitals, and other healthcare facilities. Occupancy and utilization data inform healthcare providers, managers, and consumers of how much and how effectively resources are being used to serve the needs of patients. This chapter presents several key measures of occupancy and utilization data, including bed count, bed and bassinet occupancy rates, bed turnover rate, length of stay, and discharge days. The exercises in this chapter will help familiarize you with the purpose and computation of each of these measures. Additionally, the exercises will show how to visually represent data using applications such as Wordle and PowerPoint. Finally, the chapter will introduce you to the association rules with R-Project and provide instructions on how to use the rules in data mining.

Bed Count Computation

Probably the most critical measure of a hospital's resource capacity and use is bed count. The number of areas available to attend to patient needs is critical for a medical facility. Each area requires personnel, space, and consumables for proper operation. Therefore, each medical facility receives a license from the state in which it operates for a specific number of beds. Just as elevators have maximum weight limits to prevent the motor from burning out due to excessive weight, hospitals have maximum occupancy limits to prevent them from admitting more patients than they have the resources to treat.

Definition of Inpatient Bed Count

The **bed count**—also known as inpatient bed count—is the available number of hospital inpatient beds that are either vacant or occupied on any given day. The bed count may be provided either for each unit in the hospital or for the entire hospital as a whole. This is determined by how administration defines the procedure for the count. Only beds that are set up, equipped, staffed, and ready for patients are counted.

Several types of beds are not included in this count. For example, beds located in a wing of the facility that is not in use must not be counted. Beds that are pulled into hallways and other areas in response to a natural disaster, which are known as disaster beds, are not counted in the bed count. Beds located in the physical therapy, labor and delivery, and emergency departments, are also not counted, nor are those located in examination and recovery rooms. Beds in the emergency department are typically counted as outpatient, not inpatient, beds. In some circumstances, however, beds in the emergency department may be counted as inpatient beds if certain state licensure qualifications are met. Beds used by observation patients—those who still need to be evaluated, assessed, and monitored before being admitted to an inpatient status or discharged—are also not counted. There are other information systems used to track them.

A related measure to bed count is bed count days. A **bed count day** is the attendance of one inpatient bed that is set up and staffed for a 24-hour time frame. Total bed count days are the total of inpatient bed days for each day in a specific period that you are analyzing statistically. The newborn bassinet count, which is the number of bassinets available whether occupied or vacant on any given day, is a separate count from the bed count.

Importance of Inpatient Bed Count

Bed count is used in facilities to compare and contrast past and present performances. These measures are used to guide future planning and development of patient services. The most important use of bed count is for government funding and research. Rural hospitals are categorized according to bed count so that their performance can be compared with that of other rural facilities of comparable size. Small hospitals have fewer than 50 beds, medium-sized hospitals have 50 to 99 beds, and large facilities have 100 or more beds. Bed count is used to compare the types of services that are offered and patient safety measures.

Bed Occupancy Ratios

Many of the statistical problems throughout this text can be solved by using a **ratio**, which is a comparison of two quantities. A **percentage** is a type of ratio that compares a number to 100, with 100 representing a whole. **Bed occupancy ratio**, sometimes called occupancy ratio or percentage of occupancy, is a measure of the proportion of beds that are occupied at any given time at a hospital. It is often expressed as a percentage, such as "75% of our beds are in use."

Rate, a related measure, is a type of ratio in which the two quantities being compared are of different units of measure. It is the type of measure that is used to express speed, such as 60 miles per hour. Occupancy rate is the ratio of inpatient service days to bed count days for a given period. Later we will consider bed turnover rate, which measures the number of patients who occupy a bed over a certain period.

The formula for calculating bed occupancy is as follows:

$$\frac{\text{Total inpatient service days in a period}}{\text{Total bed count days in the period}} \times 100$$

where total bed count days in the period = bed count × number of days in the period. For example, suppose that on October 1, 85 inpatient service days were provided in a 100-bed hospital. Bed occupancy ratio for October 1 is as follows:

$$\frac{85}{100} \times 100 = 85\%$$

This means that on October 1, 85% of the beds were occupied. Next we will compute bed occupancy using a spreadsheet.

Hands-on Statistics 4.1: Compute Bed Occupancy Using Excel

To calculate bed occupancy using Excel, examine **Figure 4.1** and follow these directions:

1. Open a new, blank Excel spreadsheet.
2. Type "Bed occupancy example" in cell B2, "in-patient days" in cell B3, "Beds available" in cell C3, "Percent of use" in D3, "85" in cell B4, "100" in cell C4, and the formula "=(B4/C4)*100" in cell D4, as in Figure 4.1.
3. Click out of cell D4, and the number 85 should appear in the cell, representing a bed occupancy of 85%.

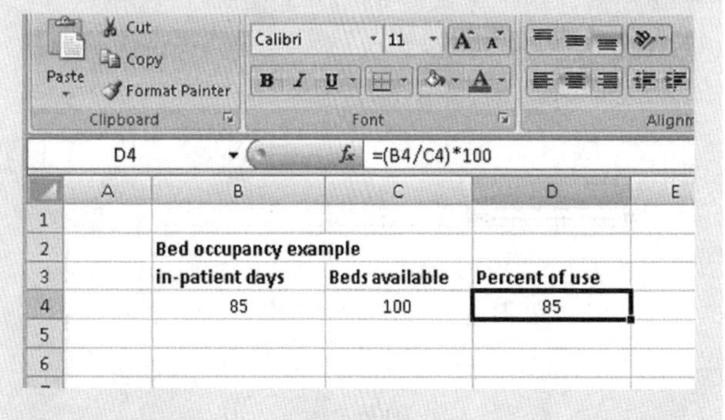

Figure 4.1 Bed occupancy.

Certificate of Need

When a facility must expand its bed count due to an increased need of services, it applies for a **certificate of need (CON)** to better serve the patient population. A CON is designed to help reduce and control healthcare facility costs and prevent any type of duplication or construction of services. In reference to construction of services, the CON prevents healthcare facilities from unnecessary expansion that involves capital spending. The CON is also used to control healthcare cost and plan any new construction or new services.

So how does the CON control healthcare cost? When a healthcare facility builds more than it needs, it tends to inflate prices to customers. That is, over-construction of beds leads to a higher cost of the beds being occupied to cover for the capital expenditures. Demographics is one of the areas reviewed before a CON is obtained by a healthcare facility. To consider it another way, local needs are compared to local medical services available (e.g., if there is already a local hospital with a neonatal intensive care unit, it would not be cost effective for a hospital 5 miles down the road to replicate the same services). If it did, costs could well be inflated for similar services.

Next you will learn how to calculate the new bed count days after expanding the bed count of a facility.

Hands-on Statistics 4.2: Calculating New Bed Count Days With Change in Certificate of Need and Bed Occupancy Ratio

Imagine that on September 20 the bed count of the hospital, at which you are employed changed from 100 to 150 and your facility had a total of 3,124 inpatient service days. It is now October, and you are calculating bed count days and bed occupancy ratio for September. How do you calculate these measures? Follow these directions:

1. Multiply the old total number of inpatient beds by the number of days in September to which this number applies:

   100 inpatient beds × 19 days of September = 1,900

2. Multiply the new total number of inpatient beds by the number of days in September to which this number applies:

   150 inpatient beds × 11 days of September = 1,650

3. Add the number of bed count days under the old total bed count to the number of bed count days under the new total bed count to find the total bed count days for September:

   1,900 + 1,650 = 3,550 bed count days

4. Now we need to calculate the bed occupancy ratio. Multiply your original inpatient service days by 100:

   3,124 × 100 = 312,400

5. Divide the result by the new bed count days, as calculated in step 3, to find the bed occupancy ratio for September:

   312,400/3,550 = 88%

It is the ultimate responsibility of health information management professionals to ensure the accuracy of statistical data. Administrative decisions rely heavily on having accurate information, so it only stands to reason that a facility's administration would have a hand in this responsibility, if not be guiding it.

Calculating Newborn Bassinet Occupancy Ratio

As we mentioned earlier, newborn bassinet occupancy ratios also are calculated separately. The formula for calculating newborn occupancy ratio is as follows:

$$\frac{\text{Total newborn inpatient service days for a period}}{\text{Total newborn bassinet count} \times \text{Number of days in the period}} \times 100$$

For example, in the month of July, Cape Hospital has a newborn bassinet count of 45 and provides 1,005 newborn inpatient service days. The formula for July would be as follows:

$$\frac{1{,}005 \times 100}{45 \times 31} = \frac{100{,}500 \times 100}{1{,}395} = 72.04\% = 72\%$$

Hands-on Statistics 4.3: Calculate Newborn Bassinet Occupancy Ratio Using Excel

To calculate newborn bassinet occupancy ratio using Excel, examine **Figure 4.2** and follow these directions:

1. Open a new, blank Excel spreadsheet.
2. Type the text and numbers in the appropriate cells, as shown in Figure 4.2.
3. In cell B7, type the following formula:

   = (B4*100)/ (B3*B5)

4. Format the cell to the decimal place of precision you desire.

	A	B	C
1			
2	**Cape Hospital**	**July**	**2013**
3	Newborn bassinet cout	45	
4	Newborn inpatient services	1005	
5	Days in Month	31	

Figure 4.2 Newborn bassinet occupancy ratio.

Now that you have learned how to calculate the occupancy ratio, we will turn to bed turnover rate.

Bed Turnover Rate

Bed turnover rate is the number of times that a bed changes occupants in a given time and/or the typical number of admissions for each bed during a period of time. For example, the emergency department may have a high bed turnover rate due to the patient length of stay being shorter, thus allowing more patients to be seen. This rate is just a calculation of how often the bed is occupied. It also shows any changes in length of stay and occupancy rate.

DID YOU KNOW? The word *bassinet* is also spelled *bassinette*. Do not confuse a baby cradle for the medieval 13th century *bascinet*, which is a metal helmet that knights used to wear; a newborn will sleep much better in the former than the latter!

Now you know!

Two Formulas for Bed Turnover

There are two formulas that can be used to calculate bed turnover rate, *direct* and *indirect*. The direct formula is as follows:

$$\frac{\text{Number of discharges (including deaths) for a period}}{\text{Average bed count during the period}}$$

The indirect formula is as follows:

$$\frac{\text{Occupancy rate} \times \text{number of days in a period}}{\text{Average length of stay}}$$

The indirect formula is used when there is a change in bed count during the period you are calculating. The direct formula works in conjunction with the total number of discharges in a given period. Bed turnover rate allows the administration to see the effect of changes when occupancy increases and/or length of stay decreases. This type of statistic permits facilities to benchmark against others or within their own facility regarding the utilization of services by patients. Length of stay is another important factor for the administration to review concerning utilization management.

Length of Stay

Length of stay (LOS) is the actual number of days that a patient is admitted in the hospital. LOS data are important for service planning, resource allocation, and bed utilization. However, they are also important for hospital efficiency. It is in the best interest of patients to leave the hospital as soon as they are able. It stands to reason that hospital personnel would need accurate data on LOS to ensure that the number of days a patient stays in the hospital is only as long as needed.

LOS is calculated from the time the patient is admitted to the time he or she is discharged. This information is important to utilization management in that it allows the facility to review its ability to provide care to the patient in the most cost-effective manner. Not only does this measure take into account the efficiency of care provided but also the underutilization and overutilization of services involving patient care.

LOS data are critical for determining methods for reducing LOS. Say, for example, that your hospital's LOS for surgical procedures was higher than the national average. In this case, you might examine what other hospitals do to reduce this time. Reducing this time benefits everyone because the hospital increases utilization and patients have recovery time decreased. Variables involved in LOS computation include average (or mean) stay, median stay, and LOS standard deviation.

The average LOS is tied to Medicare severity diagnosis-related groups (MS-DRGs) and can be used as a benchmark when measuring performance against that of other equivalent facilities as well as within the facility itself. MS-DRGs are categories used to identify the services that a hospital provides to facilitate reimbursement by Medicare. As of 2015, there were roughly 745 MS-DRGs. By having this information, facilities can easily spot whether there are any radical values or outliers. When average LOS is reviewed, each patient who falls into a particular MS-DRG is given an average LOS. This categorization is directly related to finance and reimbursement for the hospital.

For every inpatient stay, there is only one MS-DRG used. The MS-DRG includes the principal diagnosis and any additional diagnoses, the principal procedure and any additional procedures, if performed, and the patient's sex and discharge status. The diagnoses and procedures that have codes assigned through the International Classification of Diseases, 10th Revision, Clinical Modification (ICD-10-CM), starting October 1, 2014, are the driver for the MS-DRG selection. Additional diagnoses that require more utilization of services can present major complications or comorbidities (MCC) or complications/comorbidities (CC) and increase reimbursement. It is crucial for coders to select the correct diagnosis codes to achieve maximum reimbursement without committing fraud.

There are 26 major diagnostic categories (MDCs) that can relate to a specific organ system. For example, a diagnosis of osteomyelitis can be split into three MS-DRGs. MS-DRG 539 is osteomyelitis with MCC, MS-DRG 540 is osteomyelitis with CC, and MS-DRG 541 is osteomyelitis without CC/MCC. As you continue your studies covering reimbursement and coding, the importance of using the MS-DRGs for reimbursement purposes will become clearer.

Hospitals are reimbursed based on the MS-DRGs, and the actual cost of the admission has no bearing on reimbursement. For every MS-DRG, there is an assigned weight that may be adjusted according to the various resources that are utilized by the patients and their corresponding costs. These weights are updated annually to reflect changes in medical practice, hospital resources, procedural and diagnostic descriptions, and criteria for MS-DRG assignment.

By using such methods, hospitals are appropriately reimbursed for complex care cases and can prevent fraud by those trying to overcharge for services. With the ever-rising cost in care, the move to electronic health records, and the changes in the healthcare system resulting from the Affordable Care Act, the benefits of such accountability—not only for hospitals, but also for payers and patients—is evident.

For example, imagine that a patient, Sally, is admitted to Ocean View Hospital on July 6 with the status of Diabetes w/CC (diabetes with complication/comorbidity). The MS-DRG is 638 and the arithmetic mean LOS is 3.9. So, if Sally stays longer than the average LOS of 3.9, she is then an outlier due to the MS-DRG. Moreover, Sally's blood glucose level must be stable and under control before she may be discharged. Thus you can see how a patient's MS-DRG category and possible complications can affect the LOS.

Additionally, average LOS can be compared among patients with similar diagnoses but different outcomes to determine and address discrepancies and improve performance. For example, imagine that Dr. Carter on average releases patients treated for pneumonia after only 3 days of admission and Dr. Schmidt typically releases such patients after 6 days. This discrepancy should be explored by reviewing patient charts and addressing the following questions: What other conditions do Dr. Schmidt's patients have that Dr. Carter's do not? Are Dr. Carter's patients resistant to the antibiotic treatment they are receiving? Are Dr. Schmidt's patients sicker when they are admitted? The list of questions could go on.

Earlier we talked about inpatient service days and the importance of those statistics. Now we will review discharge days—what they are and how they are used.

Discharge Days and How to Calculate Them

Discharge days are a measure of the number of days a patient is hospitalized, from admission until the day the patient leaves the facility (not counting the actual discharge day). Patient deaths and patient transfers to another facility are also included in the discharge days.

Hands-on Statistics 4.4: Calculating Discharge Days

To calculate discharge days for a patient who is discharged on a day for which the count has already occurred, follow these directions:

1. First consider a patient whose stay occurred in a single month. For example, if a patient is admitted on May 14 and discharged on May 19, you would subtract 14 from 19.

   19 − 14 = 5 days

2. Now consider a patient whose stay occurred over multiple months. For example, if a patient is admitted June 28 and discharged on July 4, the discharge days would be calculated as follows:

   3 days for the month of June + 3 days for the month of July = 6 days

Total Length of Stay

Total LOS is also referred to as total discharge days, or more commonly as LOS. It represents the total LOS for a group of inpatients who were discharged during a specific period. Total LOS is not the same as inpatient service days. You will remember that discharge days are counted after discharge and inpatient service days are counted concurrently.

LOS is calculated for each type of clinical service provided, such as medicine, surgery, and obstetrics. If a patient's LOS is more than 30 days, those days are added for a complete total of LOS. Total LOS is important to administration to analyze LOS of discharged patients in groups who have similar ages, diseases, and treatments.

Average Length of Stay

Average LOS is just an average of the LOSs of a group of inpatients who were discharged in a specific period that is under review. The formula for calculating this measure is as follows:

$$\frac{\text{Total LOS (discharge days)}}{\text{Total discharges (including deaths)}}$$

The average LOS is also known as the mean LOS. Average and mean can be used interchangeably, as they define the same thing. Also remember that newborns are not included in this formula.

Hands-on Statistics 4.5: Calculating Total Length of Stay

To calculate total LOS, refer to **Tables 4.1** and **4.2**, and follow these directions:

Table 4.1 Water View Hospital Discharge List for July 27, 2017

Patient Name	Age	Clinical Service	Admission Date	Length of Stay
Adams	28	Obstetrics	7/24	3
Bolton	64	Medicine	7/15	12
Crisco	41	Surgery	7/22	5
Doran	75	Medicine	7/18	9
Evans	52	Surgery	7/10	17
Fisk	39	Obstetrics	7/25	2
Hall	83	Surgery	7/5	22
Inman	33	Surgery	7/12	15
Ingram	14	Medicine	7/1	26
Johnson	20	Medicine	7/21	6
Mills	65	Medicine	7/18	9
Nall	45	Surgery	6/2	29
Overstreet	25	Obstetrics	7/20	7
Tillison	41	Obstetrics	7/21	6
Williams	15	Medicine	7/10	17
Total:				185

(continues)

Table 4.2 Seaside Hospital Discharges in 2019

Month	Medicine Discharges	Discharge Days	Medicine ALOS
January	125	468	divide 468 by 125
February	183	525	
March	151	497	
April	162	501	
May	147	474	
June	186	510	
July	143	480	
August	138	450	
September	149	491	
October	156	505	
November	201	523	
December	196	521	

1. To calculate the LOS for medicine, surgical, and obstetrics, add the number of days from each of these categories, and insert the totals in the *Length of Stay* column.

Time and Date Admitted	Time and Date Discharged	Length of Stay
11:00 am on 1/20	1:00 pm on 1/30	
9:30 am on 5/3	10:30 am on 5/25	
2:00 pm on 7/28	3:00 pm on 8/10	
1:30 pm on 10/2	9:00 pm on 10/30	
6:00 am on 11/22	11:15 pm on 12/10	

2. Next, add the subtotals from each of these categories to find the total LOS:
   Medicine: 79
   Surgery: 88
   Obstetrics: 18
   Total: 185

Hands-on Statistics 4.6: Calculating Length of Stay Using Excel

In this example we will calculate the LOS, but we will let Excel handle the math. See **Figure 4.3** and follow these directions:

1. Create the spreadsheet shown in Figure 4.3 in Excel.
2. In cell F23, insert the sum function, selecting cells F7 through F21.
3. You should have a total of 185.

	A	B	C	D	E	F
1						
2		**Water View Hospital**				
3		Discharge List				
4		27-Jul-12				
5						
6		**Pt. Name**	**Age**	**Clinical Service**	**Admission Date**	**Length Of Stay**
7		Adams	28	OB	July 24th	3
8		Bolton	64	Medicine	July 15th	12
9		Crisco	41	Surgical	July 22nd	5
10		Doran	75	Medicine	July 18th	9
11		Evans	52	Surgical	July 10th	17
12		Fisk	39	OB	July 25th	2
13		Hall	83	Surgical	July 5th	22
14		Inman	33	Surgical	July 12th	15
15		Ingram	14	Medicine	July 1st	26
16		Johnson	20	Medicine	July 21st	6
17		Mills	65	Medicine	July 18th	9
18		Nall	45	Surgical	June 2nd	29
19		Overstreet	25	OB	July 20th	7
20		Tillison	41	OB	July 21st	6
21		Williams	15	Medicine	July 10th	17

Figure 4.3 LOS using Excel: Step 1.

Hands-on Statistics 4.7: Calculating Average Length of Stay

Table 4.3 contains a list of the number of patients by month who received medical care at Seaside Hospital in the year 2019. Using the data in that table, calculate the following measures: average LOS (ALOS), median LOS, and standard deviation. Remember to divide discharge days by the number of medicine discharges to get ALOS. You will need to round to one decimal place.

Table 4.3 Number of Patients at Seaside Hospital

Clinical Services	Admission Date	Discharge Date	Length of Stay
Medicine	8/1	8/29	
Surgery	8/20	8/20	
Obstetrics	8/25	8/26	
Cardiac	8/15	8/28	
Intensive care	8/1	8/21	

(continues)

Table 4.3 Number of Patients at Seaside Hospital (*continued*)

Total Length of Stay		
Medicine	N/A	
Surgery	N/A	
Obstetrics	N/A	
Cardiac	N/A	
Intensive care	N/A	

Hands-on Statistics 4.8: Calculating Average Length of Stay Using Excel

Let's start by filling in the spreadsheet. Follow these directions:

1. Open a blank workbook in Excel and key in the data for Seaside Hospital, as you see in **Figure 4.4**.
2. In the cell to the right of 468 for the month of January, insert the following formula:

   =C5/B5

   This will divide the number of discharge days that occurred in January, 468, by the number of discharges for that month, 125, which will produce the ALOS. Note that as soon as you click out of this cell, it will calculate the ALOS for January, which is 3.744.

	Seaside Hospital		
Month	Medicine Discharges	Discharge Days	Medicine ALOS
Jan.	125	468	
Feb.	183	525	
March	151	497	
April	162	501	
May	147	474	
June	186	510	
July	143	480	
August	138	450	
September	149	491	
October	156	505	
November	201	523	
December	196	521	

Figure 4.4 Seaside Hospital data in Excel sheet.

	Seaside Hospital		
Month	Medicine Discharges	Discharge Days	Medicine ALOS
Jan.	125	468	3.744
Feb.	183	525	
March	151	497	
April	162	501	
May	147	474	
June	186	510	
July	143	480	
August	138	450	
September	149	491	
October	156	505	
November	201	523	
December	196	521	

Figure 4.5 Dragging and autofilling in a spreadsheet.

3. Hover your cursor over the lower right corner of the cell, then click and drag the contents down to the last cell in the column, as shown in **Figure 4.5**; doing so will autofill each cell with the same formula. Excel automatically changes the row numbers within the formula in each cell that you copy.

Next, we will adjust the precision of the data displayed in the ALOS column.

1. Click on the column head, labeled *D* in this example, for ALOS, which will highlight the entire column.
2. Right-click and select *Format Cells* from the pop-up menu. The *Format Cells* window will display.
3. Click on the *Number* tab of this window then on *Number* in the *Category* field.
4. Highlight the *Decimal places* field, type "0," and then click on *OK*. The *Format Cells* window will close.
5. While the column is still highlighted, click the center button on your toolbar to center your data.

Lastly, we will view these data graphically.

1. Click on the *A* column head to highlight this column.
2. Hold down the right CTRL key and highlight the last column by left-clicking on the *D* column head.
3. Click on the *Insert* tab in your toolbar, then on *Bar* for bar chart.
4. Then click on the upper leftmost *2D* bar icon, which should appear in a pop-up window. A bar chart like the one in **Figure 4.6** should appear.

Note the following regarding the graphic that appears. All months are not displayed initially; just use your mouse to extend the graphic in Excel and you will see that they appear. Also, note that we have whole numbers displayed in the last column, yet in the bar chart the data appear more precise. In fact, Excel uses the more precise numbers that result from the calculation using the formula to create the graphic, rather than the rounded numbers that are displayed in the cells of column D. Follow step 4, only this time set the decimal places to 1. You will see that, for example, February is actually 2.9 and not 3.

(continues)

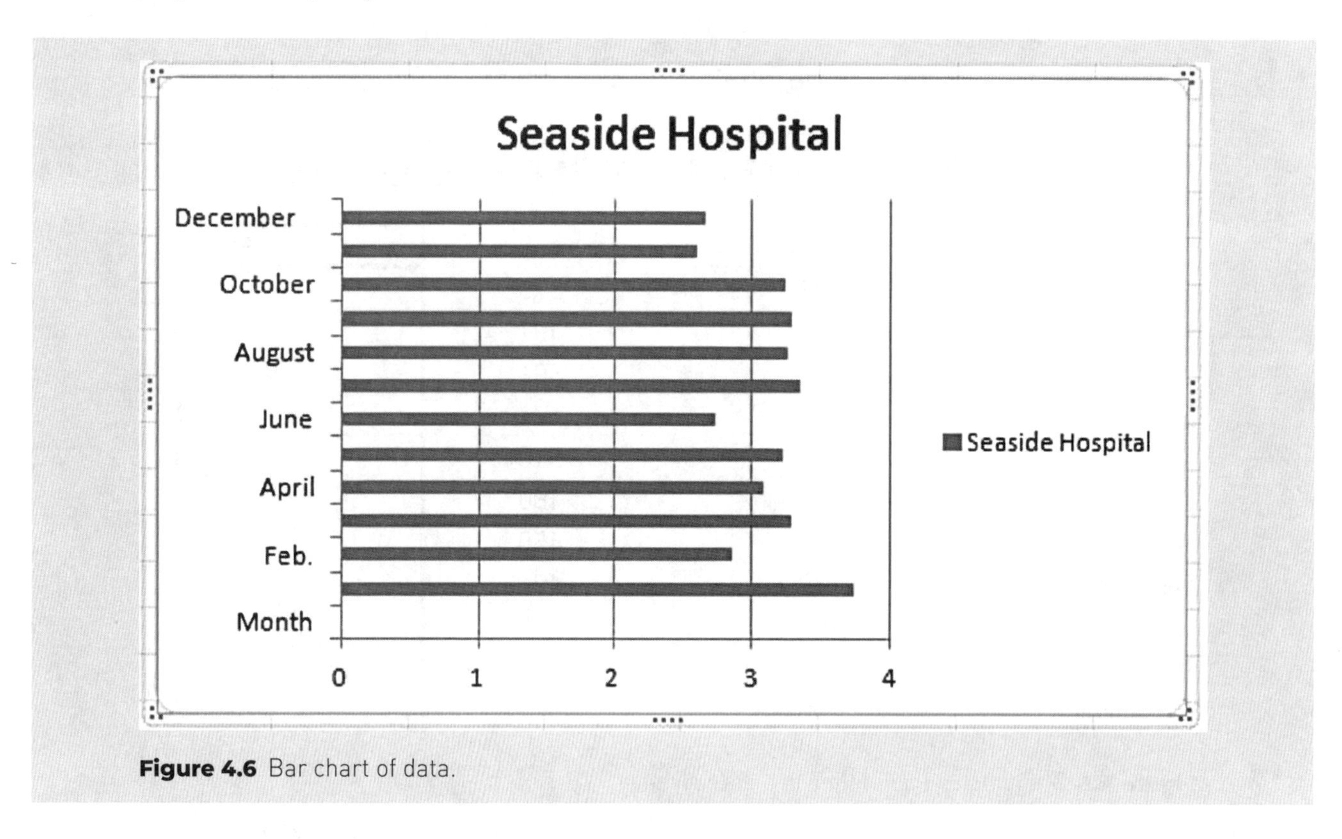

Figure 4.6 Bar chart of data.

Median and Standard Deviation for Length of Stay

Median Length of Stay

Unlike the mean (average), which is the average of all the values, median is the number that separates the upper half of your data from the lower half, such that there are exactly as many numbers greater than the median number as there are less than it. To compute median, arrange the data (sort) from lowest value to highest and pick the middle value. For example, of the following sorted numbers, 7 is the median value: 2, 3, 3, 4, 7, 7, 8, 9, 12.

For an odd number of values, the center value is the median. For an even number of values, the median is the midpoint value between the two center values (i.e., the average of those two values).

Why Might You Use Median Instead of Mean?

Mean is sensitive to data that are highly skewed. Data such as LOS could very well be skewed. For example, you may have a few patients who stayed in the hospital for months, whereas most stayed for only one or two nights. Means LOS, in this case, would not give an accurate overall picture of usage for the majority of patients. In this situation, median would likely give a more accurate representation of LOS.

Hands-on Statistics 4.9: Compute Median Length of Stay Using Excel

To find the median LOS and median discharge days with Excel, see **Figure 4.7** and follow these directions:

1. Open the "Seaside Hospital" Excel workbook that you created in the previous Hands-on exercise.
2. Type "Median Length of Stay" and "Median Discharges" in two different columns to the right of column D, as show in Figure 4.7.
3. In a cell beneath "Median Length of Stay," type the following formula:

   =median(D4:D15)

   When you click out of this cell, the median LOS should display, which should be 3.2.

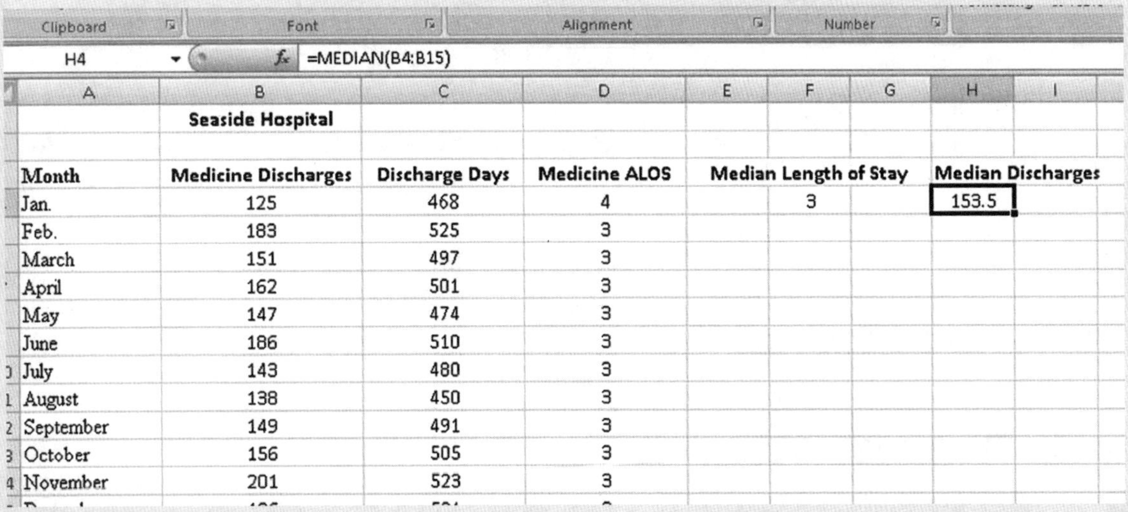

Clipboard | Font | Alignment | Number

H4 =MEDIAN(B4:B15)

Month	Medicine Discharges	Discharge Days	Medicine ALOS	Median Length of Stay	Median Discharges
	Seaside Hospital				
Jan.	125	468	4	3	153.5
Feb.	183	525	3		
March	151	497	3		
April	162	501	3		
May	147	474	3		
June	186	510	3		
July	143	480	3		
August	138	450	3		
September	149	491	3		
October	156	505	3		
November	201	523	3		

Figure 4.7 Median LOS.

4. In a cell beneath *Median Discharges*, type the following formula:

=median(B4:B15)

When you click out of this cell, the median discharges should display, which should be 153.5.

Standard Deviation of Length of Stay

Standard deviation is a measure of how spread out (or close together) a group of numbers are. For example, the dataset (1, 3, 3, 4, 5, 6, 12) features numbers that are close together, whereas the dataset (2, 3, 4, 56, 77, 1000) features numbers that are far apart. The standard deviation might inform us of skewing or outliers in the dataset, which would not be apparent by just considering the mean.

Standard deviation is measured as the square root of the variance of the values, or sigma. We will now see how to compute standard deviation of LOS with Excel.

Hands-on Statistics 4.10: Compute Length of Stay Standard Deviation Using Excel

In this example, we will compute the standard deviation of discharge days per month, which is a measure of how much one month differs from another with regard to discharge days. See **Figure 4.8** and follow these directions:

1. Open the "Seaside Hospital" Excel workbook that you used in the previous Hands-on exercise.
2. Add the heading "Standard Deviation of Discharges" in the column to the right of the column titled "Median Discharges," as shown in Figure 4.8.
3. In a cell beneath *Standard Deviation of Discharges*, type the following formula:

   =STDEV(B4:B15)

   This formula assumes your discharges data start in B4 and end in B15. When you click out of this cell, the standard deviation of discharges should display, which should be approximately 24.
4. Next, highlight the column and select the center icon to center the data. You may also change the number of decimal places, if you would like, as described in Hands-on Statistics 4.8.

(continues)

=STDEV(B4:B15)

	C	D	E–F	G–H	I	J	K
:pital							
:harges	Discharge Days	Medicine ALOS	Median Length of Stay	Median Discharges			Standard Deviation of Discharges
	468	4	3	153.5			24.3961932
	525	3					
	497	3					
	501	3					
	474	3					

Figure 4.8 Seaside Hospital with STDEV added.

Visually Representing Data

In addition to presenting data in numeric form, as in the Excel spreadsheets we created, sometimes it is effective to represent data visually, either in place of or in addition to a numeric representation. People often gravitate toward visual representations of numeric data, whereas lists of numbers often put them off. One popular way to visually represent data is via the website Wordle, as described next.

Hands-on Statistics 4.11: Use of Wordle to Visually Represent Data

Wordle is an online application that displays words that you enter as "word clouds," in which words appear in different sizes based on numeric weighting. To create a Wordle graphic to visually represent numeric data, see **Figure 4.9A** and follow these directions:

1. Download and install either the Windows or Mac version of Wordle from http://www.wordle.net/. (You will need rights to install software if in a lab.)
2. Key in words and a weight for each, as shown in Figure 4.9. The weight is simply a number, with greater numbers representing larger font sizes and smaller numbers representing smaller font sizes. In this example, we have fabricated some discharge months.
3. Click the *Go* button and your graphic is created (**Figure 4.9B**). If you do not like the layout that Wordle randomly chooses, click the *Randomize* button below the graphic and a different design will be displayed.
4. At this point you could press the Print Screen button (PRTSC) on your computer, then go to Paint (or any other graphics program), paste your screen capture in a blank document, clean it up, and be ready to use the image in a program such as PowerPoint.

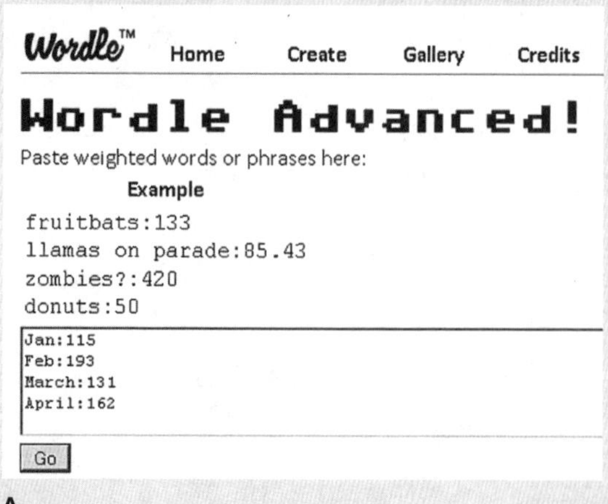

A

B

Figure 4.9 Wordle design page (A) and graphic (B).

Diagram was made using Wordle, which is available at: www.wordle.net.

PowerPoint Presentation

After compiling your data, performing necessary calculations to produce desired measures, and creating graphics, you may need to present your data to a group. If so, you will likely place your results in a word processing application, such as Word, or in a presentation application, such as PowerPoint. In the following sections, we will practice inserting statistical data into a PowerPoint presentation and into a Word document.

Hands-on Statistics 4.12: Insert a Graphic Into a PowerPoint Presentation—Static Chart

In this example, you will learn how to insert a static graphic into PowerPoint. It could be a screen capture from anything, but it must be a static graphic. Static means that if you change the data in the original application in which you created the graphic, the image as embedded in another application will not be affected, such as when inserting clip art. To insert a static image into a PowerPoint presentation, and follow these directions:

1. Open the "Seaside Hospital" Excel workbook that you created in Hands-on exercise 4.8 (**Figure 4.10**).
2. Press the PRTSC button to grab a screen capture of the chart. Note that some computers may require a key combination such as FN+PRTSC, depending on the keyboard configuration.
3. Open Windows Paint, or a different editing application of your choice.
4. Select the *Edit* tab from the toolbar and then *Paste* from the drop-down menu; your screen capture will appear, as in **Figure 4.11**.
5. Using the crop tool on the left, edit everything out of the screen capture except what you need. Select *Cut* from the *Edit* menu to cut what you want, then *File* and *New* to open a new document; then paste your chart into it.
6. Adjust the margins as needed.
7. Save the edited picture to the same folder where your PowerPoint presentation is, then insert it into your presentation like you would a piece of clip art; just select *Insert Image* from the dropdown menu of the *File* tab instead of clip art.

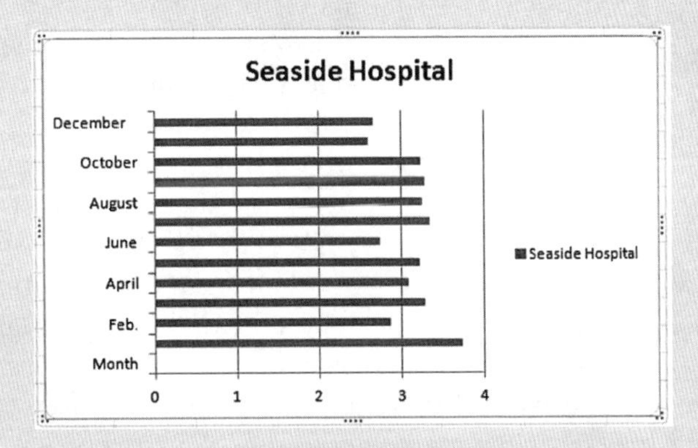

Figure 4.10 Seaside Hospital data from Excel.

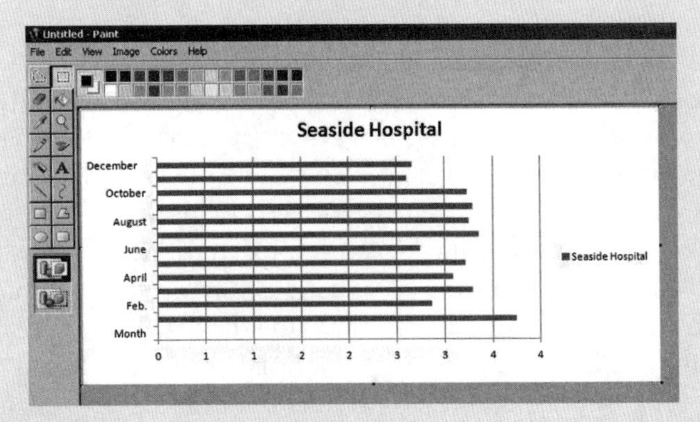

Figure 4.11 Seaside Hospital chart pasted into Paint for editing.

Hands-on Statistics 4.13: Insert Your Analysis Results in a Word Document—OLE-Linked Chart

In this exercise, you will learn how to insert a dynamic graphic into a Word document by using **object linking and embedding (OLE)**. We will cover this concept in future chapters, but to introduce it briefly for now, OLE is a method whereby an object created in one application is embedded in another but remains linked to the original application such that changes to the object in the original application will affect the image as embedded in another application. To practice using OLE, follow these directions:

1. Open a new, blank Excel spreadsheet. Create a new chart in the spreadsheet, as you did in the Seaside Hospital spreadsheet earlier. Highlight the chart and select *Copy*.
2. Open a new, blank Word document or existing one. Click in the document where you want to place your chart, and select *Paste*.
3. To prove that OLE is working, change the original data in Excel and note that the chart changes in Word.
4. Make sure both documents are in the same folder, and save both after making changes.

Data Mining: Association Rules With R-Project

We reviewed how to install R-Project and use it for basic data analysis in the previous chapter. Now we will work with R as a data mining tool. But a bit of background is in order before we start.

If you perform an internet search on phrases such as "popular data mining software" or "best free data mining software" or "free and open-source data mining software," chances are you will find R-Project listed as one of the top choices. This is a statistics package, but it can also be used for data mining. By using some of the many freely available add-on packages, you can increase the capabilities of R-Project for data mining. Next you will examine how to use R-Project for association rule data mining.

To build on what you have learned previously about data mining, we will examine several data mining **algorithms**, which are formulas for solving a problem. There are three major association rules-based data mining algorithms: Apriori, AprioriTid, and AprioriHybrid. But before we get into them, let's see what the term *a priori* really means. ***A priori*** is a Latin term that means "before the fact" and is often used to refer to knowledge that is obtained without observation or experience, typically through deductive reasoning, such as is used in solving a math equation.

In contrast to *a priori* is ***posteriori***, which means "after the fact." This approach is **empirical**, meaning that knowledge is gained through scientific observation and experience. Empirical philosophers John Locke and David Hume felt that most knowledge obtained was posteriori and that *a priori* knowledge was not even possible.

The Apriori algorithm is designed to work with association rule learning over frequently occurring items in a database. This algorithm looks for repeating items or common combinations of items in a dataset and creates linkages to other related patterns in the dataset so long as the similarly repeating items appear enough. As such, it is really looking for patterns in the data, and the algorithm makes multiple passes over the data to find these patterns, which is one of the key differentiators for this method. For example, imagine that a store that offers free coffee finds that customers who drink a cup while shopping purchase three or more items in the store 75% of the time, whereas those who do not drink a cup purchase three or more items only 35% of the time.

The second algorithm, AprioriTid, is a bit different in that after initial candidate item sets are established, the database of items is not used for counting support after the first pass looking for patterns. The first-pass candidate item sets are used and subsequent comparisons are examined. So, AprioriTid has some scaling economies after the initial first pass.

Lastly, Agrawal and Srikant first discussed a **hybrid** (mixed) version, named AprioriHybrid, of association rules for data mining at a conference in Santiago, Chile, in 1994. In their research, these authors determined a way to predict or associate what customers were purchasing based on bar-code scanning data. These market basket data were found to be very important. Based on these findings, retail chains began cross-referencing these data with customer loyalty cards, which provided customers with a discount or other incentive. Another important fact they found was that the AprioriHybrid algorithm continues to work well as a database's size increases.

For a large retail business, AprioriHybrid is an algorithm to seriously consider for predictive analysis (analyzing current and historical data to make a prediction about the future) of consumer shopping patterns. However, it is not just a tool for business, as the concept may be applied to many different areas, such as health care. It is used in health care for patient care and for business operations of the hospital. Usage as of 2019 involves text and numerical data, but image databases are also growing in use. As 2019 data-mining examples, Microsoft and Dartmouth College have developed PhotoDNA software to help thwart child pornography, and Microsoft Azure Content Moderator allows for machine-assisted content moderation of social media to help minimize cyberterrorism and other criminal activities. Regardless of the type of database, the volume of data is radically increasing. To support this growth, many more hospitals have adopted electronic health records (EHRs). As Paddock (2012) notes, "The proportion of US hospitals that now use EHRs went up from 16% in 2009 to 35% in 2011." Considering that as of 2014, most medical facilities in the United States will be required by law to use an electronic health record, the amount of data currently, or that will soon be, available for analysis is immense.

For an example related to patient care, consider data mining that uncovered some patterns of an adverse reaction to a drug, given that the patient had a unique set of preexisting conditions. These uncovered data could then be passed on to medical professionals when considering what treatment to recommend to patients. Another example would be images from a pill-sized camera that you swallow that takes many thousands of images and transmits them to a wireless storage device the patient wears temporarily on a belt. The resultant images are then "mined" for certain anomalies such as dark spots, which might indicate a need for further investigation. From a business operations perspective, data mining could perhaps help one make better choices on when to purchase consumable items the hospital requires or make better staffing decisions. Doddi, Marathe, Ravi, and Torney (2001) note that in data mining of medical records, examining claims and procedure codes using association rules discovery is "one common method for discovering such relationships."

Hands-on Statistics 4.14: Install arulesViz Package for R-Project

To install the arulesViz package for R-Project, see **Figure 4.12** and follow these directions:

1. Start R-Project and make sure you are connected to the internet.
2. From the R menu, select *Packages*, and then *Install Package(s)*, as in Figure 4.12.

Figure 4.12 Install menu in R-Project.

3. Choose *0-Cloud* as a site from the pop-up window that appears.
4. Select the *arulesViz* package from the pop-up window that appears and click *OK*, as in the figure.
5. If asked if you want to create a personal library, select *Yes*.

At this point, the installer will download the required library files, in zip (compressed) format, from the mirror download site (0-Cloud) you selected. Of course you could have chosen a different site, as they all "mirror" each other. Whenever you choose a mirror download site (for any software application that is distributed in this manner), you have options of where to download from since one may be busy.

In the next example, you will examine some built-in sample data based on grocery store receipts. In many explanations of the classic **market basket analysis**, you will see reference to an eggs milk bread database. In this type of analysis, grocery store receipts are examined to find patterns, such as the likelihood that patrons who purchase eggs will also purchase milk and bread. With this knowledge, a store could do many things, such as targeted marketing or locating items near each other.

Before we try the Hands-on exercise, you should be aware of six important concepts that are related to Apriori data mining: support, confidence, benchmark, lift, left-hand side, and right-hand side. Herrick provides descriptions for these key data mining terms. He defines **support** as the "number of data points that meet a set of rules and/or assumptions" (Herrick, 2008). These data points might be customers, transactions, visits, etc. Using the classic market basket analysis, if you found that customers who purchased milk also typically purchased bread, then the support value would be in how many of the transactions this proved true.

Confidence is defined by Herrick as "a ratio that takes the support number and divides it by the number of instances where the rule may hold true" (Herrick, 2008). Returning to our previous example, the rule would be that those who purchased milk also purchased bread. This value then is the total number of customers who purchased milk and bread divided by those who purchased only milk. Of course in data mining, as in classical statistics, the size of the sample is very important because a larger sample typically yields more accurate assumptions.

A **benchmark**, or benchmark confidence, is "the total number of items that meet an outcome divided by the total number of items in the database" (Herrick, 2008). If bread purchases were what you were trying to analyze, then the benchmark would be the total number of purchases in which bread was purchased divided by the total number of purchases.

Lift, or lift ratio, is a measure of the importance of a data mining association rule. As such, it is equal to the confidence divided by the benchmark. It is often represented visually in a **lift chart**. A lift ratio greater than 1.0 suggests support for the association rule you discovered. According to Hahsler and Chelluboina (2013), "Measures like support, confidence and lift are generally called interest measures because they help with focusing on potentially more interesting rules."

Lastly, **left-hand side (LHS)** and **right-hand side (RHS)** might be associated as "bread purchases lead to milk purchases" or by similar associations. LHS is referred to as the *antecedent* and RHS as the *consequent*.

Hands-on Statistics 4.15: Use R-Project for Apriori-Generated Association Rules

To learn how to use R-Project to generate Apriori association rules, see **Figures 4.13** and **4.14** and follow these directions:

1. Start R-Project.
2. From the command prompt, enter the following:

   library("arulesViz")

   You should see from the output screen that it loads and attaches other packages that are required.
3. Type the following:

   data("Groceries")
4. Type the following:

   summary(Groceries)

   This will show a summary of statistics on the data. Note that no quotes were used around *Groceries* this time. You will see a list of frequently purchased items, the number of item sets (32), etc.
5. Next we will mine some association rules using Apriori. Type the following at the prompt:

   rules <- apriori(Groceries, parameter =list(support=0.001, confidence=0.5))
6. Since we have generated some association rules now, let's examine them by typing "rules."

```
> inspect(head(sort(rules, by ="lift"),2))
  lhs                          rhs                  support confidence     lift
1 {Instant food products,
   soda}                    => {hamburger meat} 0.001220132  0.6315789 18.99565
2 {soda,
   popcorn}                 => {salty snack}    0.001220132  0.6315789 16.69779
> |
```

Figure 4.13 Top two highest rules according to lift.

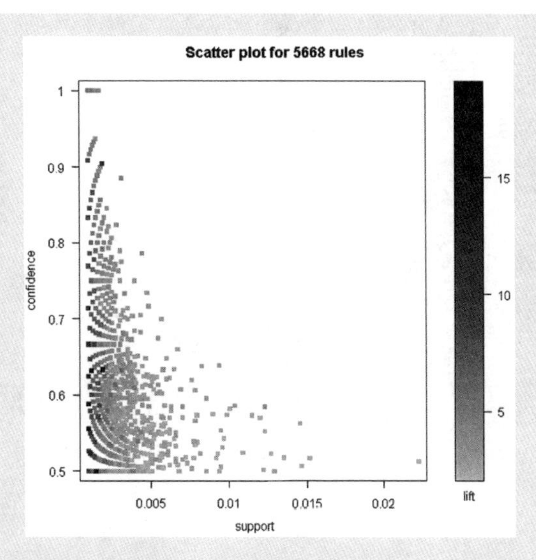

Figure 4.14 The plot(rules) command.

7. You should see that we have 5668 rules, based on Apriori association. Next let's find the top two rules based on the strength of the lift by typing the following, as shown in Figure 4.13:

   inspect(head(sort(rules, by ="lift"),2))

8. In Figure 4.13, you can see that we discovered an association between instant food products, soda, and hamburger meat, as well as soda, popcorn, and a salty snack.
9. To visualize your data, the arulesViz library offers the plot command. In the next two examples, you will see that some higher levels of lift can be found all along the range of confidence but only at lower levels of support. Also, darker blocks and circles represent higher lift, which means a stronger association between item sets.
10. Key in "plot(rules)" to show a scatter plot of all the rules discovered, as shown in Figure 4.14. Close the graphics window when done.
11. Upon returning to the command prompt window, key in "plot(rules,method="grouped")" to display a bit more information on the item sets (**Figure 4.15**). Note that you will see the LHS and RHS sides along the axis. Close the graphics window when done.

This hands-on example of Apriori association rule data mining with R-Project is very simple and certainly does not cover all of this program's capabilities, but we discovered the associations we were after.

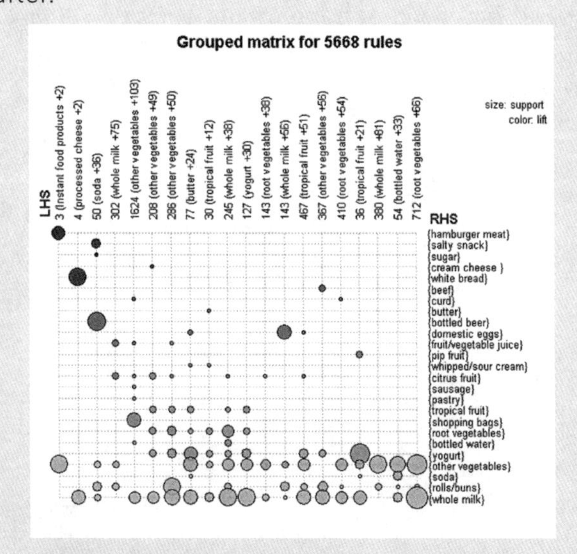

Figure 4.15 Grouped matrix for rules.

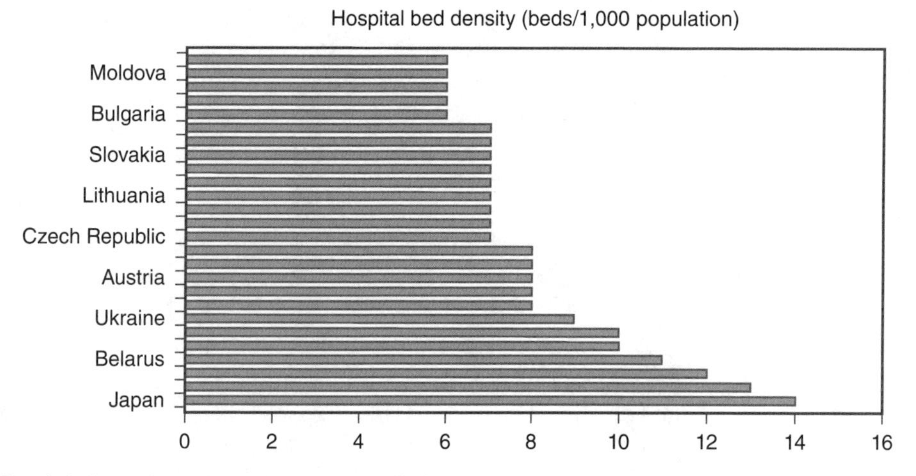

Figure 4.16 Partial view of number of beds per 1,000 people.

Global Perspective

There is no global standard for how many hospital beds per thousand people a country should have. This lack of a standard makes sense from several perspectives, as countries vary greatly in many factors that would affect the need for and supply of beds, including the burden of disease in the country (some places may need more beds) as well as the quality and type of care available. A third-world country with fewer medical funds available might have to put more patients in communal wards than a richer country. Also, the patient-to-doctor ratio would be higher in low-wealth areas.

Considering this variability, **Figure 4.16** shows the number of hospital beds available per 1,000 people for nine nations. Keep in mind that a bed does not necessarily mean a private room. In many situations, this may way be a bed in a ward or open area with many beds.

Hands-on Statistics 4.16: Import Raw Hospital Bed Density Data Using Excel

Comma-separated value (CSV) files are those in which each field, or "value," is separated from another by a comma. In this Hands-on exercise, we will take raw CSV data, in a text file, and import it into Excel so we can examine the data better.

1. Open a new, blank Excel workbook.
2. Click on the *Data* tab of the toolbar in Excel and then on the *From Text* icon.
3. In the *Import Text File* pop-up window that appears, as shown in **Figure 4.17**, select the CSV file "CH04_CIA_Factbook_Hospital_Data" in Chapter 4 of the eBook and click on *Import*. Data are also available online from http://www.indexmundi.com/map/?v=2227.

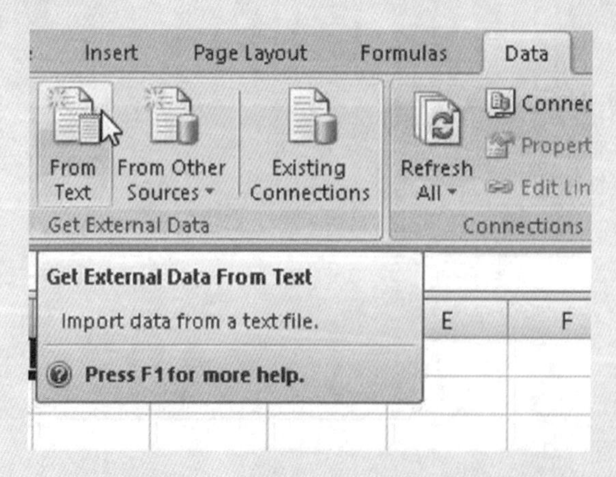

Figure 4.17 Raw CSV data import menu in Excel.

4. In the pop-up box that appears, select *Next* and make sure *Delimited* only is checked and that *Fixed* is not checked (**Figure 4.18**). Click *Next* and make sure only the *Comma* box is checked. Click *Next* one more time and select *General* format. One more click and you can select the existing Excel spreadsheet to hold your data.
5. Select *Next* and *Finish*, and your raw data will be displayed in Excel, as shown in **Figure 4.19**. Note the same data displayed in a bar chart.

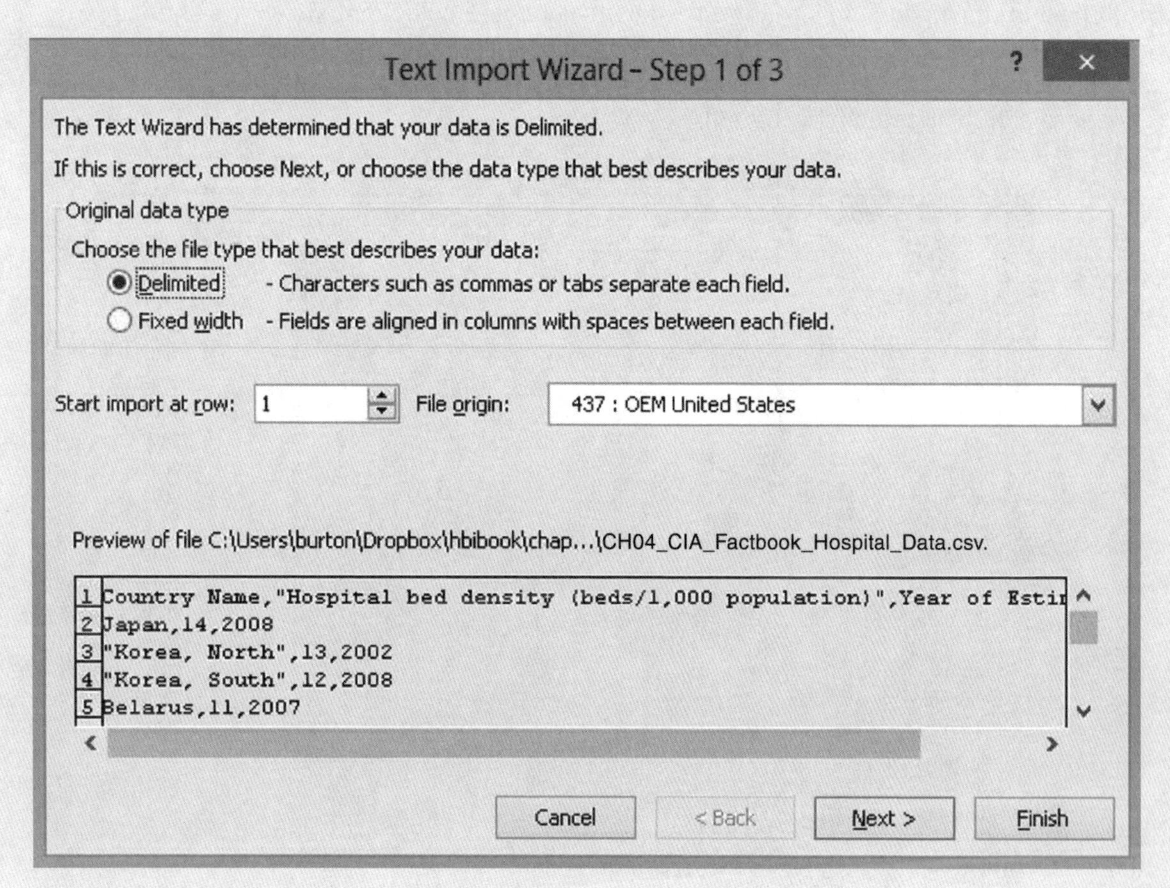

Figure 4.18 Text import wizard in Excel.

	A	B	C
1	Country Name	Hospital bed density (beds/1,000 population)	Year of Estimate
2	Japan	14	2008
3	Korea, North	13	2002
4	Korea, South	12	2008
5	Belarus	11	2007
6	Russia	10	2006
7	Russia	10	2006
8	Ukraine	9	2006
9	Germany	8	2008
10	Azerbaijan	8	2007
11	Austria	8	2008
12	Barbados	8	2008
13	Kazakhstan	8	2009

Figure 4.19 Data displayed in Excel.

Hands-on Statistics 4.17: Visually Display Raw Data Using Excel

Now let's take the raw data we imported in the previous Hands-on exercise and view it in a chart.

1. Click the A column to highlight it. Then hold down the CTRL key and highlight the B column. Partial selection of A and B is shown in **Figure 4.20**.
2. Select *Insert, Bar Chart, 2D Bar Chart, Clustered Bar* (hovering over the icons will display a screen tag showing types of Bar charts). After selecting it, you will see your chart, as shown in **Figure 4.21**.
3. Stretch your chart, keeping similar dimensions if wanted by holding down the shift key while resizing. Hover your cursor over the lower right corner of the graphic to resize so all countries can be seen. Note that you will need to stretch it down a bit to see all of the countries; otherwise it appears a bit truncated (shortened).

B1 | *fx* | Hospital bed density (beds/1,000 population)

	A	B	C
1	Country Name	Hospital bed density (beds/1,000 population)	Year of Estimate
2	Japan	14	2008
3	Korea, North	13	2002
4	Korea, South	12	2008
5	Belarus	11	2007
6	Russia	10	2006
7	Russia	10	2006
8	Ukraine	9	2006
9	Germany	8	2008
10	Azerbaijan	8	2007
11	Austria	8	2008
12	Barbados	8	2008
13	Kazakhstan	8	2009
14	Czech Republic	7	2008
15	France	7	2008
16	Hungary	7	2008
17	Lithuania	7	2008
18	Poland	7	2008
19	Belgium	7	2009
20	Slovakia	7	2008
21	Romania	7	2006

Figure 4.20 Highlighting data in columns A and B.

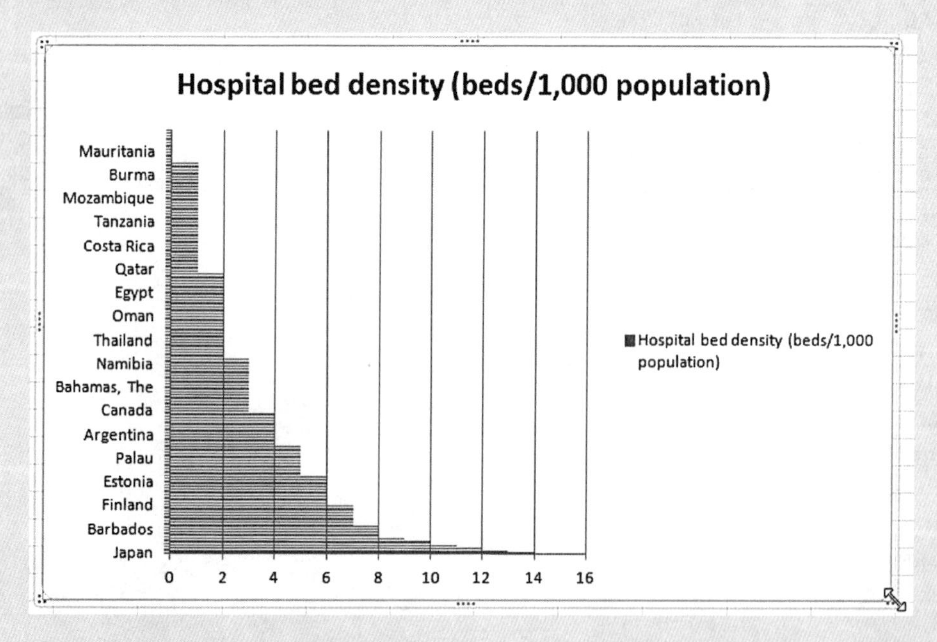

Figure 4.21 Clustered bar chart in Excel.

Chapter Summary

In this chapter you have examined many statistical measures required by hospitals such as bed count, occupancy rates, and LOS. Knowledge not only of how these measures are computed but of how they are used, in a clinical setting, is critical for all health information management and health information technology professionals. For example, data-driven decisions help hospital administrators determine if the number of beds in a given service area needs to be increased. Knowledge of these measures also facilitates **forecasting and trend analysis** of data required to predict when to increase staff availability based on high peak use times. Forecasting is a prediction of future outcomes using statistical models. It is also used in financial markets. Along with forecasting, trend analysis shows the general direction in which something is changing or developing. For example, a facility may have an increase in patients who need a particular service and a decrease in deliveries. Performing a trend analysis allows the healthcare facility to adjust budgets and staffing needs accordingly.

Data mining of association rules can also be used in forecasting and discovering relationships between these items. The large amount of healthcare data being accumulated—such as that in medical claims—requires specialized data mining tools for users to make sense of the data. Lastly, visually representing data with scatter charts and bar charts offers an efficient method to convey complex data to users in a simple visual fashion.

Apply Your Knowledge

1. Contact your state health department and learn which infectious diseases are and are not required to be reported. Does the state's list match the national infectious disease list specified by the Centers for Disease Control and Prevention?
2. What are two new facts that you learned from the Global Perspective section? Where did the United States fall in the ranking? Can you explain this outcome?
3. Using data of your own choosing (make sure you have at least 10 data elements), create two types of charts in an Excel spreadsheet.
4. From the data in the Global Perspective section, what is the number of beds per 1,000 people for the United States? How might you explain your findings?
5. Create an Excel spreadsheet using data from the PDF file obtained from the Pennsylvania Department of Health (https://www.health.pa.gov/). A PDF file containing these data is available in the Chapter 4 files in the eBook. Use the formulas provided in this chapter to compute occupancy and other data. Make sure you adjust your columns for the same precision (number of decimal places) as on the PDF.
6. Use Wordle to create an infographic image that visually describes obesity percentages by age group. You decide which age groups (e.g., students on a college campus, employees at your inner-city factory) might be appropriate. Use the downloadable version of Wordle to avoid Flash issues with the online version.
7. Create an Excel chart to show to your administrator. Using the spreadsheet you created in question 5, create a chart to display the eight highest usage amounts. Insert the result in a PowerPoint presentation.
8. Research your state healthcare website and determine the LOS for three diseases of your choice, then compare your findings to two other states using the same disease. Create a PowerPoint presentation and graph to show your findings. Also insert a short narrative in the PowerPoint presentation about the differences between hospitals.
9. For the month of June, Island Hospital reported the following: 795 inpatient service days for the 850-bed hospital. What is the bed occupancy for the month of June?
10. Define the following terms and explain why they are important: *bed turnover rate* and *observation patient*.
11. Calculate the total LOS for **Table 4.4**.
12. Define *certificate of need (CON)* and describe what is required in your state to obtain a CON.
13. Consulting the data generated in Hands-on Exercise 4.15, involving Apriori with R-Project, determine and list the top five associations.
14. St. Bart's Hospital, a 50-bed facility, reported the information in **Table 4.5** for the year 2012. Calculate the percentage of occupancy for each month and enter it in column 3. Assume that it is not a leap year.

Table 4.4

Clinical Services	Admission Date	Discharge Date	Number of Patients Discharged	Length of Stay
Medicine	8/1	8/6	150	
Surgery	8/20	8/24	35	
Obstetrics	8/25	8/28	28	
Cardiac	8/15	8/25	15	
Intensive care	8/1	8/4	8	

Table 4.5

Month	Inpatient Service Days	Percentage of Occupancy
January	568	
February	492	
March	602	
April	398	
May	468	
June	592	
July	446	
August	528	
September	588	
October	600	
November	598	
December	620	

15. Using the information provided in question 15, compute the following:
    - Percentage of occupancy for the year
    - Percentage of occupancy for each quarter
16. Using the spreadsheet on 2017 hospital data* found in Chapter 4 of the eBook, find the following information:
    a. Using the hospital census and charges by diagnosis-related group (DRG), find the range of discharges of Medicare severity DRG (MS-DRG) for the following hospitals: Auburn Regional Medical Center, Saint Joseph Medical Center, Harborview Medical Center, and Valley Medical Center.
    b. What is the mean patient days for MS-DRG 020?
    c. Calculate the frequency of swing bed admissions for MS-DRG 057.

References

Doddi, S., Marathe, A., Ravi, S., & Torney, D. (2001). Discovery of association rules in medical data. *Medical Informatics and the Internet in Medicine, 26*(1), 25–33.

Herrick, C. (2008, July 22). Defining success: Lift, support, and confidence (blog post).

Hahsler, M., & Chelluboina, S. (2013). Visualizing association rules: Introduction to the R-extension package arulesViz. Retrieved July 2014, from https://cran.r-project.org/web/packages/arulesViz/vignettes/arulesViz.pdf

Paddock, C. (2012). Electronic health record use in US hospitals has doubled in last two years. *Medical News Today*. Retrieved May 2013, from http://www.medicalnewstoday.com/articles/241871.php

Web Links

Using Excel for Statistical Data Analysis—Caveats: http://people.umass.edu/evagold/excel.html

Country Comparison—Hospital Bed Density: http://www.indexmundi.com/g/r.aspx?v=2227

Online Pattern Generator—20+ Tools for Designers: http://www.1stwebdesigner.com/freebies/free-online-tools-create-diagrams/

Sub-Saharan Africa Hospital Beds per 1000 People: http://www.tradingeconomics.com/sub-saharan-africa/hospital-beds-per-1-000-people-wb-data.html

Best Practices for an Insecticide-Treated Bed Net Distribution Programme in Sub-Saharan Eastern Africa: http://www.ncbi.nlm.nih.gov/pmc/articles/PMC3121652/ or http://www.malariajournal.com/content/10/1/157

Effects of Insecticide-Treated Bednets During Early Infancy in an African Area of Intense Malaria Transmission: A Randomized Controlled Trial: http://www.who.int/bulletin/volumes/84/2/120.pdf

Using Excel for Statistical Data Analysis—Caveats: http://people.umass.edu/evagold/excel.html

Hospital Utilization (in Non-federal Short-Stay Hospitals): http://www.cdc.gov/nchs/fastats/hospital.htm

Comparing Patient Safety in Rural Hospitals by Bed Count: http://www.ncbi.nlm.nih.gov/books/NBK20441/

Certificate of Need: http://www.ncdhhs.gov/dhsr/coneed/index.html

CON—Certificate of Need State Laws: http://www.ncsl.org/research/health/con-certificate-of-need-state-laws.aspx

Length of Stay Following Primary Total Hip Replacement: http://www.ncbi.nlm.nih.gov/pmc/articles/PMC2966203/

*Note that the spreadsheet data were obtained from the following website: https://www.doh.wa.gov/DataandStatisticalReports/HealthcareinWashington/HospitalandPatientData/HospitalDischargeDataCHARS/CHARSReports. Click on: *2017*, then *Hospital Census and Charges by DRG* under *Excel*. The file name will be 2017FYHospitalCensusandChargesbyDRG.xls.

CHAPTER 5

Morbidity and Mortality Data

Happiness depends upon ourselves.

—Aristotle

CHAPTER OUTLINE

Introduction
Morbidity Rates
 Incidence
 Prevalence
Mortality Rates
 Gross Mortality Rate
 Fatality Rate
 Net Mortality Rate
 Postoperative Mortality Rate
 Maternal Mortality Rate
 Maternal Mortality Rate as the Number of Deaths per 100,000 or 10,000 Births
 Anesthesia Mortality Rate
 Newborn Mortality Rate
 Fetal Mortality Rate
 Mortality Rates for Cancer
 Mortality-Adjusted Rates
 Interpreting Mortality Rate Results
Conducting Formal Research
 Research Design
 The Hypothesis and Null Hypothesis Statements
Statistical Measures
 P Values and Significance
 Type I Errors
 Type II Errors
 Either End of the Curve: Tails
 The Normal Distribution of Data
 Parametric and Non-parametric Tests
 z Score
 t Test
Infant Mortality by Race and County
Global Perspective
Chapter Summary
Apply Your Knowledge
References
Web Links

LEARNING OUTCOMES

After completing this chapter, you should be able to do the following:

1. Calculate the different types of morbidity and mortality rates.
2. Explain the purpose for calculating each of the different types of morbidity and mortality rates.
3. Describe the information used in calculating the different types of morbidity and mortality rates.

4. Explain the importance of the cancer registrar.
5. Define and explain hypothesis.
6. Define and explain significance.
7. Identify and describe Type 1 and Type II errors.
8. Define the normal distribution of data.

KEY TERMS

Alternative hypothesis
Cancer registrars
Degree of freedom
Early fetal death
Error
Expected mortality rate
False negative
False positive
Fatality rate
Fetal death
Hospital standardized mortality ratio (HSMR)
Hypothesis
Indirect obstetric death
Intermediate fetal death
Late fetal death
Morbidity
Mortality rate
Mortality ratio
Neonatal death
Newborn death
Nonparametric tests
Normal distribution
Null hypothesis
Observed mortality rate
Operation
P value
Parametric tests
Perinatal death
Postneonatal death
Risk-adjusted mortality rate
Significance
Surgical procedure
t test
Tail
Type I error
Type II error
z score

How Does Your Hospital Rate?

Barbara is 40 years old and expecting her fourth child. Her doctors have advised her that, due to her age, she is at high risk for possible fetal death. Barbara decides to conduct some research on the internet. She finds statistics for fetal deaths in the state of North Carolina. The statistics are based on age and race. She notes that the African American Non-Hispanic group in all age groups (15–19, 20–24, 25–29, 30–34, and 35–44 years) consistently had the highest rate of fetal deaths (North Carolina State Center for Health Statistics, 2014). The statistics also indicate that starting in the age group 30–34 years, the rate of infant mortality increased slightly among all races.

Consider the following:

1. What factors other than age and race might account for an increase in infant mortality?
2. If you were Barbara, what further research might you conduct?
3. What measures would you take, based on the statistics identified?

Introduction

It is said that there are three certainties in this world: birth, death, and taxes. Hospitals, along with many other organizations, work with data that concern two of these certainties, birth and death. So far, we have examined data for births in a medical facility (i.e., data at the beginning of the life cycle). We will now discuss morbidity rates and key aspects of death, or mortality, rates (i.e., the end of the life cycle). Mortality statistics include gross, fatality, and net mortality rates; postoperative and anesthesia mortality rates; maternal, newborn, and fetal mortality rates; and cancer-related mortality rates. You will also learn what morbidity means and how morbidity rates are measured. Finally, you will continue to learn about statistics and data mining concepts. These concepts include research design, hypothesis and null hypothesis, *P* values, significance, errors, tails, normal distribution, and parametric and nonparametric tests.

Morbidity Rates

Morbidity is illness. This term is commonly used in reference to the incidence of illness in a population, where it is usually combined with "rate"—hence, *morbidity rate*. Morbidity rates are reviewed by health insurance companies. These reports aid insurance companies in deciding the premiums that customers should pay for long-term care insurance and life insurance. The morbidity rates also assist in forecasting how many people will get sick, what diseases they may contract, and how much money they may spend in insurance claims. Healthcare providers in the United States are required to report certain infectious illnesses to the state or local authorities, usually the county health department. These illnesses are then calculated by the county, and the information is forwarded on to the state's health department. The Centers for Disease Control and Prevention (CDC), the World Health Organization (WHO), and other agencies commonly track such conditions as antibiotic-resistant infections, cancer, infectious diseases, pneumonia, Alzheimer disease, salmonella poisoning, diabetes, and the list goes on and on. The CDC publishes a weekly report called *Morbidity and Mortality Weekly* that is compiled by WHO.

Incidence

There are two other ways to measure morbidity: incidence rate and prevalence rate. Incidence rate measures the risk associated with acquiring a new condition over a specific period of time. To measure incidence morbidity rates, consideration must be given to the amount of time that each person dedicated to the study and when individuals may have been diagnosed with a specific disease.

Hands-on Statistics 5.1: Calculate Incidence Morbidity Rate by Hand

To calculate the incidence morbidity rate by hand, use the following formula:

$$\frac{\text{Number of new cases of a disease occurring in the population during a specified period}}{\text{Number of persons exposed to risk of developing the disease during that period}}$$

For example, let's say your population is 1,000 nondiseased subjects. Over the next 3 years, 125 will develop congestive heart failure. To calculate the incidence rate, you would divide 125 by 1,000, then multiply by 100:

$$125/1{,}000 \times 100 = 12.5\%$$

Your incidence report would be 12.5% newly diagnosed cases of Congestive Heart Failure.

Hands-on Statistics 5.2: Calculate Incidence Morbidity Rate Using Excel

1. Create a sheet as shown in **Figure 5.1**.
2. Use the following formula to obtain incidence rate:

=(A5/B5)*100

D5 | fx =(A5/B5)*100

	A	B	C	D	E
1					
2	Incidence Rate				
3					
4	# of patients with Congestive Heart Failure	Population	Rate over three years		
5	125	1000		12.5	

Figure 5.1 Excel incidence rate example.

Prevalence

The second rate is prevalence, which is calculated as follows:

$$\frac{\text{Number of cases of disease present in the population at a specified time}}{\text{Number of persons at risk of having the disease at that time}}$$

Prevalence is a percentage of cases in the population at a given time, which is very different from the rate of occurrence of new cases. Prevalence can also be defined as a measure of the affliction of a particular disease on civilization without consideration to time at risk or when the subjects under study were exposed to, or could have been exposed to, the potential risk factor. Of the two rates, incidence rate is more advantageous than prevalence rate in understanding disease etiology (i.e., the study of how a disease or condition occurred).

Hands-on Statistics 5.3: Calculate Prevalence Morbidity Rate by Hand

Suppose you want to measure the low-birth-weight prevalence at a hospital, where low birth weight is defined as less than 2,500 grams. For the reporting year, this hospital had 3,025 low-birth-weight infants out of a total population of 51,819 infants. You will need to use the following calculation:

$$\frac{\text{Number of infants} < 2{,}500 \text{ grams born during a reporting period} \times 100}{\text{Population during that reporting period}}$$

Thus, your calculation is as follows:

$$3{,}025 \times 100 = \frac{302{,}500}{51{,}819} = 5.84$$

This formula can be used to calculate other specific diseases for a population.

Hands-on Statistics 5.4: Calculate Prevalence Morbidity Rate Using Excel

1. Using the preceding data, create a spreadsheet as shown in **Figure 5.2**.
2. Enter the following formula in cell D5:

   =(A5*100)/B5

3. Format the cell for two decimal places to round to 5.84.

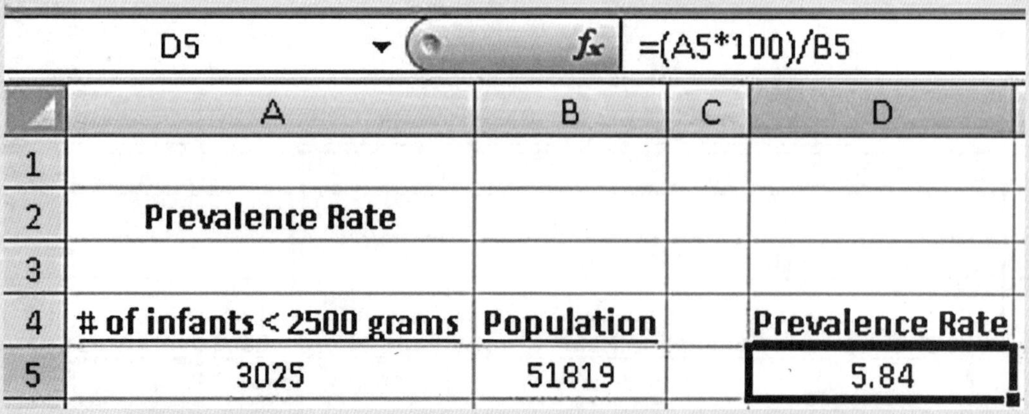

D5 | fx | =(A5*100)/B5

	A	B	C	D
1				
2	**Prevalence Rate**			
3				
4	**# of infants < 2500 grams**	**Population**		**Prevalence Rate**
5	3025	51819		5.84

Figure 5.2 Prevalence rate.

Mortality Rates

One measurement of facility quality commonly used by people outside of the hospital is mortality rate. A **mortality rate** is a measure of the number of deaths per unit of the overall population, with the unit typically being 1,000, 10,000, or 100,000. In this chapter, mortality rate will also refer to the percentage of a certain subset of people who die within a given time frame. Clearly, a high mortality rate is not a selling point for a hospital, a major concern of any facility's administration. Moreover, for internal hospital purposes, these rates and many others are calculated to assess the quality of medical care and determine whether changes are needed to operations.

Specifically, mortality rate data help organizations to plan for future and current healthcare services. Organizations such as the American Heart Association and the American Cancer Association use mortality rates to help bring awareness to diseases and the need for further research and responsiveness. Other industries, such as the automobile industry, use mortality rates to improve the safety of their products and services. We will examine several different types of mortality rates in the remainder of this chapter.

Gross Mortality Rate

Mortality data are used to track demographics relating to geography and cause of death. This information can also be compared on a global perspective. However, these statistics are somewhat difficult to gather and can be skewed due to the inability of some developing countries to accurately report deaths to the National Vital Statistics System.

Two other terms are associated with gross mortality. The first is *early mortality rate*, which is the number of deaths that occurred during the early stages of treatment. The second term is *late mortality rate*, which is the number of deaths in the late stages of ongoing treatment, or a substantial amount of time after an acute treatment. Gross mortality rate is calculated by multiplying the number of inpatient deaths, including newborns, by 100, then dividing that total by the number of discharges, including adult, child, and newborn deaths, in the same period. The formula looks like this:

$$\frac{\text{Number of inpatient deaths (including newborns) in a specified period} \times 100}{\text{Number of discharges (including adult, child, and newborn deaths) in that period}}$$

Note that the process for calculating gross mortality rates has certain rules:

- Deaths are always included in total discharges unless otherwise specified.
- Newborn deaths are added in the numerator, and discharges of newborn inpatients are included in the denominator.
- Those who are dead on arrival are not admitted, so they are not included in gross mortality rate.
- Emergency department deaths are also not included in the gross mortality rate.
- Outpatient deaths are not included in the gross mortality rate.
- Any fetal deaths are counted independently and not included in the hospital mortality rate.
- Remember to carry out to three decimal places for mortality rates, due to the number being smaller.
- When the percentage is less than 1%, add a zero before the decimal (e.g., 0.12%).

Hands-on Statistics 5.5: Calculate Gross Mortality Rate by Hand

Let's practice calculating mortality rate by hand, using the following example. Ocean Hospital has 8 deaths and 750 discharges for the month of August. Refer to the equation, and follow these directions:

$$\frac{8 \times 100}{750} = \frac{800}{750} = 1.07\%$$

1. Multiply the number of deaths by 100:

$$8 \times 100 = 800$$

2. Divide the result by the number of discharges for the period:

800/750 = 1.066666%. Round to 1.07%

Thus, approximately 1.07% of all discharges from Ocean Hospital for the month of August are deaths.

Hands-on Statistics 5.6: Calculate Gross Mortality Rate Using Excel

Next, you will practice calculating gross mortality rate using Excel. Follow these directions:

1. Open a new, blank Excel workbook.
2. Type the formula shown in **Figure 5.3** into any cell in the workbook.
3. When you click out of the cell, you should see 1.067, the result of multiplying 8 by 100 then dividing the result by 750. This result represents the gross mortality rate.
4. To display this number as a percentage, right click on the cell containing the formula and click on *Format Cells* from the pop-up menu that appears. In the *Number* tab, click on *Percentage* and then on the *OK* button at the bottom of the window. Notice that the number will now be displayed as 106.67%, as the percentage format automatically multiplies the answer from the formula by 100. To fix this, click on the cell once more and revise the formula so that it appears as follows:

   =(8*100)/750

   This should render the correct answer of 1.07%.

Figure 5.3 Formula bar in Excel, gross rate.

Fatality Rate

Another important rate to consider is the **fatality rate**, which is the number of people who die from a specific disease or accident. Facilities may calculate the fatality rate for all types of conditions, including heart transplants, influenza, AIDS, pneumonia, and heart failure, to name a few. The formula to calculate fatality rate is as follows:

$$\frac{\text{Number of people who die of a disease in a specified period} \times 100}{\text{Number of people who have the disease}}$$

Hands-on Statistics 5.7: Calculate Fatality Rate by Hand

Let's practice calculating fatality rate by hand, using the following example. Wave Hospital operated on 15 kidney transplant patients in the previous year. Of the 15 patients, 2 died. Refer to the equation, and follow these directions:

$$\frac{2 \times 100}{15} = 13.33\%$$

1. Multiply the number of deaths by 100:

   $$2 \times 100 = 200$$

2. Divide the result by the number of transplant patients:

   $$200/15 = 13.33\%$$

Thus, the fatality rate for patients with kidney transplants at Wave Hospital is 13.33%.

Hands-on Statistics 5.8: Calculate Fatality Rate Using Excel

To calculate fatality rate in Excel, follow these directions:

1. Open a new, blank Excel workbook.
2. Type the formula shown in **Figure 5.4** into any cell in the workbook.
3. When you click out of the cell, you should see 13.33333333, which represents the fatality rate.

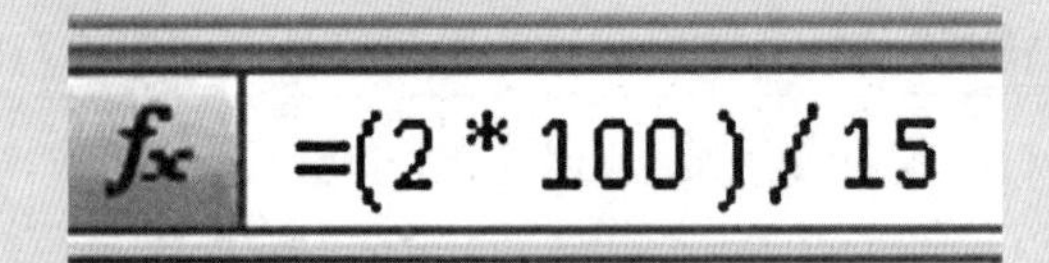

Figure 5.4 Fatality rate formula in Excel.

So, what do these calculations mean? If you were comparing hospitals to see which would be the best place to undergo kidney transplantation, you might well consider the facility with the highest survival rate (lowest fatality rate). Of course this is but one measure of quality, but it should be included in an overall total evaluation of the facility.

DID YOU KNOW? The following fascinating statistics regarding kidney transplantations in the United States were provided by the Organ Procurement and Transplantation Network as of January 2014. Visit this website (https://optn.transplant.hrsa.gov), and see if these transplantation statistics have changed.

- As of January 2014, there were 106,497 patients who were waiting for a kidney transplant.
- On any given day, there are at least 79 patients who receive organ transplants, whereas 18 persons will die waiting for a transplant.
- Five years after receiving kidney transplants, 69% of recipients are still alive.
- Sixty-two percent of living kidney donors in 2010 were women. **Figure 5.5** shows data related to patient sex and transplants by year since 1996.
- Of deceased kidney donors, 67% were white, 16% black, 13% Hispanic, and 2.3% Asian.
- If everyone in the United States would agree to become a donor, the number of lives that could be saved would be immense.

Now you know!

Transplants in the U.S. by Recipient Gender
U.S. Transplants Performed : January 1, 1988 - March 31, 2013
For Organ = Kidney, Format = Portrait
Based on OPTN data as of June 28, 2013

Change Report (Optional) :
Organ
Kidney
Go

Add Field to Report :

	To Date	2013	2012	2011	2010	2009	2008	2007	2006	2005	2004	2003	2002	2001	2000	1999	1998	1997	1996
All Genders	338,443	4,061	16,485	16,813	16,899	16,829	16,521	16,634	17,095	16,485	16,006	15,138	14,780	14,279	13,623	12,765	12,454	11,708	11,411
Male	203,777	2,479	10,057	10,237	10,282	10,146	10,060	10,213	10,479	9,941	9,573	9,056	8,757	8,482	8,076	7,590	7,354	7,027	6,804
Female	134,666	1,582	6,428	6,576	6,617	6,683	6,461	6,421	6,616	6,544	6,433	6,082	6,023	5,797	5,547	5,175	5,100	4,681	4,607

Figure 5.5 Kidney transplants from 1996 to 2013 according to the US Department of Health and Human Services.
Data retrieved 7/4/2013 from: http://optn.transplant.hrsa.gov/

Hands-on Statistics 5.9: Calculate Fatality Rate With Excel Using COUNTIF

In this example you will explore the ability of Excel to count occurrences of data. Specifically, you will use the COUNTIF Excel function to determine how many firefighters died of a heart attack. Follow these directions to examine this powerful data analysis function:

1. Visit the following website: https://apps.usfa.fema.gov/firefighter-fatalities/
2. Scroll down, and in the middle of the screen, under *Search firefighter fatality incident data*, click *Download our firefighter fatality data*. A pop-up window will ask if you want to open or save this file. Click on the down arrow next to *Save* and then on *Save as*. Another window will appear allowing you to select if you would like the calculator to save the file. Select a location and click *Save*. This is a 3-mb file, so it may take a bit of time to download, depending on your internet speed.
3. Open the file in Excel. Resize the columns as needed to be able to read some of the comments in *Column A*.
4. In a cell at the bottom of the data, type the following formula:

   =COUNTIF(L1:L3747,"Heart Attack")

   This function counts all of the cells in this column that contain "Heart Attack" as a cause of death.
5. Your result should be that 985 firefighters died of a heart attack. Note that the quality of the data is important here. For the COUNTIF function to operate properly and produce accurate results, it is critical that the data entry person spelled "heart attack" properly and did not miss capitalization, make a typo, change the case, etc.

DID YOU KNOW? The US Fire Administration (USFA) is a part of the Department of Homeland Security's Federal Emergency Management Agency. The core mission of USFA is to provide a leadership role for all firefighters (USFA, 2020). This organization responds to emergency disaster situations and also provides training to the public about home smoke alarms, cigarette safety, and many other public services. One item you would think the agency would supply is the stickers to place on the window of a child's bedroom. It does not, however—the reasoning being that families move, there is no guarantee that the child will be in that room, and some families do not want to identify the rooms in which children are sleeping.

Some of the statistics that the USFA tracks are the numbers of fires in residential and nonresidential buildings for a year, the number of deaths involved, the number of injuries, the dollar value of damaged property, and the cause of the fire, such as cooking or ignition of a mattress and bedding. The statistics that the USFA tracks are used for corrective action, to set priorities, and to provide a baseline for evaluating programs. The USFA also tracks relative risk nationally and by state. Their publications include "Sesame Street Fire Safety Program: Color and Learn," "Fire Safety Checklist for Older Adults," and "Is Your Home Fire Safe."

Now you know!

Net Mortality Rate

Net mortality rate is calculated a little differently from the regular fatality rate in that you subtract any deaths that occurred within 48 hours of admission from the total fatalities. Net mortality rates are also referred to as adjusted rates for the simple reason that certain deaths are not counted against the hospital. The rate is adjusted based on the time frame within which the death occurred. The main reason for not including these patients in the net mortality rate is that 48 hours is not a sufficient amount of time to establish if a patient's death is related to treatment given by physicians at the hospital during that time.

Certain state agencies require statistical information from hospitals on net mortality rate, so hospitals report this information to the state in which the death occurred. The CDC collects this information from all states in the United States. The information is then reviewed to identify any trends or patterns that provide information to aid in the prevention and detection of a possible outbreak or to improve overall healthcare quality. Researchers and other organizations, such as insurance companies, also use this information.

Of course there is much more to this topic, but this general overview will suffice for our purposes of performing basic calculations of net mortality rate. The formula to calculate net mortality rate is as follows. Note that the inpatient deaths and discharges include those of newborns.

$$\frac{(\text{Total number of inpatient deaths} - \text{Deaths within 48 hours of admission for specified a period}) \times 100}{\text{Total number of discharges} - \text{Deaths within 48 hours of admission for that period}}$$

Hands-on Statistics 5.10: Calculate Net Mortality Rate by Hand

Imagine that Ocean View Hospital for the month of June had 185 discharges, which included 6 deaths. Four of those deaths occurred less than 48 hours after admission. Refer to the equation, and follow these directions to calculate net mortality rate for this facility:

$$\frac{(6-4)\times 100}{185-4} = \frac{200}{181} = 1.10\%$$

1. Subtract the deaths that occurred less than 48 hours after admission from the total number of deaths, and then multiply the result by 100:

   $$(6-4)\times 100 = 200$$

2. Subtract the deaths that occurred less than 48 hours after admission from the total discharges:

   $$185 - 4 = 181$$

3. Divide the result of step 1 by the result of step 2 to get the net mortality rate as a percentage:

   $$200/181 = 1.1\%$$

Hands-on Statistics 5.11: Calculate Net Mortality Rate Using Excel

To calculate net mortality rate using Excel, follow these directions:

1. Open a new, blank Excel workbook.
2. Type the formula shown in **Figure 5.6** into any cell to calculate net mortality rate.
3. Click out of the cell, and the answer should appear; it should be something close to 1.104972, depending on the current format selected in Excel. To round the answer to 1.1%, right click on the cell and select *Format Cells* from the pop-up window. Then click on *Number* in the *Category* field of the pop-up window and select two decimal places of precision. Click *OK*, and the answer in the cell should now be 1.10.

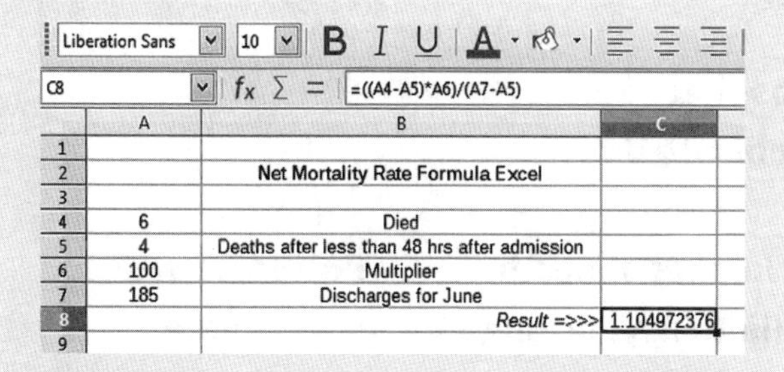

Figure 5.6 Net mortality rate formula for Excel.

Postoperative Mortality Rate

Postoperative mortality rate, also known as surgical mortality rate, is the sum of deaths that occur within 10 days after a surgical procedure, compared with the number of persons who were operated on in a given period. Remember that it is not unusual for a patient to have more than one operation during a hospital stay. Note that an **operation** is considered a surgical intervention to save a person's life, whereas a **surgical procedure** is a procedure that is performed to repair an injury to a specific body part.

Why should a hospital calculate postoperative mortality rates? Hospitals have been rating their quality of care based on postoperative mortality rates (or surgical mortality rates) for quite some time. This rate can be used to assess the surgical outcomes of individual physicians in a healthcare facility. The information allows the facility to decrease the physician's surgical caseload if there is an increase in surgical mortality rates.

The formula for postoperative mortality rate is as follows:

$$\frac{\text{Total number of deaths (within 10 days after surgery)} \times 100}{\text{Total number of patients who were operated on for a given period}}$$

Hands-on Statistics 5.12: Calculate Postoperative Mortality Rate by Hand

To calculate the postoperative mortality rate by hand, follow these directions:

1. For each of the physicians listed in **Figure 5.7**, multiply the number of deaths within 10 days after surgery by 100.
2. Divide each of the results of step 1 by the corresponding number of patients who were operated on by each physician. Each result is the postoperative mortality rate for the individual physician.

Tidal Hospital July-September, 2012		
Physician Name	Number of surgery patients	Number of deaths within 10 days of surgery
Dr. Wind	136	2
Dr. Sims	56	1
Dr. Blue	82	6
Dr. Piddler	62	3
Dr. Wilde	76	4

Figure 5.7 Tidal Hospital postoperative mortality death rates.

Hands-on Statistics 5.13: Calculate Postoperative Mortality Rate Using Excel

To calculate the postoperative mortality rate using Excel, follow these directions:

1. Open a new, blank Excel workbook.
2. Except for the last column, create a spreadsheet with data shown for Tidal Hospital in **Figure 5.8**. Note that Dr. Wind's statistics are given in row 10, the number of deaths within 10 days after surgery for each physician is shown in column F, and the total number of patients for each physician is shown in column D.
3. In the cell under *Postoperative Mortality Rate* in the row for Dr. Wind, insert the following function:

   =(F10*100)/D10

   Click out of the cell, and the rate should appear.
4. Now drag the cell containing the formula down to Dr. Wilde's row; the postoperative mortality rates for each physician will appear automatically.
5. Lastly, change your precision to two decimal places by right clicking on the column containing the postoperative mortality rates, selecting *Format Cells*, then *Number*. Set the decimal places to "2," then click the *OK* button.

Tidal Hospital July-September, 2012			
Physician Name	Number of surgery patients	Number of deaths within 10 days of surgery	Postoperative Death Rate
Dr. Wind	136	2	1.47
Dr. Sims	56	1	1.79
Dr. Blue	82	6	7.32
Dr. Piddler	62	3	4.84
Dr. Wilde	76	4	5.26

Figure 5.8 Tidal Hospital postoperative mortality using Excel.

How Does Your Hospital Rate?

Barbara is really worried her baby could die after evaluating the infant mortality statistics. Barbara is in good health and is eating the proper diet. Will this be enough to have a healthy baby?

Visit the following website and review the fetal death rates by specific age groups and by race: https://schs.dph.ncdhhs.gov/interactive/query/*Data*.

Consider the following:

1. Which race had the highest fetal death rate?
2. At what maternal age did the fetal death rates start to increase?
3. Could there have been other causes for the fetal deaths, such as other conditions that affected the mother or poor prenatal care?

Maternal Mortality Rate

When we calculate maternal mortality rate, we are considering any death of a pregnant woman or of a woman who has recently given birth that is linked to or provoked by the pregnancy. Other causes could be due to the health care the mother receives or did not receive. However, accidental deaths (such as from a boating accident or a fall from a ladder) and incidental deaths (such as murder or suicide) are *not* included in this rate.

In some cases, hospitals may want to keep separate statistics for direct and indirect obstetric deaths. In a **direct obstetric death**, the mother dies from complications associated with a procedure that is directly related to pregnancy or birth, such as incidental damage to blood vessels during a C-section. **Indirect obstetric death** is caused by some underlying condition, such as cancer or other conditions that could aggravate the pregnancy. When calculating the maternal mortality rate, use only the direct obstetric mortality rate. Included are only those mortality rates of patients who are hospitalized at the time.

Note in the following example that you will use the term *prepartum*, which means before birth. In the case of calculating maternal mortality rates, prepartum means the mother died before giving birth. *Postpartum* is after childbirth. Again, in the calculation of maternal mortality rate, the mother would have died after giving birth.

The formula for maternal mortality rate as a percentage is as follows:

$$\frac{\text{Number of direct maternal deaths for a period} \times 100}{\text{Number of obstetrical discharges (including deaths) for the period}}$$

Hands-on Statistics 5.14: Calculate Maternal Mortality Rate as a Percentage by Hand

Imagine that the Seaside Women's Center had the following statistics for the year 2002:

3,500 babies delivered, 35 aborted births, 83 prepartum maternal deaths, and 6 postpartum maternal deaths, 5 of which occurred shortly after delivery. Refer to the equation, and follow these directions to calculate maternal mortality rate for this facility:

$$\frac{5 \times 100}{(3{,}500 + 35 + 83 + 6)} = \frac{500}{3{,}624} = 0.1379 = 0.14$$

1. Multiply the number of direct maternal deaths by 100. Note that in this case the maternal deaths that occurred shortly after delivery are considered direct, whereas the one maternal death that occurred later is not direct:

   $$5 \times 100 = 500$$

2. Add together the number of deliveries, aborted births, prepartum maternal deaths, and postpartum maternal deaths:

   $$3{,}500 + 35 + 83 + 6 = 3{,}624$$

3. Divide the result of step 1 by the result of step 2 to determine the maternal mortality rate. Remember to round to two decimal places:

   $$500/3{,}624 = 0.1379 = 0.14$$

Hands-on Statistics 5.15: Calculate Maternal Mortality Rate as a Percentage Using Excel

To calculate maternal mortality rate for this facility using Excel, follow these directions:

1. Open a new, blank Excel workbook.

(continues)

2. Type the data shown in **Figure 5.9** into the cells indicated, but leave cell F10 blank.
3. In cell F10, insert the following formula:

=(F8*100)/(F4+F5+F6+F7)

4. Right click on the cell and set the precision to two decimal places. The result should be 0.14%.

F10 | fx =(F8*100)/(F4+F5+F6+F7)

	A	B	C	D	E	F
1						
2		Seaside Women's Center for the year 2002				
3						
4		Delivered				3500
5		Aborted				35
6		Prepartum				83
7		Postpartum				6
8		Mothers who died shortly after deliver				5
9						
10						0.13797

Figure 5.9 Maternal mortality rate using Excel.

Maternal Mortality Rate as the Number of Deaths per 100,000 or 10,000 Births

In addition to presenting maternal mortality rate as a percentage, researchers often calculate it as the number of maternal deaths associated with pregnancy per 100,000 or per 10,000 live births during a specific time frame. For example, for the year 2013, a hospital may have experienced 112 maternal deaths per 100,000 live births, which would be its maternal mortality rate for that year.

The formula for maternal mortality rate as the number of maternal deaths per 100,000 is as follows:

$$\frac{\text{Number of deaths credited to maternal conditions during a given time} \times 100{,}000}{\text{Number of live births during the given time}}$$

Calculating the maternal deaths per 10,000 works the same way, but you would use 10,000 as the multiplier instead of 100,000.

Hands-on Statistics 5.16: Calculate Maternal Mortality Rate as the Number of Deaths per 100,000 or 10,000 Births by Hand

Sea View Hospital recorded 1,423 live births for the year 2018 and 19 maternal deaths following abortions in the same period. Refer to the equation, and follow these directions to calculate maternal mortality rate per 100,000:

$$\frac{19 \times 100{,}000}{1{,}423} = \frac{1{,}900{,}000}{1{,}423} = 1{,}335.207 \text{ per } 100{,}000 \text{ live births}$$

1. Multiply the number of maternal deaths by 100,000:

$$19 \times 100{,}000 = 1{,}900{,}000$$

2. Divide this total by the number of live births during this period:

$$1{,}900{,}000/1{,}423 = 1{,}335.27 \text{ per } 100{,}000 \text{ live births}$$

To calculate for every 10,000 live births, you would just replace the 100,000 with 10,000.

Anesthesia Mortality Rate

Anesthesia mortality rate is a measure that compares the deaths that occur due to anesthesia-related issues during a given period with the number of patients to whom anesthesia was administered during the same period who survived. Anesthetic agents may be classified as local, regional, or general. The formula represents deaths caused by the total of the three agents.

The formula for anesthesia mortality rate is as follows:

$$\frac{\text{Total number of deaths caused by anesthetic agents} \times 100}{\text{Total number of anesthetics administered to patients}}$$

Hands-on Statistics 5.17: Calculate Anesthesia Mortality Rate by Hand

During the past year, Sea View Hospital administered anesthesia 4,200 different times to patients. Three deaths occurred due to anesthesia during that time. Refer to the equation, and follow these directions to calculate anesthesia mortality rate for this facility:

$$\frac{3 \times 100}{4,200} = \frac{300}{4,200} = 0.07\%$$

1. Multiply the number of deaths caused by anesthesia that occurred during the past year by 100:

$$3 \times 100 = 300$$

2. Divide the result of step 1 by the total number of times patients were administered anesthesia during that period:

$$300/4,200 = 0.0714\%$$

3. Round your percentage two decimal places: 0.07%.

Hands-on Statistics 5.18: Calculate Anesthesia Mortality Rate Using Excel

To calculate anesthesia mortality rate using Excel, follow these directions:

1. Open a new, blank Excel workbook.
2. Type the data shown in **Figure 5.10** into the cells indicated, but leave cell F8 blank.
3. Type the following formula in cell F8:

   =(F5*100)/F4

   When you click out of the cell, the anesthesia mortality rate should appear there.
4. Set the decimal places by right clicking and selecting two decimal places of precision.

	A	B	C	D	E	F
1						
2		**Sea-View Hospital**				
3						
4		Patients adminstered anesthesia				4200
5		Deaths due to anesthesia				3
6						
7						
8				Anesthesia death rate		0.07143
9						

Figure 5.10 Anesthesia mortality rate using Excel.

Newborn Mortality Rate

Newborn mortality rates are used to follow national trends in newborn mortality and to prevent future newborn deaths. Some of the terms used in discussing this measure are **perinatal deaths**, which are stillborn infants, and **neonatal deaths**, which are deaths that occur up to 28 days following birth. **Newborn deaths** are those in which the infant is born but dies during the same hospital admission. Infant deaths include those of children up to 1 year old. **Postneonatal deaths** refer to those that occur from 28 days after birth to 1 year of life.

The formula for calculating newborn mortality rate is as follows:

$$\frac{\text{Total number of newborn deaths for a period} \times 100}{\text{Total number of newborn discharges (including deaths) for that period}}$$

Perinatal death, or perinatal mortality, refers to the death of a fetus and is the measure that serves as the basis to calculate the perinatal mortality rate. The precise definition of perinatal mortality varies, specifically concerning the issue of *inclusion* or *exclusion* of early fetal and late neonatal fatalities.

Hands-on Statistics 5.19: Calculate Newborn Mortality Rate by Hand

Sea Breeze Hospital had 3,300 newborn discharges and 5 newborn deaths in 1 year. Refer to the equation, and follow these directions to calculate newborn mortality rate for this facility:

$$\frac{5 \times 100}{3{,}300} = \frac{500}{3{,}300} = 0.15\%$$

1. Multiply the number of newborn deaths by 100:

   $$5 \times 100 = 500$$

2. Divide the result by the number of newborn discharges:

   $$500/3{,}300 = 0.1515$$

3. Round your percentage two decimal places: 0.15%.

Hands-on Statistics 5.20: Calculate Newborn Mortality Rate Using Excel

To calculate newborn mortality rate using Excel, follow these directions:

1. Open a new, blank Excel workbook.
2. Type the data shown in **Figure 5.11** into the cells indicated, but leave cell B7 blank.
3. Type the following formula in cell B7:

   =(E5*100)/E4

   When you click out of the cell, the newborn mortality rate should appear there.
4. Set the number of decimal places of precision in cell B7 for two.

B7 | fx =(E5*100)/E4

	A	B	C	D	E
1					
2		**Sea Breeze Hospital**			
3					
4		Newborn Discharges			3300
5		Newborn Deaths			5
6					
7		0.15152			

Figure 5.11 Newborn mortality rate.

Fetal Mortality Rate

Fetal deaths occur when a fetus is expelled and does not breathe or show any evidence of life through pulsations of the umbilical cord, heartbeat, or muscle movement. Fetal deaths are required to be reported to state agencies, just as instances of sexually transmitted infections and gunshot wounds are required to be reported. Fetal deaths are not included in patient deaths; they are counted separately.

There are three classifications for fetal deaths: early, intermediate, and late. These classifications are based on gestation period and fetal weight. Follow along as we examine the details associated with each classification.

Early fetal deaths are those that occur at less than 20 weeks of gestation with a fetal weight of 500 grams or less. **Intermediate fetal deaths** occur at or following 20 weeks of gestation but before 28 weeks, with a fetal weight between 501 and 1,000 grams. **Late fetal deaths** occur after 28 weeks of gestation, with a fetal weight of 1,001 or more grams. Intermediate and late fetal deaths are considered stillbirths. All three types are standardized nationally, so you must clearly understand the delineation between them.

State and federal laws require that fetal deaths be tracked and reported. The CDC and other research

Fetal Mortality Rates 1985-2006	Total	20-27 Weeks	28 Weeks or more
2006	6.05	3.1	2.97
2005	6.22	3.21	3.03
2004	6.28	3.17	3.14
2003	6.32	3.25	3.08
2002	6.41	3.24	3.19
2001	6.51	3.25	3.28
2000	6.61	3.31	3.32
1999	6.74	3.39	3.38
1998	6.73	3.35	3.41
1997	6.78	3.29	3.51
1996	6.91	3.33	3.6
1995	6.95	3.33	3.64
1990	7.49	3.22	4.3
1985	7.83	2.91	4.95

Figure 5.12 Fetal mortality rate.

groups examine trends and patterns in fetal deaths in different states and across the county. Data tracked by the CDC include the relationship between fetal mortality and maternal age, effect of twins and multiple pregnancies on mortality rates, and mortality rates of ethnic groups. For example, note in **Figure 5.12** that late fetal deaths decreased from a high of 4.95 per 100,000 births in 1985 to 2.97 per 100,000 births in 2006. You can also review other disease data from the website. More information on fetal deaths by state is available at http://statehealthfacts.org.

The formula for calculating fetal mortality rate is as follows:

$$\frac{\text{Total number of intermediate and/or late fetal deaths for a period} \times 100}{\text{Total number of live births + Intermediate and late fetal deaths for that period}}$$

Hands-on Statistics 5.21: Calculate Fetal Mortality Rate by Hand

During December, Coastal Women's Center had 323 live births, 4 intermediate fetal deaths, and 6 late fetal deaths. Refer to the equation, and follow these directions to calculate fetal mortality rate for this facility:

$$\frac{(4+6)\times 100}{323+10} = \frac{10\times 100}{333} = \frac{1000}{333} = 3.0\%$$

1. Add the number of intermediate fetal deaths to the late fetal deaths and multiply the result by 100:

   $$(4 + 6) \times 100 = 1{,}000$$

2. Add the number of live births to the numbers of intermediate and late fetal deaths:

   $$323 + 10 = 333$$

3. Divide the result of step 1 (the numerator) by result of step 2 (the denominator):

   $$\frac{1000}{333} = 3.0\%$$

DID YOU KNOW? **Cancer registrars** are collections of data on cancer, such as cancer mortality rates. Cancer *registrars* are the healthcare professionals who capture these data by working closely with physicians, administrators, and researchers. They also provide support for cancer programs and ensure compliance.

Data for cancer deaths are classified by age, sex, race, and cancer site for a specific period. This information can be used to track trends in certain types of cancer for specific geographical areas or nationwide. These registries are used

(continues)

to make essential healthcare decisions to take full advantage of limited public health funds, such as screening funds. Cancer registries are another essential tool to aid those who look at etiology, diagnosis, and treatment of cancers.

The registries are also used for studying the causes and spread of cancer and for follow-up and tracking of patients who are currently under treatment or have completed treatment for a type of cancer. The follow-up is lifelong. It includes reminders to patients and doctors of appointments and provides crucial survival information. The data that are collected are used to assess patient outcomes; to provide information on cancer surveillance, quality of life, cancer program activities, and referral patterns; and to help determine the allocation of resources to healthcare facilities at the community and state levels. The data are also used to assist in reporting cancer incidence, to develop educational programs, and, most importantly, to evaluate the effectiveness of treatment modalities.

Now You Know!

Mortality Rates for Cancer

The National Center for Health Statistics, a division of the CDC, collects data on cancer, including cancer-related mortality rates. These data can be searched on the basis of any given time frame, such as a year or a quarter, for the entire United States or for smaller regions. The formula for the cancer mortality rate is as follows:

$$\frac{\text{Number of cancer deaths during a period} \times 1{,}000}{\text{Total number in population at risk}}$$

Hands-on Statistics 5.22: Calculate Mortality Rates for Cancer by Hand

According to the American Cancer Society, in 2010 there were 569,490 deaths from all types of cancer. The population according to the Census Bureau for 2010 was 308,745,538. Refer to the equation, and follow these directions to calculate the overall cancer mortality rate for the United States:

$$\frac{569{,}490}{308{,}745{,}538} \times 1000 = 1.84\%$$

1. Multiply the number of cancer deaths for that period by 1,000:

$$569{,}490 \times 1{,}000 = 569{,}490{,}000$$

2. Divide the result by the number in the US population at risk:

$$569{,}490{,}000/308{,}745{,}538 = 1.84\%$$

Hands-on Statistics 5.23: Calculate Mortality Rates Using Excel

To calculate mortality rate for cancer using Excel, follow these directions:

1. Open a new, blank Excel workbook.
2. Type the data shown in **Figure 5.13** into the cells indicated.
3. In cell C5, enter the following formula:

   =(A5*1000)/B5

   Press Enter; cell C5 should display a cancer mortality rate of 1.844528681.
4. To round to two decimal places (1.84), right click on the cell, select *Format Cells*, then *Number*, and change the decimal places to "2."

C5 | f_x =(A5*1000)/B5

	A	B	C
1			
2	Calculating Mortality Rates for Cancer with Excel		
3			
4	Cancer Deaths	Population	Cancer Mortality Rate
5	569490	308745538	1.844528681

Figure 5.13 Cancer mortality rate using Excel.

Mortality-Adjusted Rates

As you have seen, hospital mortality rates show the percentage of patients who die in the hospital as a result of a specific condition. For example, if 100 people were admitted to the hospital for drug overdose and 50 of them died, then the mortality rate for drug overdose patients would be 50%.

The problem with a hospital mortality rate is that it does not allow for mitigating issues. The process of risk adjustment allows us to include other factors in our assessment to compensate for these issues. These mitigating factors, all of which relate to survival, include the age of the patient, preexisting conditions, and severity of the illness. For example, an older patient who has a drug overdose might not have the same odds of surviving as a younger patient due to previous complications. Thus, hospital administrators often adjust hospital mortality rate to provide a more accurate benchmark for internal hospital improvement and to be used as a way to promote hospital care. In this way, it serves both internal and external customers.

There are many terms for adjusted mortality rates, including **risk-adjusted mortality rate** and **hospital standardized mortality ratio (HSMR)**. Regardless of what it is called, the computation is an adjusted hospital mortality rate, based on the mortality rate of other facilities. The **mortality ratio**, noted as a decimal value, is simply the observed number of actual deaths divided by the number of expected deaths in a certain population. A 1.0 rating means the survival rate of the hospital is as expected when compared with other facilities. Of course, a lower value means the hospital has a lower survival rate.

For an HSMR, multiply the mortality ratio result by 100. The formula for HSMR is as follows:

$$\frac{\text{Number of observed deaths} \times 100}{\text{Number of expected deaths}}$$

An **observed mortality rate** is the number of patient deaths that occur in the hospital for any reason. The **expected mortality rate** is the predicted number of deaths in the hospital based on the various levels of illness of the patients in the hospital, with very sick or complicated patients having a higher expected mortality rate. Note that the expected mortality rate does not necessarily reflect a hospital's quality of care. Other factors affect this rate, including the size of the facility, hospital and/or doctor specialization, and the equipment available. For example, patients who have a terminal illness may prefer to go to a large and prestigious hospital as opposed to a local hospital to gain access to a specialized procedure but then die, not as a result of low quality of care at the hospital, but due to the nature of their illness. The end result is that the prestigious facility has a higher mortality rate than the smaller local one, yet it is no reflection on the quality of the facility's care.

Interpreting Mortality Rate Results

An HSMR equal to 100 suggests that there is no difference between the hospital's mortality rate and the overall expected average rate. Less than 100 HSMR suggests that the local mortality rate is lower than the overall expected average rate. Greater than 100 HSMR suggests that the calculated mortality rate is higher than the overall average.

However, not everyone agrees that this measure is helpful. Research by Thomas and Hofer in 1999, based on six parameters, found that risk-adjusted mortality was not an accurate measure of hospital care. The parameters that were used by Thomas and Hofer were (1) patients who received high-quality care, (2) those who received poor-quality care, (3) proportion of patients who received poor-quality care among all hospitals, (4) hospitals that were already considered to give poor-quality care, (5) patient relative risk of getting poor-quality care in high- and poor-quality care hospitals, and (6) the number of patients who were treated in those hospitals. Of those hospitals that delivered poor-quality care, less than 12% were recognized as high mortality rate outliers. More than 60% of the hospitals provided high-quality performance. Thomas and Hofer (1999) report, "Under virtually all realistic assumptions for model parameter values, sensitivity was less than 20% and predictive error was greater than 50%. Reports that measure quality using risk-adjusted mortality rates misinform the public about their performance." Other researchers, such as Pitches, Mohammed, and Lilford (2007), have come to the same conclusion as Thomas and Hofer—that hospitals that provided poor-quality care would potentially have higher risk-adjusted mortality rates, while other hospitals with higher than likely risk-adjusted mortality rates do not deliver poor-quality care to patients. The overall conclusion is that risk-adjusted mortality rates are unreliable.

Conducting Formal Research

Although in previous chapters you have considered many methods that are related to examining data that the facility collects, in this chapter we will discuss more thoroughly the topic of formal research. Note that merely collecting and reporting data do not constitute a study. A research study, for health care purposes, should be for the improvement of a situation and diagnostic in nature.

Research Design

The design of a research study is largely determined by the type of study. Research may be classified as either experimental or descriptive. Experimental studies generally include an experimental group of subjects, which undergoes a specified intervention, and a control group, which unknowingly does not undergo the intervention but instead is given a placebo treatment. Such a design allows for comparison of the two groups and the elimination of extraneous factors that may also affect the outcome. Descriptive research—and thus descriptive statistics—in contrast, usually does not include a control group. This type of research describes patterns that are seen in data, perhaps survey data or even historical data.

Also, your research study design should be based around an important and specific research question you are trying to answer. Examples of effective research questions are, "Does an aspirin a day lower blood pressure in people identified as prehypertensive?" and "If support staff is increased by a certain factor, do patient satisfaction survey results improve?" Moreover, the research design should be such that you can obtain measurable results that help answer the question at hand. Time spent early on developing a good research question to examine is paramount to the success of the study.

When you are working on a design and refining your questions, remember to keep it simple and clear. Ask yourself the following questions: What exactly is my question? What factors would affect the research findings? What other data might I want to know? Should I focus on baby aspirin or regular aspirin? How will the weight, age, race, sex, and family history of high blood pressure of respondents affect the results? Other aspects of experimental design that you should consider include selection of statistical tests, data collection methods, determination of the significance of findings, and reporting of your findings.

The Hypothesis and Null Hypothesis Statements

Keep in mind when creating your research question that you will either need to prove or disprove the premise or concept behind it. That is, your research study can have only one of two possible outcomes: either the **hypothesis** will be true or the **null hypothesis** will be true. The hypothesis statement—also known as the **alternative hypothesis**—is a guess at the what the answer could be and answers your research question in the affirmative. In contrast, the null hypothesis statement answers your research question in the negative. Your goal, ultimately, is the reject the null hypothesis. For example, consider the following research question: "Is there a difference in blood pressure with the consumption of an aspirin a day versus not taking an aspirin?" The null hypothesis to this question, typically denoted as H_0, would be, "An aspirin a day has no significant effect on blood pressure." An alternative hypothesis to this question, in this case denoted as H_1, would be as follows: "Taking an aspirin a day has significant effect on blood pressure: it is significantly lower."

Let's take a closer look at the null hypothesis for this study. Suppose you have three groups of subjects participating in the study, designated as u1, u2, and u3. You would write the null hypothesis as follows: "H_0 u1=u2=u3 Taking an aspirin a day has no significant effect on blood pressure: it does not significantly lower blood pressure." When the null hypothesis is shown to be true, as in this case, it indicates that there is not enough significance in the research findings to prove the hypothesis (remember *significance*, as we will visit that important word soon). All three groups have the same end result as far as changes in blood pressure, as indicated by the equal signs. The values may not be exactly the same, but based on the *confidence level* (another term we will visit soon) of the value we derived, we say that there is no significant or notable difference in the findings for the groups. The alternative hypothesis, on the other hand, we would write as follows: "H_1 Not H_0. The use of aspirin does significantly affect blood pressure."

Now let's cover statistical significance and how to measure it in the next section.

Statistical Measures

P Values and Significance

In a previous chapter, you were introduced to the concept of variance with regard to measures of central tendency. Now you will learn how to compare the variances between two or more datasets to determine whether there is a significant difference between them. It is the amount of the difference between the datasets compared with the significance level that determines whether it is significant.

Significance is a percentage measure of the odds that the result of a research study did not happen by statistical accident. For example, on analysis of a given set of data, you might determine that there was a 95% chance that dataset A varied significantly from dataset B.

Significance is often measured by ***P* value**, which is an estimate of the probability that the result happened by accident and the datasets really are not significantly different. Put another way, it is an estimate of the probability that the change or difference between one dataset and the other dataset under observation is attributable only to chance.

P value is important in expressing whether there is significant variation or difference between the datasets. Again, there may well be a numeric difference, but that does not mean it is significant.

A larger *P* value shows a small level of statistical significance between datasets. A smaller *P* value shows a greater level of statistical significance between datasets. The *P* value is a standard significance measure and is derived from a table of values, based on the significance or confidence level and whether it is one- or two-sided. The value from the table is what you will compare with your findings from your analysis. In the table, you will have a choice of precision levels, with different predicted values for each level of confidence, such as .05, .01, etc.

To interpret .05 and other levels, think of percentages, with 100% being complete certainty. For a *P* value of .05, there is 95% certainty that the difference did not occur by chance. There is no more than a 5%, or 1 in 20, probability of observing such a result solely due to chance. In this case, the results are significant.

A *P* value of .01 would equate to a 99% probability that the difference is not due to chance. Thus, a significance level of $P < .01$ is more precise than $P < .05$, although many texts and researchers use the latter as a valid level. If you see a significance level higher than $P < .05$, however, examine the results closely to determine why it is set so high. Although the significance level can be as precise as the researcher wishes it to be, one that is too high can lead to a **false positive** or **false negative** result, which is known as an interpretation **error**, discussed below.

Type I Errors

Type I errors, or false positives, occur when you reject the null hypothesis when it is true; that is, you think there is some statistical significance in the results when there is not. In a healthcare test setting, for example, you might think the patient tested positive for a condition that the patient does not, in fact, have. An overly sensitive test can lead to false positives. The probability of a type I error is the level of significance of the test of hypothesis and is denoted by the lowercase Greek letter *alpha* (α).

For example, we diagnose a person with a disease based on certain criteria. The more sensitive or exact the confidence range, the more likely one will get a false positive; in other words, the test is too sensitive. You will see α used in other contexts and with different meanings, but the use here is specific to hypothesis testing only.

Type II Errors

Type II errors, or false negatives, occur when you fail to reject a false null hypothesis. You accept the null hypothesis when you should have accepted the hypothesis. Put another way, you did not think there was anything significant, but there was. The probability of a type II error in the hypothesis results is denoted by the lowercase Greek letter *beta* (β). If β is found, then you have determined that there was no statistical significance when there actually was.

Next, we will consider how probability might appear on a normal bell curve, whether significant or not.

Either End of the Curve: Tails

A **tail** is either one or both extreme ends of a bell curve. With a one-tailed distribution, all of the significance level is represented on only one tail or side—either upper or lower—with the rest of the curve added to it, equaling 100%. In this case, your result is significant only if it falls into the one side where the α falls. With a two-tailed distribution, half of the significance level is represented on each side of the curve, with the rest of the curve added to it, equaling 100% (**Figure 5.14**). In this case, your result is significant if it falls in either tail. For example, for a two-tailed distribution with a significance level of $P = .05$, the tail on each side would equal .025 and the rest of the curve would add up to 95%. *P* is really then the probability that you will get a

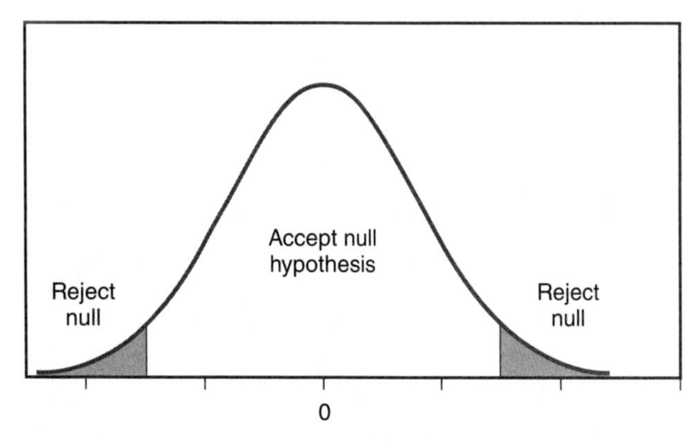

Figure 5.14 Two-tails, with each having half of the *P* value (e.g., if .05, each is .025).

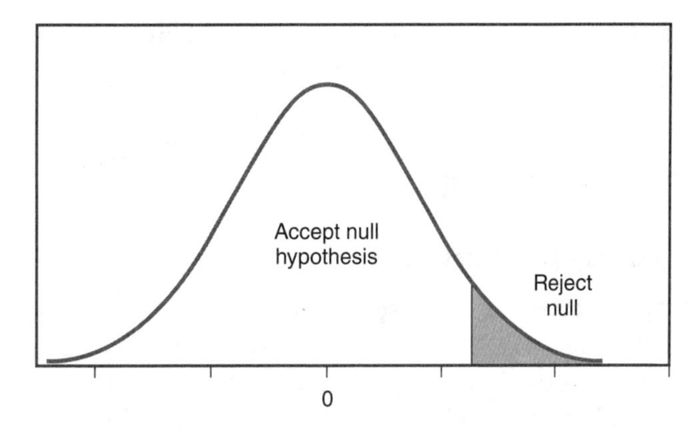

Figure 5.15 One-tailed distribution, with $P = .05$.

result in an extreme, small or limited area, which is represented by the tail or tails. Most tests are two-tail.

A one-tailed test (**Figure 5.15**) is better at detecting significance because you are examining only one direction. However, you would need to be sure your findings would fall into the range of only one tail; if they do not, you should use a two-tailed test. For example, imagine your values span a continuum that ranges from –5 to +5, such as –4.02 and +4.2. In this case, you would need to select a two-tailed test. However, if you know your values would all fall close to one end of the continuum, then you might consider a one-tailed test.

The Normal Distribution of Data

A **normal distribution**, or average distribution, of data looks like a bell when graphically depicted—hence the name "bell curve" for this classic form. You may also see it referred to as a Gaussian distribution. If you have a reasonable amount of sample data, you can safely assume that a normal distribution of data represents an entire population. For example, if you find that in a normally distributed sample of reasonable size, 25% of the people have a certain trait you are testing for, you could assume that 25% of the entire population would have that trait.

For now we will assume the data we are examining fit a normal distribution and, more importantly, that any measures we use are appropriate. Also, sample size will determine the type of test used, so that must also be considered by the researcher. To that end, you should be aware of two general types of tests we will consider in this text: parametric tests for normally distributed data and nonparametric tests for data that are not normally distributed.

Parametric and Nonparametric Tests

You must choose the type of test that fits your data. Most tests you use will be parametric, as most of your data will likely be normally distributed. **Parametric tests** assume certain things about the data under observation, such as a normal distribution and equal variance within each dataset (homogeneity of variance). Pearson product moment correlation, *t* test, chi-square test of independence, and analysis of variance (ANOVA) are all parametric tests.

Nonparametric tests, or distribution-free tests, are not based on the assumption of a normal distribution or of homogeneity of variance. Nonparametric tests compare medians rather than means because the latter would not be valid due to the non-normally distributed data.

Rank-ordered (ordinal) data would call for nonparametric tests as well. Nonparametric tests include the chi-square goodness-of-fit test, Mann-Whitney test (also called the Wilcoxon rank sum test or *U* test), and Spearman rank correlation. You will examine these tests later in the text.

z Score

A ***z* score**, or standard or standardized score, is a measure of the deviation of a value (in standard deviations) from the mean of a dataset. This commonly used descriptive statistical method is often visually represented on a bell curve chart via a percentile ranking of the area to which it refers. The *z* score can let you know whether a certain value is greater than, equal to, or less than the mean of all the values in the dataset under consideration. This handy function is not hard to calculate, as you can see from the following formula:

$$z = \frac{x - M}{SD}$$

where x is the value in the dataset to be compared against the mean, M is the mean, and SD is the standard deviation.

At a glance, z can be described as follows:

- Greater than the mean if positive
- Less than the mean if negative
- Equal to the mean if 0

Depending on the value, a value of +1 = one standard deviation beyond the mean, –1 is one standard deviation below the mean, and so on. Assuming a normal distribution of data (a critical assumption), 68% of

Hands-on Statistics 5.24: Use the STANDARDIZE Function in Excel

The STANDARDIZE function calculates a standard for a given z score, mean, and standard deviation on a set of data under consideration. Follow these directions to see how the STANDARDIZE function in Excel works with regard to the exam grades of a class of students:

1. Open a new, blank Excel workbook.
2. Type the data shown in **Figure 5.16** into the cells indicated, but leave cell A24 blank.
3. Type the following formula for standard deviation into cell A24:

   =stdev (A4:A22)

   Click out of the cell, and the standard deviation should appear (Figure 5.16).
4. Type the following formula for mean into cell A26:

   =average (A4:A22)

   Click out of the cell, and the mean should appear (**Figure 5.17**).
5. Lastly, type the following STANDARDIZE formula into cell D24:

   =STANDARDIZE (89,A26,A24)

In this case, we are trying to see where a score of 89 falls on a standard curve. Click out of the cell, and the z score should appear (**Figure 5.18**).

3	Exam Scores	Student
4	68	A
5	79	B
6	97	C
7	89	D
8	67	E
9	84	F
10	98	G
11	95	H
12	89	I
13	90	J
14	79	K
15	89	L
16	85	M
17	79	N
18	59	O
19	91	P
20	72	Q
21	99	R
22	86	S
23		
24	11.21741545	<=S.D.

Figure 5.16 Part 1 of z score using Excel.

3	Exam Scores	Student
4	68	A
5	79	B
6	97	C
7	89	D
8	67	E
9	84	F
10	98	G
11	95	H
12	89	I
13	90	J
14	79	K
15	89	L
16	85	M
17	79	N
18	59	O
19	91	P
20	72	Q
21	99	R
22	86	S
23		
24	11.21741545	<=S.D.
25		
26	83.94736842	<=Mean

Figure 5.17 Part 2 of z score using Excel.

D24 | =STANDARDIZE(89,A26,A24)

	A	B	C	D	E	F
1						
2						
3	Exam Scores	Student				
4	68	A				
5	79	B				
6	97	C				
7	89	D				
8	67	E				
9	84	F				
10	98	G				
11	95	H				
12	89	I				
13	90	J				
14	79	K				
15	89	L				
16	85	M				
17	79	N				
18	59	O				
19	91	P				
20	72	Q				
21	99	R				
22	86	S				
23						
24	11.21741545	<=S.D.		0.450427	z score	
25						
26	83.94736842	<=Mean				

Figure 5.18 Part 3 of z score using Excel.

your data should fall between −1 and +1, 95% of your data should fall between −2 and +2, and about 99.7% should fall between −3 and +3.

Read on to learn how to use Excel to calculate a z score.

To interpret this result, look at the z table shown in **Figure 5.19**, find 0.4 in the z (first) column on the left, then come down the column headed 0.05 until you reach the row corresponding to the z score you just found (we are trying to match 0.45 as closely as

Z	0	0.01	0.02	0.03	0.04	0.05	0.06
0.0	0.5000	0.5040	0.5080	0.5120	0.5160	0.5199	0.5239
0.1	0.5398	0.5438	0.5478	0.5517	0.5557	0.5596	0.5636
0.2	0.5793	0.5832	0.5871	0.5910	0.5948	0.5987	0.6026
0.3	0.6179	0.6217	0.6255	0.6293	0.6331	0.6368	0.6406
0.4	0.6554	0.6591	0.6628	0.6664	0.6700	> 0.6736	0.6772
0.5	0.6915	0.6950	0.6985	0.7019	0.7054	0.7088	0.7123

Figure 5.19 Partial data of a *z* table showing .05 significance. Note that the arrow points to .4 at .05 significance.

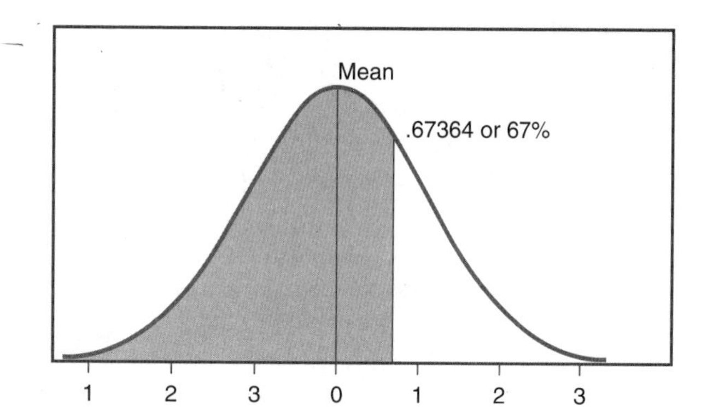

Figure 5.20 Graph of where item of interest falls on curve at 67%.

we can), and you should find 0.67364. This means that about 67% of the students' scores were at least an 89, as shown in **Figure 5.20**. The *z* table shown in Figure 5.19 can be referenced in texts and on the internet or can be created in Open Office, Excel, or other software applications.

You can search the web for PDF files and spreadsheet files with *z* table scores. Some are free for personal and other use. You might start with the following website: https://www.ztable.net/.

Next you will examine another classic and handy significance test, the *t* test for comparing two groups.

t Test

The ***t* test** is a test of significance that is used to compare the means of two groups to determine whether they are statistically different from each other, in contrast to the *z* score, which compares one item of interest against the mean. It is a standard inferential test, and one you should be familiar with. It is also known as the Student *t* test. Student is the pen name William Sealy Gosset, inventor of the *t* statistic, used as he developed and wrote about this test while working with the Guinness Brewery company in the early 1900s. He did not use his real name because of the brewery's desire that their competitors not know that the secret to good stout was based on statistics! Gosset was not satisfied with the current system used for estimation, so he devised a correction to the *z* deviate.

The *t* test is appropriate only for situations in which there are two groups of only one independent variable, the sample size is appropriate (not too small), and there is a normal distribution. This test is important in that from a small number of items you can estimate conclusions about an entire population, but if the sample is too small, the results might not be generalizable to a larger group. You are inferring conclusions about a larger population based on findings from a smaller sample of that population.

The distribution curve is wider than a normal distribution with this test. With a *t* test, you will have degrees of freedom, which are based on sample size. If the sample size increases, then *t* tends to mimic the findings of a *z* score.

The *t* test measures the difference between two means in relation to the variability of the sets. This is a classic test to use when trying to determine the outcomes of an experiment. For example, you have a control group and an experimental group that are being used to determine whether a treatment helps cure a certain condition. The control group does not receive the treatment, but the experimental group does. Researchers then tabulate and examine the results via a *t* test to determine whether there is a significant improvement. Significance is then defined by the *P* value.

There are many different types of *t* tests, and they are used for different types of data being examined. Factors affecting choice of test would include whether both samples had the same or different variance or whether there was only one sample and the experiment was run on it twice.

You will examine three types of t tests that Excel supports: type 1, type 2, and type 3. A type 1 test in Excel is a paired or dependent t test, which has one measurement variable and two nominal variables. For example, consider a study in which the scores from a group of students who took a dexterity test are compared with scores from the same group of students on a dexterity test after drinking a cup of strong caffeinated coffee.

Type 2 and type 3 tests are used when there are two groups of subjects, either independent or unpaired, not one group of subjects with two sets of measurements, as in the previous example. For example, imagine a study in which you are trying to determine whether business students in a school drink more energy drinks than science students drink. If the standard deviations for both groups are equal, you would use a type 2 test; if they are not, you would use a type 3 test. Again, note that the size and type of distribution of data in a study should influence the test you choose.

With a t test, you could examine a one-sided or two-sided t distribution (two-tailed) along the bell curve. Two-sided is more common, and in such cases you would halve the P value, although the P value will be listed for you in a standard t table for one- or two-sided tests.

When interpreting your results, your t value will be positive if the first mean is larger than the second and will be negative if the first mean is smaller. To determine whether your findings are significant, you must determine your α level. In most cases, an α level of .05 would be appropriate, which simply means that in 5 cases out of 100 you might determine a difference between the means under examination. If you find significance, you can reasonably say that your findings are not due to chance. Considering our preceding example that includes both a control group and an experimental group, if you find significance, you could say that the energy drinks did help the students' dexterity.

If you are using a significance table, you only need to determine the degrees of freedom for your test and then α level, and then you can find in the table what values would be significant. You can compare those values to your t test result and know whether you have significance.

The **degree of freedom** (df) is the number of values in a calculation that can vary. It is used to find a significance level in a table in many situations. For example, if n = the sum of the number of observations, then df = $n - 1$. To further refine this, with a single dataset of n observations, you are trying to estimate the mean, so you have one parameter you are trying to determine, and the result is $n - 1$ df. If you have two datasets, then you have two means you are estimating, and this would be $n1 + n2$.

Alpha values of .05 for a one-tailed test and .1 for a two-tailed test are shown in **Figure 5.21**.

You now have the theory behind you. It is time to use Excel to make sense of some data using a t test.

1 tail	**0.05**	**0.025**
2 tail	**0.1**	**0.05**
df		
1	6.3138	12.7065
2	2.92	4.3026
3	2.3534	3.1824
4	2.1319	2.7764
5	2.015	2.5706
6	1.9432	2.4469
7	1.8946	2.3646
8	1.8595	2.306
9	1.8331	2.2621
10	1.8124	2.2282
11	1.7959	2.201
12	1.7823	2.1788
13	1.7709	2.1604
14	1.7613	2.1448
15	1.753	2.1314
16	1.7459	2.1199
17	1.7396	2.1098
18	1.7341	2.1009
19	1.7291	2.093
20	1.7247	2.086
21	1.7207	2.0796
22	1.7172	2.0739
23	1.7139	2.0686
24	1.7109	2.0639
25	1.7081	2.0596
26	1.7056	2.0555
27	1.7033	2.0518
28	1.7011	2.0484
29	1.6991	2.0452
30	1.6973	2.0423

Figure 5.21 Table of critical t values.

Hands-on Statistics 5.25: Perform a *t* Test Using Excel, Two-Tailed

In this example, we will perform a *t* test with Excel for data for which it is appropriate to use a two-tailed test.

1. Open a new, blank Excel workbook.
2. Type the data shown in **Figure 5.22** into the cells indicated, but leave cells B14, C14, and D15 blank.
3. Type the following formula to find variance in cell B14:

   =var(B3:B13)

   Recall that variance is a measure of the spread of observations in a distribution of data. Click out of the cell, and the variance should appear, as in Figure 5.22.
4. Copy the formula from cell B14 to cell C14, and then click out of the cell. The variance for this column should now appear, as in Figure 5.22.
5. Type the following Excel formula for *t* test in cell D15:

   =TTEST(B3:B13,C3:C13,2,3)

   Note that after you type "C13," into the formula, Excel will prompt you to select either one-tailed or two-tailed distribution. This is why you enter the "2" in the preceding formula, as you will want to use the two-tailed test. After you type the "2," in the formula, Excel will prompt you to select one of the following three *t* test types:

   1—Paired
   2—Two-sample equal variance (homoscedastic)
   3—Two-sample unequal variance (heteroscedastic)

   We select *3* in this case because the two variances shown in cells B14 and C14 are unequal.
6. Click out of the cell, and the *P* value will be displayed, as in Figure 5.22. Selecting a confidence value of .025, you see that our *P* value (about .22) is greater than .025 (or .05 for a one-tailed test): $P > .025$. This suggests that there is no significant difference between the means of our datasets and that we would not reject our null hypothesis.

   If $P < .025$ (say, for example, $P = .002$), we would say this finding suggests that we do have a significant difference, because *P* falls into the range of one of the two tails. In fact, it falls into the extremely narrow range we specified. Again, what you are trying to determine is whether your *P* value falls in the range of a tail area; if so, you have found significance. For this example, there is no significant difference between the means of the datasets.

D15 *fx* =TTEST(B3:B13,C3:C13,2,3)

	A	B	C	D	E
1					
2		Set A	Set B		
3		4	5		
4		3	19		
5		5	4		
6		8	2		
7		10	13		
8		2	3		
9		2	5		
10		3	15		
11		5	1		
12		4	5		
13		1	3		
14	Variance	7.218182	35.76364		
15			P-value	0.218904	

Figure 5.22 *t* test in Excel.

How Does Your Hospital Rate?

Well, Barbara did not become a statistic for infant mortality. She made it to full term and delivered a healthy baby girl. Although all women are not a statistic for having children in their late 30s or early 40s, they are at a higher risk for infant mortality. Infant mortality rates for the state of North Carolina for 2013 were high compared with the national average, according to the North Carolina Healthy Start Foundation (visit http://www.nchealthystart.org/infant_mortality/, then click *Statistics* on the left). Key findings include the following:

- 7 babies died for every 1,000 live births
- 5.5 white babies died for every 1,000 live births
- 3.7 Latino babies died for every 1,000 live births
- 12.9 African American Non-Hispanic babies died for every 1,000 live births
- 832 total babies died in North Carolina during the year

By comparison, the overall infant mortality rate for the United States in 2013 was 6 deaths for every 1,000 live births. Since 1988, the overall US infant mortality rate has dropped 43%. On a county level for the state of North Carolina, Brunswick County had 8 infant mortality cases reported for the year 2011, compared with only 3 in Bladen County. See **Table 5.1** for a breakdown of infant deaths, births, and infant mortality rates in these two counties by race. Although Brunswick County has two hospitals, only one delivers babies.

Consider the following:

1. How does your county hospital rate compared to your neighboring county?
2. How does it compare to the two North Carolina counties mentioned here?
3. How does it compare to the national averages for infant mortality?

Table 5.1 Birth Statistics in Two North Carolina Counties

Statistic	Race of Infant	Brunswick County, NC	Bladen County, NC
Infant deaths	White, Non-Hispanic	7	1
	African American, Non-Hispanic	1	1
	Other Non-Hispanic	0	0
	Hispanic	0	1
	Total	**8**	**3**
Births	White, Non-Hispanic	795	147
	African American, Non-Hispanic	140	131
	Other Non-Hispanic	14	12
	Hispanic	98	52
	Total	**1,047**	**342**
Infant mortality rates	White, Non-Hispanic	8.8	6.8
	African American, Non-Hispanic	7.1	7.6
	Other Non-Hispanic	0.0	0.0
	Hispanic	0.0	19.2
	Total	**7.6**	**8.8**

Infant Mortality by Race and County

The CDC was formed in 1946 as part of the Public Health Service by Dr. Joseph Mountin of Atlanta, Georgia. Engineers and entomologists (specialists in the study of insects) made up most of the 400 initial employees of the CDC. Their main objective at that time was to keep the southeastern states free of malaria and murine typhus fever. The CDC has grown tremendously over the years.

The purpose of the CDC now is to aid in the prevention and control of infectious and chronic diseases and to protect the public's health and safety by delivering information regarding health decisions. The CDC provides this information through group partnerships with state health departments and other associations. It seeks to improve US citizens' health through promotion of education programs related to injury prevention, environmental health, prevention of infectious diseases, and many other aspects related to health. The CDC is the world's most notable epidemiological center.

The CDC has even expanded its mission to include preparation for a zombie outbreak! One of their more recent marketing strategies has been to provide a series of materials that would prepare people for a hypothetical zombie outbreak. With the recent popularity of zombie movies, strategists at the CDC thought that younger members of the population who were interested in this genre of movies might pay attention to the steps for preparation for a zombie outbreak, which are almost the same as those for hurricanes, tornadoes, and other disease outbreaks. In fact, it has worked wonderfully. This approach and its success make sense when you consider that zombie outbreaks in the popular culture are often linked to the outbreak of an infectious disease or an epidemic. The CDC promotes readiness for the impending zombie outbreak via posters, their website, and blogging. They have even produced a graphic novel, titled *Preparedness 101: Zombie Pandemic* (https://www.cdc.gov/cpr/zombie/novel.htm). More information on how they are preparing for the hypothetical outbreak is on the CDC website.

Global Perspective

Globally, more than 3.3 million babies died in their first month of life in 2009. One-quarter to almost one-half of those deaths occurred within the first 24 hours of life, and three-quarters within the first week. Another 2.6 million babies were stillborn. These numbers can be overwhelming. Why are there so many deaths among babies so early? Two major reasons are preterm births and asphyxia. In urban areas of Africa, Asia, and the Americas, the poorest 20% are more than twice as likely to die before their first birthday than are the richest 20% (WHO, 2016).

Chapter Summary

In this chapter we discussed many statistical methods required by hospitals, such as gross mortality rate, net mortality rate, postoperative mortality rate, fetal mortality rate, and cancer mortality rate. We examined research design, hypothesis testing, significance, and types of error. These processes can assist in improving the quality of care that patients receive, while contributing to the betterment of existing policies and procedures at the facility and adding to the common body of knowledge. The material in this chapter is foundational. We will build on these topics in later chapters.

Apply Your Knowledge

1. Summerville Hospital reported the statistical information shown in **Table 5.2** for discharges and deaths in the second quarter of 2018. Calculate the gross mortality rates of patients treated by each physician. Enter the gross mortality rate for each physician in the column provided.
2. Try some other demo features of R. Which one did you find of most interest? Which one was of least interest?
3. Using the spreadsheet you worked on for Hands-on Statistics 5.9, use the COUNTIF function to find summary data for causes of death other than heart attack. Also, allow for no listed cause of death, and display your findings graphically, such as in a bar chart.
4. List a pro and a con for using Excel and for using R to report or analyze your data findings.

Table 5.2

Physician Name	Discharges	Deaths	Gross Mortality Rate
Ian	35	2	
Jacob	54	4	
Jones	49	3	
Landry	58	6	
Munn	41	5	
Neil	62	9	

5. Using the following link from the CDC, compare 20 of the towns listed in the table and write a brief description on your findings: http://wonder.cdc.gov/mmwr/mmwr_reps.asp?mmwr_table=4A&mmwr_year=2013&mmwr_week=07&mmwr_location=Click+here+for+all+Locations
6. Why is it important to calculate mortality rates and mortality?
7. The Supervisor of Anesthesia at Sunnyside Hospital would like to review the anesthesia mortality rates for the months January to June of 2012. Use the information provided in **Table 5.3** to calculate the following anesthesia mortality rates:

Table 5.3

Inpatient discharges	1,841
Operations	1,432
Patients who underwent an operation	1,400
Anesthetics given	1,426
General anesthetics given	668
Regional anesthetics given	238
Local anesthetics given	520
Total inpatient deaths	68
< 10 days after surgery	42
> 10 days after surgery	22
Inpatient deaths due to anesthesia	18
Calculate the following:	
Deaths due to general anesthesia	
Deaths due to regional	
Deaths due to local anesthesia	
Anesthesia death rate	

- Deaths due to general anesthesia
- Deaths due to regional anesthesia
- Deaths due to local anesthesia
- Anesthesia death rate

8. Look at your state's general anesthesia mortality rate and compare it with those of two other states. Put your information in a graph and write a brief description of your findings.
9. Madison Hospital has calculated their yearly statistics for maternal mortality rate and abortion mortality rate. See **Table 5.4**, and use the direct maternal death rate to calculate the maternal mortality rate and the abortion mortality rate.
10. Go to the Kaiser Family Foundation's webpage on state health facts at https://www.kff.org/statedata/?ind=47&cat=2&rgnhl=35, click on your state, and compare infant mortality rates with five other states. Now compare those same mortality rates by race and ethnicity. How does your state rank on each of these? Write a brief description of your findings.

Table 5.4

Inpatient discharges	Delivered	848
	Aborted	92
	Undelivered, prepartum	208
	Undelivered, postpartum	126
Deaths	Delivered	10
	Aborted	8
	Undelivered, prepartum	6
	Undelivered, postpartum	3
Calculate the following:		
Maternal mortality rate		
Abortion mortality rate		

Table 5.5	
Newborn discharges	398
Births	380
Newborn deaths	18
Newborn mortality rate	

11. Sunnyside Hospital is reporting their yearly newborn mortality rate. Deaths will be included in the discharges. Calculate the newborn mortality rate based on the data in **Table 5.5**.
12. Review the cancer statistics provided by the National Cancer Institute's Surveillance, Epidemiology, and End Results Program by visiting the following webpage: http://seer.cancer.gov/statfacts/index.html. Select a cancer of your choice. After you have chosen a cancer, review the number of new cases and deaths. Was there a greater incidence of cancer among men or women? Which race had the highest cancer rates? Write a brief summary of your findings.
13. Sea Breeze Hospital has reported the information shown in **Table 5.6** from the hospital's cancer registry for the year. Calculate the mortality rate associated with each type of cancer. Remember that deaths are included in the discharge total.
14. Using the data in **Table 5.7**, create a graph. Then write a brief description regarding the type of postoperative death that had the highest ranking. What questions could you pose about the cause of the deaths? For example, do the patients who die from postoperative respiratory failure have other comorbid conditions that factor in?

    Now locate postoperative death data from two other countries and compare them with those of the United States. For example, you may want to locate data on accidental puncture or lacerations. How does the United States compare with other countries? Create a graph using these data and write a brief report.
15. Wave Hospital reported the data shown in **Table 5.8** for deaths in the month of September. Using these data, calculate the net mortality rate. In this report, the deaths are not included in the discharges.

Table 5.6

Cancer Type	Discharges	Deaths	Mortality Rate
Uterus	59	5	
Breast	92	10	
Lung and bronchus	168	22	
Ovary	34	2	
Melanoma	124	22	
Colon	44	4	
Kidney/renal pelvis	21	2	

Table 5.7 US Postoperative Deaths, 2018–2020	
Patients with foreign body retained after a procedure	6.25%
Accidental puncture or laceration	7.65%
Complications of anesthesia	1.44%
Postoperative respiratory failure	30.97%
Wound dehiscence	9.63%
Transfusion reactions	8.95%

Table 5.8	
Adult and children discharges	538
Adult and children deaths	15
Deaths < 48 hours after discharge	8
Deaths > 48 hours after discharge	7
Newborn discharges	41
Newborn deaths	5
Deaths < 48 hours after discharge	3
Deaths > 48 hours after discharge	2
Net mortality rate	

References

North Carolina State Center for Health Statistics. (2014). Reported pregnancies 2011. Retrieved August 20, 2015, from http://www.schs.state.nc.us/schs/data/pregnancies/2011/

Pitches, D. W., Mohammed, M. A., & Lilford, R. J. (2007). What is the empirical evidence that hospitals with higher risk adjusted mortality rates provide poorer quality care? A systematic review of the literature. Retrieved June 1, 2009, from http://www.ncbi.nlm.nih.gov/pmc/articles/PMC1924858/

Thomas, J. W., & Hofer, T. P. (1999). Accuracy of risk-adjusted mortality rate as a measure of hospital quality of care. *Medical Care, 37*(1), 83–92.

US Fire Administration. (2020). About the US Fire Administration. Retrieved February 6, 2020, from http://www.usfa.fema.gov/about/

World Health Organization. (2016). Global Health Observatory data repository. Retrieved March 10, 2017, from http://apps.who.int/gho/data/view.main.vEQINFANTMORTTOTv?lang=en

Web Links

Economic Data Freely Available Online: http://economicsnetwork.ac.uk/links/data_free

Basic Statistical Data Used in Acute Care Facilities: http://www.jblearning.com/samples/0763750344/45561_ch01.pdf

Population Reference Bureau: http://www.prb.org/

Maternal Mortality in Vietnam, 2000–2001: http://www.wpro.who.int/publications/docs/Maternal_Mortality_in_VietNam.pdf

An Introduction to Maternal Mortality: http://www.ncbi.nlm.nih.gov/pmc/articles/PMC2505173/

Organ Procurement and Transplant Network: https://optn.transplant.hrsa.gov/data/

Deaths From Surgical Errors/Complications: http://www.rightdiagnosis.com/s/surgical_errors_complications/deaths.htm

Infant Mortality in North Carolina: http://www.nchealthystart.org/infant_mortality/statistics.htm

2011 North Carolina Infant Mortality Report, Table 1: http://www.schs.state.nc.us/SCHS/deaths/ims/2011/2011rpt.html

Historical Perspectives History of CDC: http://www.cdc.gov/mmwr/preview/mmwrhtml/00042732.htm

Mortality Rate (Wikipedia): http://en.wikipedia.org/wiki/Mortality_rate

National Vital Statistics System—Mortality Data: http://www.cdc.gov/nchs/deaths.htm

Morbidity Rate (Investopedia): http://www.investopedia.com/terms/m/morbidity-rate.asp

Division of Nutrition, Physical Activity, and Obesity: http://www.cdc.gov/pednss/how_to/read_a_data_table/calculating_prevalence.htm

Cancer Stat Facts: http://seer.cancer.gov/statfacts/index.html

CHAPTER 6

Autopsy Data

Get correct views of life, and learn to see the world in its true light. It will enable you to live pleasantly, to do good, and, when summoned away, to leave without regret.

—General Robert E. Lee

CHAPTER OUTLINE

Introduction
What Is an Autopsy and Why Is It Important?
Centers for Disease Control and Prevention Data
Types of Autopsy Data
- Autopsy Rate
- Net Autopsy Rates
- Inpatient Hospital Autopsies
- Adjusted Hospital Autopsy Rate
- Autopsy Rate for Newborns
- Fetal Autopsy Rate

Statistical Measures
- *F* Test: Comparison of Two Variances
- Analysis of Variance
- One-Way Analysis of Variance
- Two-Way Analysis of Variance

Global Perspective
Chapter Summary
Apply Your Knowledge
References
Web Links

LEARNING OUTCOMES

After completing this chapter, you should be able to do the following:

1. Describe and calculate gross and net autopsy rates.
2. Discuss inpatient hospital autopsies.
3. Describe and calculate the adjusted hospital autopsy rate.
4. Describe and calculate the autopsy rate for newborns.
5. Describe and calculate the fetal autopsy rate.
6. Define *F* test and calculate it using Excel.
7. Define analysis of variance (ANOVA) and calculate one-way and two-way ANOVA using Excel and R-Project.
8. Harvest data from the US Census Bureau.

KEY TERMS

Adjusted hospital autopsy rate
Analysis of variance (ANOVA)
Arrays
Autopsy
Balanced design
Clinical autopsy
Coroner
F test
F value

Fetal autopsy rate
Forensic autopsy
Gross autopsy rate
Hospital autopsy
Inferential statistics
Inpatient hospital autopsy
National Center for Health Statistics
Net autopsy rate
Newborn autopsy rate
One-tailed test
One-way ANOVA
Pathologist
Postmortem
Two-tailed test
Two-way ANOVA
Unbalanced design
Verbal autopsy

How Does Your Hospital Rate?

Leah is a nurse at a hospital in Brunswick County in North Carolina. She has been assigned the responsibility of researching the number of autopsies performed in her county in the past 5 years and learning what the primary causes of death were determined to be in these cases. She finds that an average of 54% of the deaths between 2003 and 2019 were autopsied. The reasons for autopsies included motor vehicle accidents, falls, homicide, and suicide. Pick a county in the state in which you live and conduct your own research.

Consider the following:

1. How many autopsies have been performed in your county in the past 5 years?
2. What percentage of deaths in the county were autopsied during this time?
3. What were the primary causes of death?

Introduction

Hospitals deal with both life and death every day. A wealth of information about those who have died can often be gained through autopsy. For instance, medical professionals learn valuable diagnostic information from the deceased to help the living. Certainly legal and insurance representatives gain important data from the examination of the bodies of the deceased.

The person in charge of performing autopsies is a **coroner**, who collects much data for statistical analysis, including net death rates and other rates pertaining to death to look for patterns in deaths from certain diseases. These figures also allow the facility to determine what cases have been sent to the county coroner, as only certain types of cases require an autopsy according to state regulations, such as those in which criminal activity might be involved.

This chapter opens with a discussion of data gathered by the Centers for Disease Control and Prevention (CDC) and an examination of the National Center for Health Statistics. The chapter next introduces you to the various types of autopsy data commonly tracked and used by healthcare professionals, including gross autopsy rate, net autopsy rate, inpatient hospital autopsies, adjusted hospital autopsies, newborn autopsy rate, and fetal autopsy rate. Hands-on Statistics sections instruct you on how to calculate these rates both by hand and using Microsoft Excel. Finally, statistical measures are presented, including the *F* test and analysis of variance, along with corresponding Hands-on Statistics sections.

What Is an Autopsy and Why Is It Important?

Autopsy is an ancient practice, undertaken as long ago as the 4th century BCE by Aristotle and the 3rd century BCE by Herophilus of Chalcedon, the latter of whom performed it in public, explaining the process to bystanders (Gulczynski, Iżycka-Swieszewska, & Grzybiak, 2009).

An **autopsy**, also known as a postmortem examination or necropsy, is typically a surgical procedure performed on patients who have died to verify the exact cause of death. It can also be a nonsurgical procedure in which various imaging technologies are used to examine the body (known as a virtual autopsy) or in which people who knew the deceased are interviewed about details related to the death (a **verbal autopsy**). Either a virtual or a verbal autopsy may be ordered by a family who does not wish the body to be desecrated by surgical techniques due to religious or other beliefs. Governments of states or countries may conduct a verbal autopsy, especially in developing countries, where mortality reporting may go unrecorded unless the central government conducts such verbal autopsies to collect valuable national data. In fact, according to an article in the *BBC World News*, two-thirds of deaths that occur around the world are not recorded (Soy & Lacey, 2013). The article discusses an innovative use of smartphone technology whereby field workers can enter verbal autopsy data into the device and later upload it to a central government database. With the increase in quality data, the country can determine, among other things, whether healthcare money is being spent properly.

Regardless of whether it is a traditional autopsy or a nonsurgical one, the underlying goal is the same: to determine the cause of death, whether disease or injury, natural or related to criminal activity. The two general reasons for autopsy are forensic and clinical. A **forensic autopsy** is conducted for legal purposes and to determine the cause and manner of death, such as when criminal activity is suspected in the person's death, as in homicide. A **clinical autopsy**, on the other hand, is conducted for medical purposes. It helps caregivers better understand the conditions that caused the patient's death and helps hospitals ensure and improve quality of care at the facility. Whereas forensic autopsies are often required by state agencies, clinical autopsies are for medical benefit, are not required but elective, and typically require permission of the patient's next of kin or someone else who can legally authorize the procedure.

There are two types of hospital autopsy, based on the location of the patient: a hospital autopsy and an inpatient hospital autopsy. A **hospital autopsy** is a postmortem exam of a person who was a hospital patient, an emergency department patient, a walk-in outpatient, or a home-care patient. A dead fetus would not be included in this group because a fetus is not considered a living patient. **Inpatient hospital autopsies** are performed only on patients who died during hospitalization after being formally admitted. The medical procedure for both types of hospital autopsy is the same. It includes an internal and external examination of the body and a toxicology screening.

Unless it is a verbal autopsy, autopsies are generally performed in a morgue by either a hospital **pathologist**, who is a physician who specializes in the study and diagnosis of diseases, or another physician.

In areas where people often die at home and there is no efficient way to obtain the data gained through autopsy, which are so important into supporting and improving the healthcare system, other means must be used to gather this information (Baiden et al., 2007). For instance, the US Department of Veterans Affairs has developed a questionnaire that can be used to gather these valuable statistical data.

According to *Frontline*, there is a decline in the use of autopsies in the United States, with statistical data showing that, as a national average, only 8.5% of deaths are followed by autopsy, and only 4.3% of those who are autopsied died from a disease (i.e., natural causes) (Breslow, 2012). In fact, through the years, this figure has declined, since in the past there was less value put on the data that could be obtained and reported to clinicians to improve health care, diagnosis, and prevention. Not only is there less value, but there are fewer experts trained to determine the cause of death. Currently, there is no federal oversight of death investigators, and accreditation is voluntary. The Criminal Justice and Forensic Science Reform Act, proposed in 2014, would require investigative units receiving federal funding to have certifications from national organizations.

In a 2012 interview on National Public Radio with Habiba Nosheen on the decline in autopsies, George Lundberg, MD, stated that autopsies used to be the standard in hospitals, but not anymore. According to Lundberg, the number of autopsies has "gone down so far now that there are large numbers of hospitals that do almost no autopsies at all" (Nosheen, 2012). Whereas the Joint Commission required a certain percentage of autopsies to accredit hospitals in the past, this is no longer has not been a stipulation for accreditation since 1971. Nosheen notes that, "Until 1971, hospitals had to autopsy at least 20% of patients who died in their care, but they are not required to meet a quota." (Nosheen 2012). In Nosheen's interview Dr. Lundberg explains that, "The people who pay the bills say, 'Well, why would I want to pay money for an autopsy on a patient who's already dead? My money is supposed to be spent on people while they're alive.'" (Nosheen, 2012).

? DID YOU KNOW? "Gangrenous pus does not taste good," states Antonio Valsalva (Freedman, 2012). Valsalva further notes the result of his tasting as "leaving the tongue tingling unpleasantly for the better part of the day" (Freedman, 2012). The Italian physician Antonio Maria Valsalva recorded these observations in his early 18th century book *De Aure Humana Tractatus*. In the 1700s, they did not have the sophisticated chemical analysis techniques we have today. As a result, tasting was a valid technique promoted by expert physicians of the day.

Of course, this was before confirmed evidence of routes of disease transmission. Today, we know that such a technique would not be safe. Valsalva is probably most famous, however, for the Valsalva maneuver, which is a technique to clear the auditory canal that involves bearing down, or closing the throat and contracting the abdominal muscles to increase intrathoracic pressure.

Now you know!

How Does Your Hospital Rate?

Leah, the nurse we met earlier in the chapter, has been assigned to conduct research on the number of forensic pathologists in the state of North Carolina, how many autopsies they performed in the preceding year, and the budget for these autopsies. She finds that North Carolina has 51 part-time provisional forensic pathologists and 3 full-time forensic pathologists, who together performed 4,119 autopsies in the previous year, with a budget of $5,584,140. Using the *Frontline* "Post Mortem Death Investigation in America" webpage provided by PBS.org (http://www.pbs.org/wgbh/pages/frontline/post-mortem/map-death-in-america/), see how your state compares with North Carolina. If data are not listed for your state, choose a state for which data are available.

Consider the following:

1. How many full- and part-time forensic pathologists work in your state?
2. How many autopsies were performed in your state in a recent year?
3. What is the budget for autopsies for 2010?

So, why has the number of autopsies performed declined in the United States? Hospitals are not required to have an autopsy performed, and insurance companies do not pay for them. Moreover, with the average cost of an autopsy being around $2,000 or more, many families are unable to afford them. Added to this expense is a fear that an autopsy will lead to a malpractice suit against the physician. All of these factors contribute to the low number of autopsies. The consequences of this lack of autopsies are significant. Errors in medical care go unnoticed, opportunities to evaluate the effectiveness of medical treatments and the progression of diseases are lost, and inaccurate information is placed on death certificates and can skew healthcare statistics. For instance, the number of people who die from certain diseases will be inaccurate if autopsies are not performed on those who actually have the disease but have not been diagnosed at the time of death.

Because autopsies provide many benefits, they can, potentially improve the quality of medicine and contribute considerably to medical education. For these reasons, the Institute for Healthcare Improvement, the National Quality Forum, and The Joint Commission all support an increase in the number of autopsies being performed.

So how is it determined who needs to be autopsied? Patients who do not have a physician and die will need someone to sign a death certificate as to the cause of death. The bodies of individuals who have died due to suspicious circumstances and, possibly as a result of violence, such as (e.g., homicide, suicide, or accidental death), are sent to the medical examiner for an autopsy as standard practice.

The authors interviewed Greg White, the Brunswick County Coroner, on July 15, 2013, Mr. White explained just exactly what the coroner's duties are with regards to autopsies. As an elected official, he has been coroner for the county since 1985. This particular coroner's office responds to approximately 200 to 250 cases a year, a number that has increased in recent years due to a rise in population. Mr. White oversees cases involving suicides, unattended deaths, drug overdoses, and unexpected deaths. When asked how he determines what cases to follow up on, he replied that there are guidelines in place to help him determine which cases warrant further action based on the suspected cause of death, such as homicide, questionable suicide, drug overdose, or suspicious deaths.

Primary responsibilities of the coroner include communicating and coordinating with the families of the deceased and arranging for transport of the body. Although they use autopsy data, Brunswick County Hospitals do not perform autopsies on-site, as not all facilities have the capability to do so. For this reason, when an autopsy must be performed, the body is sent to a hospital in another county, such as Onslow Memorial, Pitt County Memorial Hospital, or several others, all of which are several hours' drive away. If the body has significantly decomposed, making autopsy more difficult, it could be sent to Chapel Hill or Raleigh, North Carolina, for autopsy. Wake Forest Baptist Hospital serves as the regional forensic pathology center for 30 of North Carolina's 100 counties.

Centers for Disease Control and Prevention Data

We considered the mission and roles of the CDC in the previous chapter. Now, we will explore a useful tool for harvesting data that the CDC offers.

The CDC's **National Center for Health Statistics** is a powerful online search tool to harvest care data from many areas. This site presents data tables on various US health trends, including crude death rates, infant mortality rates, and other types of data. You can view and search the respective tables of data separately or use the search box to search all the tables simultaneously for the criteria you are trying to find.

To visit the National Center for Health Statistics, use the following URL: https://www.cdc.gov/nchs/hus/contents2018.htm.

Types of Autopsy Data

Autopsy Rate

Gross autopsy rate is a measure of all autopsies, regardless of who performed them, on inpatient deaths for a period compared with the total inpatient deaths for the same period. It is typically reported as a percentage. Newborn autopsies may be incorporated in the gross autopsy rate or calculated separately, depending on the hospital's policy.

The formula for gross autopsy rate is as follows:

$$\frac{\text{All autopsies on inpatient deaths during a specified period} \times 100}{\text{All inpatient deaths for that period}}$$

Hands-on Statistics 6.1: Calculate Gross Autopsy Rate by Hand

Let's practice calculating the gross autopsy rate by hand, using the following example. Sunny Hospital for the month of November discharged 798 patients. Of those discharges, 14 were deaths, including newborns, 9 of which were autopsied. Follow these directions:

1. Multiply the number of autopsies by 100:

$$9 \times 100 = 900$$

2. Divide the result by the number of deaths, including newborns:

$$900/14 = 64.28$$

Thus, after rounding, you would report a 64% gross autopsy rate for Sunny Hospital for the month of November.

Hands-on Statistics 6.2: Calculate Gross Autopsy Rate Using Excel

To calculate the gross autopsy rate in Excel, follow these directions:

1. Open a new, blank Excel workbook.
2. Enter the data as shown in **Figure 6.1**, but leave cells D4 and E7 blank.
3. In cell D4 enter the following function:

=B4*C4

Click out of the cell, and 900 should appear, as in Figure 6.1.

	A	B	C	D	E
1					
2	Gross Autopsy Rate				
3		Autopsies		Result	
4		9	100	900	
5					
6		Deaths			Gross Autopsy Rate
7		14			64.28571429

Figure 6.1 Gross autopsy rate using Excel.

4. In cell E7, enter the following function:

=D4/B7

Click out of the cell, and the gross autopsy rate should appear, as in Figure 6.1.
5. Format the result cell for two decimal places of precision at the minimum.

This gross autopsy rate is high; usually a high rate is due to a relatively low number of deaths in a given period.

Net Autopsy Rates

Net autopsy rate is a ratio of all inpatient autopsies to all inpatient deaths minus cases that were released to the coroner or medical examiner, and thus not autopsied in the hospital. Each state appoints a coroner who is responsible for establishing the time and cause of death. Certain cases of death are reportable to the coroner.

In the state of North Carolina, the following types of cases are reported to the coroner: homicide, suicide, accidental death, death from trauma, death from natural disaster, death from violence, death while in jail or in protective custody, death from poisoning or suspected poisoning, death from a public health hazard, death from surgery or anesthesia, sudden unexpected deaths, and death without medical attendance. When a body has been sent for autopsy and the coroner is not available to perform the autopsy, a hospital pathologist may report initial autopsy-related findings.

The net autopsy rate is as follows:

$$\frac{\text{All autopsies on inpatient deaths during a specified period} \times 100}{\text{All inpatient deaths} - \text{Cases released to the coroner or medical examiner}}$$

Hands-on Statistics 6.3: Calculate Net Autopsy Rate by Hand

Let's practice calculating the net autopsy rate by hand, using the following example. Ebb Tide Hospital had 19 patient deaths and performed 8 autopsies. Three bodies were released to the coroner. Follow these directions:

1. Multiply the number of autopsies by 100:

$$8 \times 100 = 800$$

2. Subtract the bodies released to the coroner from the total number of deaths:

$$19 - 3 = 16$$

3. Divide the result of step 1 by the result of step 2 to determine the net autopsy rate as a percentage:

$$800/16 = 50$$

Hands-on Statistics 6.4: Calculate Net Autopsy Rate Using Excel

To calculate the net autopsy rate in Excel, follow these directions:

1. Open a new, blank Excel workbook.
2. Enter the data as shown in **Figure 6.2**, but leave cell B5 blank.

B5 | fx =(C3*100)/(B3-D3)

	A	B	C	D
1	Net Autopsy Rate			
2		Deaths	Autopsies	Bodies released to coroner
3		19	8	3
4				
5	Rate=	50		

Figure 6.2 Net autopsy rate using Excel.

(continues)

3. In cell B5, enter the following function:

   (C3*100)/(B3-D3)

4. Click out of the cell, and 50 should appear, as in Figure 6.2. This is the net autopsy rate.

Inpatient Hospital Autopsies

Postmortem, or after death, examinations of patients who die during their admission to the hospital are normally performed at the facility at which the patients were admitted. However, if the hospital does not have the capability to perform the autopsy, the patient may be transferred to an outside laboratory, possibly at a different local hospital, or funeral home for this procedure. Timing of the autopsy is also important; it should be performed within 24 hours after death, before the internal organs begin to deteriorate. Second autopsies may be performed if there are any questions related to findings from the first autopsy. In some instances, the second autopsy reveals information not found in the first autopsy.

Inpatient hospital autopsies should not be confused with hospital autopsies, as previously noted earlier in the chapter. Hospital autopsies are performed on patients who have previously been in the hospital for various services, including outpatient services. Inpatient hospital autopsies, in contrast, are performed on patients who died *during admission* to the hospital as an inpatient.

Some guidelines for autopsies include the following:

- All homicide victims must have an autopsy.
- Hospitals usually perform the autopsy unless they do not have a morgue, in which case the patient may be transferred to a funeral home or other hospital. The next of kin must give legal consent for the autopsy, except in certain cases when the autopsy is required.
- Fetal autopsies are not included in the hospital autopsy rate, as fetuses are not considered inpatients.

DID YOU KNOW? The CDC has taken many innovative approaches to educating citizens on health care and related issues. One of these initiatives is Solve the Outbreak, which allows you to act in an investigative capacity, similar to a medical examiner. Visit the following link and see how good your medical sleuthing skills are: www.cdc.gov/mobile/applications/sto/web-app.html.

Now you know!

Adjusted Hospital Autopsy Rate

To get a more precise percentage of hospital and physician resources that are used for autopsies, the **adjusted hospital autopsy rate** is used. Inpatients, outpatients, prior patients who died elsewhere, and those bodies available for a hospital autopsy are included in the calculation of this rate. Patients who are autopsied by a physician or pathologist who is acting as an agent for the coroner or medical examiner are also included in this rate.

The formula for the adjusted hospital autopsy rate is as follows:

$$\frac{\text{All hospital autopsies for a specified period} \times 100}{\text{All hospital patients who died and whose bodies are accessible for autopsy}}$$

Hands-on Statistics 6.5: Calculate Adjusted Hospital Autopsy Rate by Hand

Let's practice calculating the adjusted hospital autopsy rate by hand, using the following example. In the month of January, Sea Breeze Hospital had 378 discharges, of which 10 were deaths. The pathologist autopsied 4 of those bodies. Also in the month of January, 1 patient who was receiving home care died and then received an autopsy. Note that the number of discharges has no effect on autopsy rate. Follow these directions:

1. Add the number of autopsies performed by the pathologist on inpatients to the number of autopsies performed on other types of patients and multiply the sum by 100:

   $(4 + 1) \times 100 = 500$

2. Add the total number of deaths of inpatients to the number of non-inpatients who died and whose bodies are to be autopsied:

$$10 + 1 = 11$$

3. Divide the result of step 1 by the result of step 2; the result is the adjusted hospital autopsy rate for Sea Breeze Hospital for the month of January, as a percentage:

$$500/11 = 45.45$$

Hands-on Statistics 6.6: Calculate Adjusted Hospital Autopsy Rate Using Excel

To calculate the adjusted hospital autopsy rate in Excel, follow these directions:

1. Open a new, blank Excel workbook.
2. Enter the data as shown in **Figure 6.3**, but leave cell C6 blank.
3. In cell C6, enter the following function:

   =((B4+B5)*100)/(B3+B4)

   Click out of the cell, and 45.45455 should appear, as in Figure 6.3. This is the adjusted hospital autopsy rate.
4. Format cell C6 so that zero decimal places are displayed.

Clipboard | Font

C6 | fx | =((B4+B5)*100)/(B3+B4)

	A	B	C	D
1	Adjusted Hospital Autopsy Rate			
2				
3	10 deaths	10		
4	1 died while receiving home care	1		
5	4 patients autopsied	4		
6			45.45455	

Figure 6.3 Adjusted hospital autopsy rate using Excel.

Autopsy Rate for Newborns

As we have stated previously, newborn autopsies may be calculated separately from other autopsies. This decision is made by the hospital administration. When newborn autopsies are calculated independently of the adults and children, it is called the **newborn autopsy rate**.

The formula for the newborn autopsy rate is as follows:

$$\frac{\text{Newborn autopsies performed during a specified period} \times 100}{\text{All newborn deaths for that period}}$$

Hands-on Statistics 6.7: Calculate Newborn Autopsy Rate by Hand

Let's practice calculating the newborn autopsy rate by hand, using the following example. In the month of March, Women's Hospital had 68 births. Of these 68 births, 4 newborns died and were autopsied. Follow these directions:

1. Multiply the number of newborns autopsied by 100:

$$4 \times 100 = 400$$

(continues)

2. Divide the result by the number of newborn deaths for the period to determine the newborn autopsy rate for Women's Hospital for the month of March:

$$400/4 = 100\%$$

Hands-on Statistics 6.8: Calculate Newborn Autopsy Rate Using Excel

To calculate the newborn autopsy rate in Excel, follow these directions:

1. Open a new, blank Excel workbook.
2. Enter the data as shown in **Figure 6.4**, but leave cell C9 blank.
3. In cell C9, enter the following function:

   =B7*100/B6

   Click out of the cell, and 100 should appear, as in Figure 6.4. This is the newborn autopsy rate.

Note: In this example we did not use parentheses because the order of operations dictates that, given multiplication and division, they are taken in order left to right as opposed to addition or subtraction, which would need to be forced, via parentheses, to take place first.

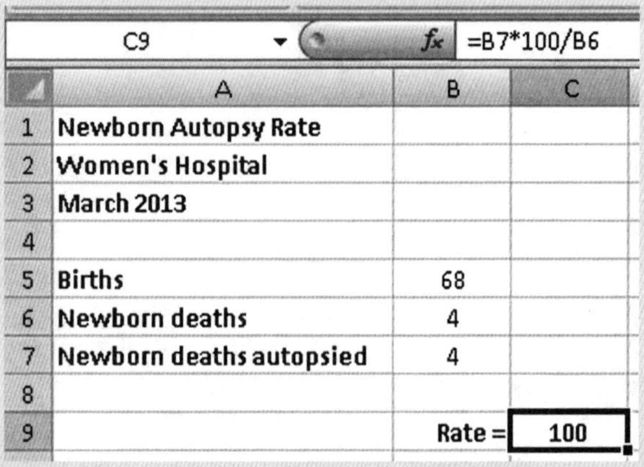

C9 | fx =B7*100/B6

	A	B	C
1	Newborn Autopsy Rate		
2	Women's Hospital		
3	March 2013		
4			
5	Births	68	
6	Newborn deaths	4	
7	Newborn deaths autopsied	4	
8			
9		Rate =	100

Figure 6.4 Newborn autopsy rate using Excel.

Fetal Autopsy Rate

As mentioned earlier, fetal autopsies are not included in the inpatient autopsy count. Instead, they are counted separately in the **fetal autopsy rate**. Note that only intermediate and late fetal deaths are included in this rate.

The formula for the fetal autopsy rate is as follows:

$$\frac{\text{Autopsies conducted on intermediate and late fetal deaths for a specified period} \times 100}{\text{All fetal deaths (intermediate late) for that period}}$$

Hands-on Statistics 6.9: Calculate Fetal Autopsy Rate by Hand

Let's practice calculating the newborn autopsy rate by hand, using the following example. For the months of January through March, Brook Hospital reported 18 fetal deaths, 4 of which were early, 8 intermediate, and 6 late. Of these fetal deaths, 4 late fetal deaths and 6 intermediate deaths were autopsied. Follow these directions:

1. Add the number of late fetal deaths to the number of intermediate fetal deaths, and multiply the sum by 100:

   $$(4 + 6) \times 100 = 1{,}000$$

2. Add the total number of intermediate fetal deaths to the number of late fetal deaths:

   $$8 + 6 = 14$$

3. Divide the result of step 1 by the result of step 2 to determine the fetal autopsy rate for Brook Hospital for the period of January through March:

   $$1{,}000/14 = 71.42\%$$

DID YOU KNOW? Some people equate a *z* score with a sigma level (as related to Six Sigma quality control); however, they are not the same thing, depending on how the Six Sigma defect rate is calculated. For short-term calculation, a *z* score and a sigma level would be the same, but for a long-term calculation there is a conversion factor, so they are not the same.

Now you know!

Statistical Measures

F Test: Comparison of Two Variances

Named in honor of Sir R. A. Fisher, the *F* test statistic, Fisher test, or ***F* test** is a ratio of two sample variances or standard deviations that was developed by

George W. Snedecor. An *F* test, ***F* value**, or *F* calculation compares two variances or standard deviations to see whether they are equal. As a parametric test, it assumes certain conditions about the data under analysis, such as a normal distribution. In a later chapter, we will see tests for non-normally distributed data.

The *F* test is key to **analysis of variance (ANOVA)** and gives a value for statistical significance that is found by dividing the between-group variance by the within-group variance. If the variances are equal, the result will be 1. The degrees of freedom for the numerator are the degrees of freedom for the between group ($k - 1$), and the degrees of freedom for the denominator are the degrees of freedom for the within group ($N - k$) (note that capital N is an entire group and lowercase n is a sample). The *F* test should not be confused with *F* statistics, which deals with population genetics and is a different subject altogether (and not one covered in this text). To use the result of this test, compare the calculated *F* value to a table of values aligned with your significance level to find the result. If you use Excel, R-Project, or another software application, then the software will do this for you.

The *F* test can be a one-tailed or a two-tailed test. A **one-tailed test** examines one direction or end of the curve. An example might be that the group under examination is either greater than or less than the second group's variance or standard deviation under examination. A **two-tailed test** examines the hypothesis versus the null hypothesis to see whether they are equal. So, considering the hypothesis or alternate hypothesis and the null hypothesis, you would find the following:

- If group A = group B, then you would accept the null hypothesis (H_0; what you are attempting to reject).
- If group A < group B, then you would accept the alternate hypothesis (H_a; what appears to be true or what you expect) and reject the null hypothesis.

Certainly you must be intrigued at the power of the test now. Now, to get more detailed:

- If group A > group B for an *upper* one-tailed test, then you would accept the null hypothesis.
- If group A < group B for a *lower* one-tailed test, then you would accept the null hypothesis.
- If group A is *not equal* to group B for a two-tailed test, then you would reject the null hypothesis.

Hands-on Statistics 6.10: *F* Test Using Excel

Now we will practice using the *F* test in Excel. Note that Apache Open Office uses the same syntax, so feel free to use either application. Although you may use the FTEST function in Excel for this exercise, it is recommended that you use the *F* test function that is included with the free Microsoft Office Excel add-in program Analysis ToolPak, as it offers more features and is more accurate, according to Microsoft. Apache Open Office has the FTEST function built in if you are not using Excel. If using the Analysis ToolPak in Excel, just make sure it is installed first. You may install it easily by selecting *Add-ins* from the *Options* menu on the *File* tab. Use the *F test for two samples function*, which provides much more information and accuracy.

In this example, we have two columns of data representing hypothetical scores from two groups. In Excel they are referred to as **arrays**, which in this context simply means groups of the same type of data.

1. Open a new, blank Excel workbook.
2. Enter the data as shown in **Figure 6.5**.
3. Next, assuming you have properly installed Analysis ToolPak, select the *Data* menu in the toolbar in Excel, then *Data Analysis* on the right, then *F test Two-Sample for Variances* in the pop-up window, and then *OK*.
4. In the pop-up window that appears, click on the small icon with the red arrow next to *Variable 1 Range* in the *Input* section. This will cause the pop-up window to minimize.
5. Highlight all of the cells containing numerical data under Set 1 (cells A5 to A10); do not select the cell containing the title *Set 1*.
6. Click again on the icon in the pop-up window to return your selection box.
7. Repeat steps 4 through 6 for *Variable 2 Range*, but highlight the data under *Set 2* this time. Both text fields should be populated now, as in **Figure 6.6**.

(continues)

	A	B
1	**F-test with Excel**	
2	*Two groups with various scores for each group*	
3		
4	**Set1**	**Set2**
5	3201	3066
6	2770	2731
7	2724	2839
8	3045	2912
9	2789	2788
10	2844	2841

Figure 6.5 *F* test data setup in Excel.

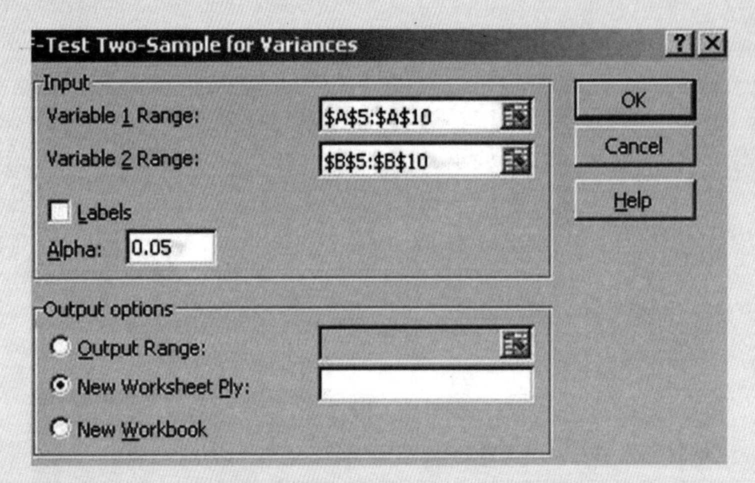

Figure 6.6 Step 7 of *F* test using Excel.

8. For the next step, you may either put your results in a different sheet or in the sheet you are working on. For the sake of this example, we will put the results further down in the same spreadsheet. Select the radio button *Output Range* in the pop-up window, then click on the icon with the red arrow in the field next to this button, and then highlight cells A13 to C22 in the Excel spreadsheet, then click on the icon in the pop-up window again, and then click *OK*. The data analysis results should appear in the cells you selected, as shown in **Figure 6.7**.
9. Analysis of results: In this example, we find that the *F* value is less than the *F* critical value, and the *P* value is greater than our alpha of .05, so we cannot reject the null hypothesis. This means that the variances are not significantly different. Note that a standard alpha of .05 was chosen.

F-Test Two-Sample for Variances		
	Variable 1	*Variable 2*
Mean	2895.5	2862.833333
Variance	34967.5	13543.76667
Observations	6	6
df	5	5
F	2.581815005	
P(F<=f) one-tail	0.160597217	
F Critical one-tail	5.050329058	

Figure 6.7 Results of *F* test using Excel.

F is of course the value you generated. *F* critical is the value of the *F* statistic at which it is probable, at your alpha level, of accidentally rejecting a true null hypothesis, which would constitute a type I error. It is computed in part by the degrees of freedom, but for our purposes we will let Excel handle the computations. Recall that degrees of freedom are the number of data values in the set of data for a sample that are free to vary.

If *F* is less than *F* critical (as it is here, with 2.58 [*F*] < 5.05 [*F* critical]), the difference in variance between the two sets of data is statistically significant. If this were the next step, this finding might lead you to perform a *t* test based on *unequal* variances in our datasets. So, a summary of your findings would be:

- If *F* statistic > critical value (i.e., *F* > *F* critical), then reject the null hypothesis.
- If *P* value < alpha, then reject the null hypothesis.
- If *F* statistic < critical value (i.e., *F* < *F* critical), then accept the null hypothesis.
- If *P* value > alpha, then accept the null hypothesis.

Now let's examine another handy test, ANOVA.

Analysis of Variance

So far we have been working with descriptive statistics. Descriptive statistics, also known as summary statistics, explain or summarize existing data such as mean, median, mode, and standard deviation. With these tools, you assume the data are all known to you and you are describing what you see in summary form, or at least you have a valid sample from the population. For example, you would need all of the data to determine that the average of all student grades in a class (i.e., the mean) is 85 and that the most frequently occurring grade in a class (i.e., the mode) is 90.

Inferential statistics, in contrast, are used to make predictions—or inferences—about future events or trends based on past data. For example, if for the past 10 years a fast-food restaurant sold four times as many hamburgers on the Friday after Thanksgiving as on any other day of the year, it would be safe to predict, or forecast, that it will do the same this year as well. ANOVA, which is a type of inferential statistics, produces two estimates of variance between two or more groups (Batten, 1986). The first estimate is the variance among the means, and the second is the variance among data elements in the groups.

After determining that an ANOVA is appropriate to examine your data, you must select either a one-way ANOVA (one-factor of interest ANOVA) or a two-way ANOVA (two-factor of interest ANOVA) test, each of which is described in the following discussion. If you are considering the means of three or more independent (unrelated) groups of data, use a one-way test. If you are considering two independent groups of data, use a two-way test. In either case, if the means are the same (or fall within your confidence range), then you accept the null hypothesis. Conversely, if you find significant variance in the means, you accept the alternative hypothesis. Regardless of which you choose, try for a **balanced design**, in which each group has the same number of data, as opposed to an **unbalanced design**, in which the groups have differing numbers of data, which calls for other statistical considerations.

One-Way Analysis of Variance

One-way ANOVA should be used when you have one categorical variable (for example, blood type—A, B, AB, or O) and one independent variable (the variable that is manipulated by the researcher, such as the amount of aspirin needed to lower blood pressure), with more than two groups of data. For example, say you have three groups—A, B, and C—and each group, composed of an equal number of people, has received increasing amounts of a different drug treatment. Using one-way ANOVA, you could examine the data to see whether the outcome for each of the three groups was the same. If all three outcomes were equal, you would accept the null hypothesis; if not, you would reject the null hypothesis in favor of the alternative hypothesis, which indicates that the treatment worked to some degree.

Two-way Analysis of Variance

Two-way ANOVA (also known as two-way factor ANOVA and factorial ANOVA, with two factors) should be used when you have two nominal variables (also known as effects, measures, or factors) and one measurement variable. For example, consider that you are examining whether cats with and without chronic diarrhea have the same insulin levels in their blood and you are concerned that the differences between male and female cats may affect the results. The two nominal variables are the presence or absence of chronic diarrhea and sex (male vs. female). Note the contrast here between two nominal variables in this example and the one nominal variable (drug type) in the preceding example of one-way ANOVA. With the two-way ANOVA, both the group with diarrhea and the one without have male and female cats.

There are two types of two-way ANOVA tests in Excel: *with replication* and *without replication*. An Excel two-way ANOVA with replication (or replicates) is

Subjects	Treatment 1	Treatment 2	Treatment 3
Male	7	8	4
	5	10	6
	0	5	8
Female	15	3	15
	11	3	12
	7	1	10

Figure 6.8 Two-way ANOVA with replication.

used when there is more than one observation for each combination of nominal variables. For example, imagine you are examining the effects of three different drug concentrations on men and women in a study.

There are two samples under consideration, males and females, and three different drug concentrations. The study has three *replications* (repetitions of an experiment), three men and three women with three treatments for each, as shown in **Figure 6.8**.

A two-way ANOVA without replication is appropriate when you have only one observation for each combination of nominal variables. For example, instead of having groups divided into males and females in the study, as in the previous example, imagine that the groups are mixed, males and females, and that the subjects' sex is irrelevant (**Figure 6.9**). All that matters is the difference in effects of the three drug treatments. What is the difference, regardless of sex, of each of the three treatments on each of the patients?

Regardless of the test, you are considering the significant differences between the means of the groups. Both tests assume that samples are independent, variances of the population are equal or homogenous, and data are normally distributed. Additionally, with a two-way test, the groups must have the same sample size.

In the next example, we examine a one-way (or one-factor) ANOVA and determine whether the means from two or more samples are equal.

Golfer	Brand X	Brand Y	Brand Z
Johnson	251	245	261
Reid	226	234	231
Dickerson	271	281	285
Thompson	235	242	245
Taylor	211	221	224

Figure 6.9 Two-way ANOVA without replication.

Hands-on Statistics 6.11: Using Excel for a One-way Analysis of Variance

To try a one-way ANOVA with Excel on three classes, follow these directions:

1. To add the Data Analysis function to Excel, open a new, blank Excel workbook and click on the *Options* button (found in the *File* menu in Excel 2010 and in the Office menu in Excel 2007 [online versions of MS Office and other versions are similar]).
2. The Excel Options window will open. In the column on the left, click on the *Add-Ins* heading. On the right side of the window, scroll down to *Inactive Application Add-Ins* and click on *Analysis ToolPack VBA* to select it. Then click the *Go* button. If the *Add-ins* window appears, click in the checkbox next to *Analysis ToolPak—VBA* and click on *OK*.
3. Click on the *Data* tab, and the *Data Analysis* button will appear in the *Analysis* group on the right side of the ribbon.

Now, to try your hand at an ANOVA, complete the following steps:

1. Key the data shown in **Figure 6.10** into a blank spreadsheet. Figure 6.10 shows a simple one-way analysis of final grades from three classes.
2. Select the cells containing data in all three columns, excluding the column headings.
3. Select *Data Analysis* to the right on the menu and then *Anova: Single Factor* from the pop-up window. Click *OK*.
4. In the next pop-up window that appears, make sure the alpha is set to .05 and then click *OK*. The ANOVA data should appear in a new workbook.

3	Class A	Class B	Class C
4	89	89	89
5	88	88	88
6	67	84	67
7	88	88	88
8	89	89	66
9	91	99	91
10	99	99	67
11	95	95	95
12	87	87	87
13	99	99	0
14	98	98	98
15	89	89	89
16	79	69	100

Figure 6.10 Data table for one-way ANOVA.

5. Look at the degrees of freedom and *P* value data in the table to check for significance. Now you can tell whether classes A, B, and C are similar or different with regard to the significance level.

Hands-on Statistics 6.12: Using Excel for a Two-Way Analysis of Variance With Replication

In the following example, we will examine a two-way ANOVA with replication to determine the number of deep-knee bends participants can perform in a given period, with and without an energy drink.

1. Open a new, blank Excel workbook.
2. Key in the data shown in **Figure 6.11**, and save the file to a location you can easily access.

	Without energy drink	With energy drink
male	16	17
male	15	20
male	12	18
male	16	22
male	19	22
male	20	24
male	28	26
male	25	27
male	18	21
male	15	21
female	12	15
female	16	19
female	17	20
female	19	24
female	20	24
female	23	26
female	23	27
female	28	28
female	21	25
female	17	20

Figure 6.11 Data table for two-way ANOVA with replication.

(continues)

3. Click in a blank cell in the sheet where you want the output analysis to appear, select the *Data Analysis* menu, and then select *Anova: Two-Factor With Replication* from the pop-up window (**Figure 6.12**). Click *OK*.

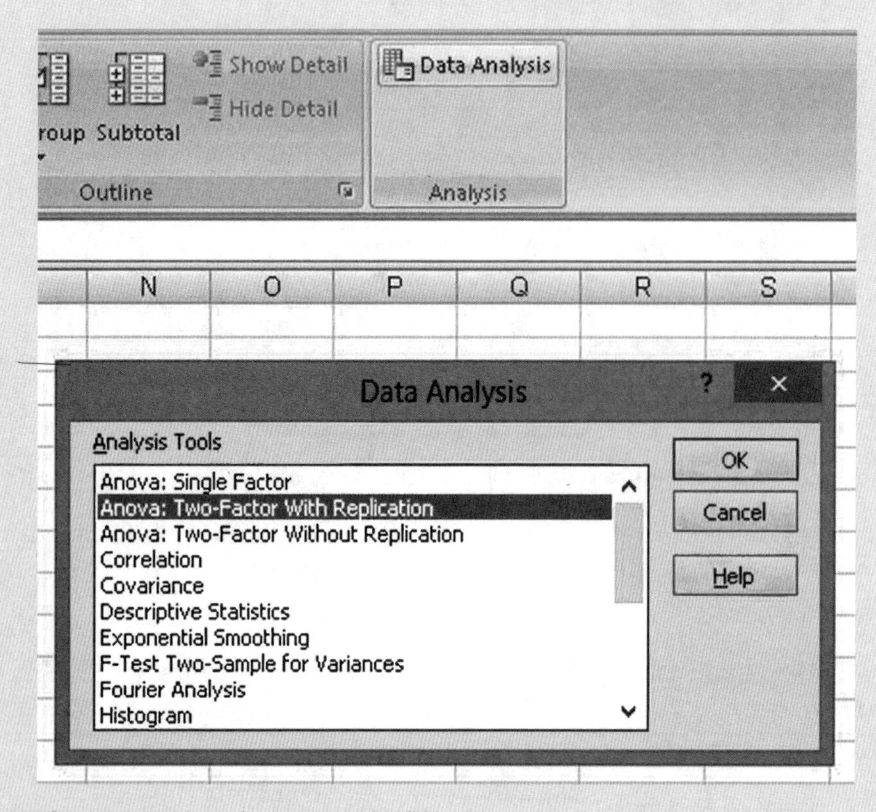

Figure 6.12 Menu for two-way ANOVA with replication.

4. Minimize the next pop-up window that appears and then select the cells in the two columns that contain numerical data (i.e., exclude the headings).
5. Return to the pop-up window and key in "20" in the *Rows per sample* field, since we have 10 males and 10 females. Note the alpha value is .05.
6. For *Output Options*, select *Output Range*, and highlight an area to the right of the data you entered on the sheet.
7. The *F* critical value is greater than the *F* statistic, and the *P* value (.016099) is less than the alpha of .05, so you could conclude that there is some performance enhancement when an energy drink is consumed (**Figure 6.13**).

ANOVA						
Source of Variation	*SS*	*df*	*MS*	*F*	*P-value*	*F crit*
Sample	0	0	65535	65535	#NUM!	#NUM!
Columns	108.9	1	108.9	6.344986	0.016099	4.098172
Interaction	0	0	65535	65535	#NUM!	#NUM!
Within	652.2	38	17.16316			
Total	761.1	39				

Figure 6.13 Results of two-way ANOVA with replication.

Next we will try an ANOVA using R-Project software.

Hands-on Statistics 6.13: Using R-Project to Import Data From a File and Perform an Analysis of Variance

Next we will try an ANOVA using R-Project software. Follow these directions:

1. Start your copy of R. From the command prompt, determine the working directory R is using (the default directory) to read or write data files by typing "getwd()" and pressing Enter. If using a Windows-based computer, it will report something like the following:

   "C:/Users/*yourname*/Documents"

   Your name (e.g., *Smith*) should appear in place of *yourname*. We will now change the working directory to your desktop.
2. In R, type the following, replacing *yourname* with the name you are signed in as on the computer:

   setwd("c:/Users/*yourname*/Desktop")

3. Next, type "getwd()"; make sure you see it is changed. Now you can create or save your data file to the desktop and R can see it properly. Press the up arrow key to display the previous commands, and use the up and down arrow keys to scroll through what you have typed during that session.
4. For our data, we will examine a two-way ANOVA with replication (males and females) for the results of a drug treatment. One group consists of 10 males and the other of 10 females, for a total of 20 participants. Each group is broken into two subgroups, one receiving treatment *a* and one receiving treatment *b*. The alpha will be .05.

   Key in the data shown in **Figure 6.14** using a text editor such as Notepad and save a file named "treatments" (Notepad will automatically append a "*.txt" extension).

```
1,m,a,8
2,m,a,12
3,m,a,13
4,m,a,12
5,m,a,6
6,m,b,7
7,m,b,23
8,m,b,14
9,m,b,15
10,m,b,12
11,f,a,22
12,f,a,14
13,f,a,15
14,f,a,12
15,f,a,18
16,f,b,22
17,f,b,23
18,f,b,14
19,f,b,15
20,f,b,18
```

Figure 6.14 Data file for use with R ANOVA.

(continues)

5. Referred to as "fixed format input," next we will import a local data file, in comma-separated values (CSV) format, and assign our column names manually. In this sample code, we tell R the file name, the names of the columns of data, and the following:

```
mydata = read.table("treatments.txt",
+ col.names = c("Observation","Gender","Dosage","Result"),
+ sep=c(","))
```

   From the R console, type in the first line, press Enter, and you will move to the next line (a + symbol will be inserted for you). Type the next line, and then the third and last.

   Assuming no errors, type "mydata" at the prompt, press Enter, and you will see the data appear, with the column headings you specified, as shown in **Figure 6.15**.

```
> mydata
   Observation Gender Dosage Result
1            1      m      a      8
2            2      m      a     12
3            3      m      a     13
4            4      m      a     12
5            5      m      a      6
6            6      m      b      7
7            7      m      b     23
8            8      m      b     14
9            9      m      b     15
10          10      m      b     12
11          11      f      a     22
12          12      f      a     14
13          13      f      a     15
14          14      f      a     12
15          15      f      a     18
16          16      f      b     22
17          17      f      b     23
18          18      f      b     14
19          19      f      b     15
20          20      f      b     18
> |
```

Figure 6.15 R output of raw data file.

   Examine this code and note that the variable with data is named *mydata*, the file *"treatments.txt"* is in the working directory (the desktop), and the + symbol means you are concatenating, or adding, the other lines to the first line to make one long computer programming statement, so the statement will not run off the page. This would be fine on a computer screen but hard for you to read in the text.

   With *col.names*, we are manually assigning column names for the data in the file. Lastly, *sep=c(","))* is saying that commas separate the field values, and the extra parenthesis is just closing the statement (e.g., if you have an element on the left side, you must have an associated one on the right, as in algebra).

   Summarize your findings by typing "summary(mydata)" and pressing Enter. You will note minimum, mean, and count values for data in each column, as shown in **Figure 6.16**.

6. Next perform an ANOVA on the data. Type the following:

```
myresult = aov(Observations~Gender*Dosage,data=mydata)
```

```
> summary(mydata)
  Observations    Gender Dosage      Result
 Min.    : 1.00   f:10   a:10   Min.    : 6.00
 1st Qu.: 5.75    m:10   b:10   1st Qu.:12.00
 Median :10.50                  Median :14.00
 Mean   :10.50                  Mean   :14.75
 3rd Qu.:15.25                  3rd Qu.:18.00
 Max.   :20.00                  Max.   :23.00
```

Figure 6.16 Summary function of R on *mydata* set.

7. Next, type "summary(myresult)" to run a summary. Results will display. Looking at the asterisk to the right and then *signif.codes:* at the bottom, you will see that three asterisks represent less than the .05 *P* value, so you would reject the null hypothesis and accept the hypothesis, indicating that there is a difference in effectiveness among drugs.
8. To visualize the findings, key in the following and press Enter:

   boxplot(Observations~Dosage*Gender,data=mydata)

The results, shown in **Figure 6.17**, indicate that treatment *b* seems to be better regardless of sex.

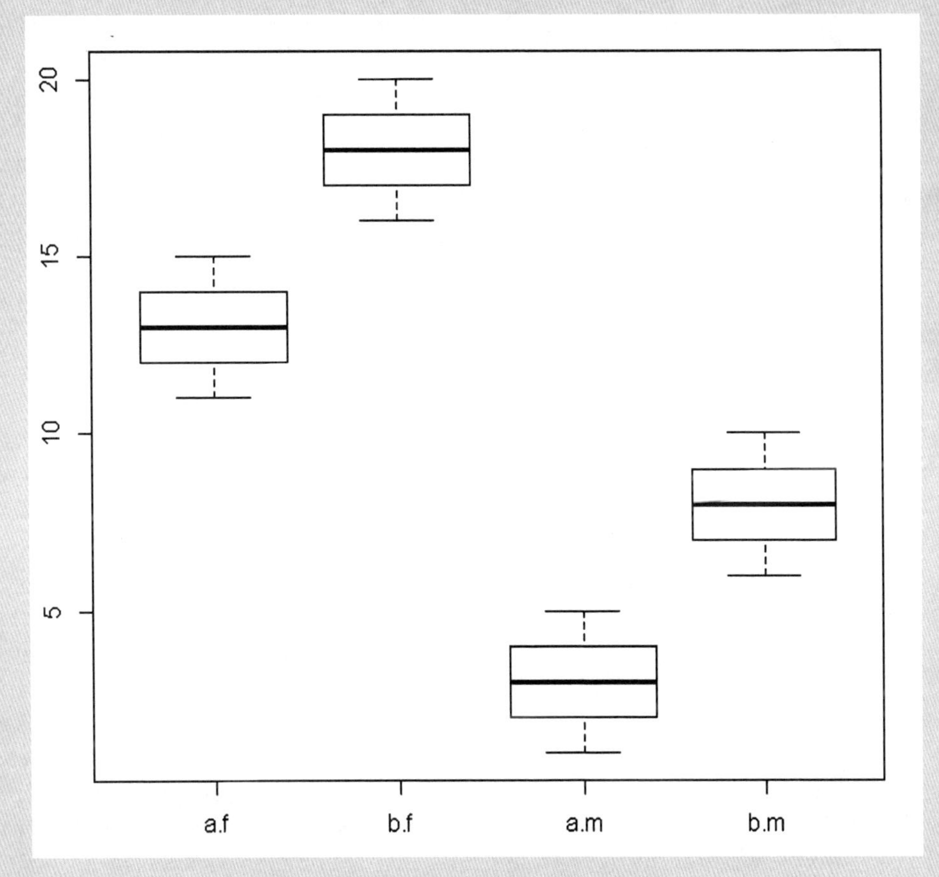

Figure 6.17 Boxplot summary showing treatment *b* has some advantages over treatment *a*.

Two-way ANOVA, like all ANOVAs, assumes that the observations within each cell are normally distributed and have equal variances. However, if the data do not fit a normal distribution or do not fit under the constraints of parametric data, such as with categorical data (e.g., *yes*, *no*, *no-answer*), consider using a nonparametric method, such as the Kruskal-Wallis test, which is covered later in the text.

How Does Your Hospital Rate?

Autopsy statistics are kept by hospitals and reported to the state. They are valuable in addressing key issues such as increases in alcohol-related deaths. For example, in Brunswick county, from 2003 to 2007, there were 265 natural deaths in which alcohol was confirmed to be involved among those 15 years or older. Contact a hospital in your area and ask an administrator for the hospital's autopsy statistics.

Consider the following:

1. Does the hospital perform autopsies?
2. If not, is there a facility nearby that performs autopsies?
3. Is there a common cause of death in the facility, such as staph infection, natural death by alcohol, or homicide deaths related to alcohol?

Global Perspective

In this section we will examine some data visualizations on census data from the US Census Bureau, download raw census data, and compare similar data from other countries' census sites. The US Census Bureau contains a wealth of data that are freely available to researchers. As data collection and analysis methods have improved, so has the quantity and quality of collected data. The US Census Bureau's tools have changed as of late 2020, so using the online tutorials is your best way to learn the new features and tools available on the site.

Follow these steps to explore data harvesting options with US Census Data:

1. Visit the US Census Bureau: https://data.census.gov/cedsci/.
2. Under *Stay Connected* to the right, select *Tutorials*.
3. Under *Data Gems*, watch the tutorial on *How to Visualize Data for Your Area*.
4. Under *Data Gems*, watch the tutorial on *How to Customize and Download Tables*.

Chapter Summary

In this chapter, we have examined the role of a coroner, autopsy facts, and various autopsy rates, including gross and net autopsy rates, adjusted hospital autopsy rate, autopsy rate for newborns, and fetal autopsy rate. The importance of each of these statistics has been discussed, as well as which stakeholders require the data and when. Common social science statistical methods were also examined, such as *F* test and one-way and two-way ANOVA. Lastly real-world data were reviewed from the Census Bureau. The methods you learned can be put to use on many datasets you may encounter.

Apply Your Knowledge

1. Sea Coast Hospital for the months October through December reported the following newborn statistics:
   - Births were 485 and discharges 474.
   - Newborn deaths were 4, and newborn autopsies were 2.
   - Early fetal deaths were 5, intermediate fetal deaths were 3, and late fetal deaths were 1.
   - Of those fetal deaths, early fetal death autopsies were 3, intermediate fetal death autopsies were 3, and late fetal death autopsies were 1.

   Using these data, calculate the following using Excel: newborn death rate, fetal death rate, newborn autopsy rate, and fetal autopsy rate.
2. **Figure 6.18** presents data reported for Sunrise Hospital. Using these data, calculate the following for the year using Excel: gross death rate, gross autopsy rate, and net autopsy rate.
3. Harrington Heights Hospital reported the following for the month of February:
   - Inpatient deaths: 15
   - Emergency service department deaths: 4
   - Home health deaths: 1
   - Inpatient autopsies: 3
   - Emergency service department autopsies: 2
   - Home health autopsies: 1
   - Coroner's cases: 1 (The coroner's cases are those that are unavailable for autopsy.)

   Referring to the data above, calculate the adjusted hospital autopsy rate.

	A	B	C	D	E
1			***Sunrise Hospital 2015***		
2	**Month**	**Discharges**	**Inpatient Deaths**	**Coroner's Cases**	**Autopsies**
3	January	685	10	2	8
4	February	708	8	1	1
5	March	691	15	2	3
6	April	700	5	0	2
7	May	650	5	1	1
8	June	676	9	3	5
9	July	681	7	2	4
10	August	715	15	4	6
11	September	711	6	0	1
12	October	686	4	0	2
13	November	690	12	2	5
14	December	654	7	1	1

Figure 6.18 Sunrise Hospital data table.

4. Valley View Hospital has just released the following information for the year 2015:
   - Inpatient discharges for adults and children: 15,360
   - Inpatient discharges for newborns: 1,325
   - Inpatient deaths for adults and children: 128
   - Inpatient deaths for newborns: 27
   - Autopsies for adults and children: 43
   - Autopsies for newborns: 11

   Referring to the data above, calculate the gross autopsy rate.
5. Scenic Hospital has just released the following information for the year 2015:
   - Adult and child deaths: 128
   - Newborn deaths: 27

   Of those deaths, there were 43 autopsies of adults and children, and 11 newborns were autopsied. Three bodies were autopsied by the hospital pathologist.

   Using these data, calculate the new autopsy rate.
6. Palm Hospital has just released the following information for the year 2015:
   - There were 15,360 adults and 1,325 newborns discharged.
   - Of those discharged, inpatient deaths were 128 for adults and children, and newborn inpatient deaths were 27.
   - Deaths of newborns included 13 early fetal deaths, 7 intermediate fetal deaths, and 2 late fetal deaths.
   - There were 43 autopsies performed on adults and children, and 11 on newborns.
   - Nine of the bodies were unavailable for autopsy, and 3 were examined by the hospital pathologist.
   - They performed autopsies on 5 intermediate and late fetal deaths, while 4 other autopsies were brought in for autopsy.

   Using these data, calculate the adjusted hospital autopsy rate.
7. Using the data from question 6, calculate the newborn autopsy rate and the fetal death autopsy rate.
8. Using information gained by searching the CDC website, graph newborn deaths from five states, including the state in which you live, and compare the statistics. Write one to two paragraphs on your findings.

9. Using the mortality rate information from the World Bank website (http://data.worldbank.org/indicator/SH.DYN.NMRT), graph and compare fetal death rates for 10 countries, including the United States. How does the United States rank in terms of fetal death rates compared with other nations? Write one to two paragraphs on your findings.
10. Using the World Bank website (https://data.worldbank.org/indicator/SP.DYN.CDRT.IN), review the crude death rate. Research by topic and choose *health*. Graph and compare the number of deaths from 10 countries, including the United States. How does the United States rank among other nations in terms of crude death rates? Write one to two paragraphs on your findings.
11. Which types of cases does your state report to the coroner? How did you identify this information?
12. Imagine that you are a pathologist. Consider the facts provided by MTSamples.com regarding autopsy cases (http://www.mtsamples.com/site/pages/browse.asp?type=94-Autopsy), and try to determine the cause of death. Complete at least three cases, and provide a short narrative (three or more sentences) about each one. Mention how many guesses it took you to determine the cause of death.
13. Apply your knowledge of the *F* test. Students took two tests. Their scores are shown in **Table 6.1**. Did they perform any better on the second test than they did on the first (e.g., Was there a difference?). Assume a .05 confidence level.

Table 6.1

Test 1	Test 2
67	99
99	78
89	90
95	79
85	82
90	69
85	82
99	100
87	77
69	86
87	81

14. Complete a one-way ANOVA using the data from **Table 6.2**. Examine three different assembly techniques used by workers to make widgets in a factory. Using a .05 confidence level, is there a difference in processes?

Table 6.2

Process 1	Process 2	Process 3
45	52	44
44	50	46
40	49	45
44	54	43
50	52	44
43	53	42
44	51	47
45	49	43
42	45	45
45	49	49

15. Use two-way ANOVA with replication to solve this problem. A group of male and female subjects were administered two different medical interventions to a problem, with the result of 100 subjects being 100% cured. Determine a .05 confidence ANOVA using the data shown in **Table 6.3**. Was one intervention different from the other?

Table 6.3

M/F	Intervention 1	Intervention 2
F	67	99
F	99	78
F	89	90
F	95	79
F	85	82
F	84	83
M	90	69
M	85	82
M	99	100
M	87	77
M	69	86
M	87	81

16. According to the World Factbook website (https://www.cia.gov/library/publications/the-world-factbook/rankorder/2223rank.html), in 2010 the United States had 21 maternal deaths for every 100,000 live births, compared to 1,100 maternal deaths per 100,000 live births in Chad. Estonia had the lowest maternal death rate, with 2 per 100,000 live births. This poses several questions: Why does Estonia have the lowest maternal death rate? Are there fewer pregnancies in Estonia, do women simply take better care of themselves, or do they have fewer preexisting conditions that would complicate their pregnancy? Create a short presentation to report some interesting comparisons of data from this site using a statistical method you have learned so far in the text.
17. Using this link for the CDC National Center for Health Statistics, https://www.cdc.gov/nchs/hus/contents2018.htm, click on the link for *Table 003 Infant Mortality for 2017*. Compare infant death rates under 28 days by ethnicity. Prepare a chart and provide five questions that you feel would answer why a certain ethnicity rate is higher than others.
18. The Healthcare Cost and Utilization Project (http://www.ahrq.gov/research/data/hcup/) is a comprehensive source for national, state, and local information on inpatient care, ambulatory care, and emergency department visits. Massive databases with this information are available to researchers. Use the following direct link: http://www.hcup-us.ahrq.gov/. Write a one-page max summary of at least four tools/links that a researcher would find useful. List at least one item related to your state.

References

Baiden, A., Bawah, A., Biai, S., Binka, F., Boerma, T., Byass, P., Yang, G. (2007). Setting international standards for verbal autopsy. *Bulletin of the World Health Organization, 85*(8), 570–571.

Batten, J. W. (1986). *Research in Education* (rev. ed.). Greenville, NC: Morgan Printers.

Breslow, J. M. (August 8, 2012). More deaths go unchecked as autopsy rate falls to "miserably low" levels. *FrontLine*. Retrieved March 1, 2013, from https://www.pbs.org/wgbh/frontline/article/more-deaths-go-unchecked-as-autopsy-rate-falls-to-miserably-low-levels/

Freedman, D. H. (2012, September). Twenty things you didn't know about . . . autopsies. *Discover*. Retrieved May 11, 2013, from https://www.discovermagazine.com/the-sciences/20-things-you-didnt-know-about-autopsies

Gulczynski, J., Iżycka-Swieszewska, E., & Grzybiak, M. (2009). Short history of the autopsy part I. From prehistory to the middle of the 16th century. *Polish Journal of Pathology, 3*, 109–114.

Nosheen, H. (2012, February 5). Fewer autopsies mean crucial info goes to the grave. *NPR*. Retrieved December 19, 2012, from http://m.npr.org/news/Health/146355717

Soy, A., & Lacey, A. (2013, September 21). Mobile phone app offers "verbal autopsies" in Malawi. *BBC World News*. Retrieved April 8, 2014, from http://www.bbc.com/news/health-24164824

Web Links

Without Autopsies, Hospitals Bury Their Mistakes: http://www.propublica.org/article/without-autopsies-hospitals-bury-their-mistakes

Map Death in America: http://www.pbs.org/wgbh/pages/frontline/post-mortem/map-death-in-america/

Autopsy 101: http://www.pbs.org/wgbh/pages/frontline/post-mortem/things-to-know/autopsy-101.html

Office of the Chief Medical Examiner, Frequently Asked Questions (New Hampshire Department of Justice): http://doj.nh.gov/medical-examiner/faq.htm

More Deaths Go Unchecked as Autopsy Rate Falls to "Miserably Low" Levels: http://www.pbs.org/wgbh/pages/frontline/criminal-justice/post-mortem/more-deaths-go-unchecked-as-autopsy-rate-falls-to-miserably-low-levels/

Inferential Statistics: http://www.socialresearchmethods.net/kb/statinf.php

Introducing R: http://data.princeton.edu/R/readingData.html

CHAPTER 7

Infection, Consultation, and Other Data

The combination of some data and an aching desire for an answer does not ensure that a reasonable answer can be extracted from a given body of data.

—John Tukey

CHAPTER OUTLINE

Introduction
Infection Rates
- Infection Control Committee
- Nosocomial Infection Rate
- Specific Infection Rate
- Postoperative Infection Rate

Complication Rates
Cesarean Section Rate
Consultation Rates
General Occurrence Rates
Statistical Measures and Tools
- R Commander Graphical User Interface
- Post-Hoc Analysis of Variance
- Tukey Honest Significant Difference
- HAI Reporting and Tracking

Text Mining and Visualization Using R-Project and Wordle
Global Perspective
Chapter Summary
Apply Your Knowledge
References
Web Links

LEARNING OUTCOMES

After completing this chapter, you should be able to do the following:

1. Define the role of an infection control committee.
2. Describe and calculate hospital infection rates.
3. Discuss the importance of postoperative infection rates to hospitals and how to calculate them.
4. Explain what complication rates are and how to calculate them.
5. Describe and calculate cesarean section rates.
6. Discuss the purpose of consultation rates, and demonstrate how to calculate them.
7. Describe the R Commander application, and demonstrate how to install and use it to perform statistical calculations.
8. Explain what an "iatrogenic" death is.
9. Demonstrate how to conduct a post-hoc analysis.
10. Create Wordle word clouds with R.
11. Conduct text data mining with R.

KEY TERMS

Complication rate
Consultation
Consultation rate
Golden hour
Graphical user interface (GUI)
Healthcare-associated infections (HAIs)
Infection control committee
Nosocomial infection
Post-hoc analysis
Postoperative infection rate
Text corpus
Text mining
Tukey honest significant difference (HSD)

How Does Your Hospital Rate?

Sally has decided to have elective surgery to correct shoulder pain. She would like to choose a hospital nearby but has heard stories from friends and read reports of patients developing infections while being treated in local facilities. Some have even died from these infections.

Consider the following:

1. When researching and comparing hospitals for her surgery, what specific statistical measures related to infection should Sally investigate?
2. What qualities should she look for in the hospital that would lessen her risks of acquiring an infection?

Introduction

From the earliest of times, caregivers of patients with injuries from warfare, animal attacks, and accidents have fought life-threatening infections. From ancient Greek surgeons to healthcare practitioners of today, all have had to deal with the same core issues of wound care: "wound management, the **golden hour** (the principle that a victim's chances of survival are greatest if he receives resuscitation within the first hour after a severe injury), and infection control" (Manring, Hawk, Calhoun, & Andersen, 2009). Today, however, healthcare workers have the benefit of applying statistical measures to infection, allowing them to better track, manage, and prevent infection.

This chapter covers how hospitals attempt to control infection and the statistical measures they use to quantify it. It also presents other critical rates hospitals measure, including complication, cesarean section, and consultation rates, as well as how all of these data are used in hospital settings to ultimately improve patient care. Common statistical techniques such as post-hoc analysis of variance and Tukey honest significant difference are explained, along with tools such as R Commander and new applications of R-Project and Wordle for text mining and visualization.

Infection Rates

One of the key statistical measures used to track hospital infections is infection rates. In this section, we will learn about the role of infection control committees, the types of infections that typically occur in hospitals, and specific infection rates that are commonly tracked.

Infection Control Committee

All hospitals have **infection control committees**, whose primary goal is to prevent or reduce the occurrence of **healthcare-associated infections (HAIs)** in the facility (**Figure 7.1**). Such committees

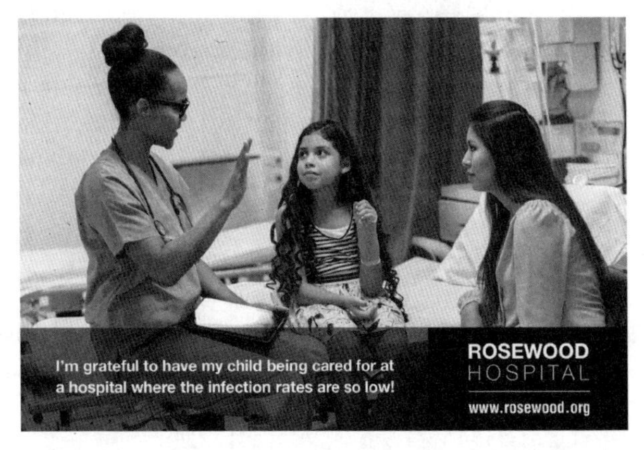

Figure 7.1 Dosher Memorial Hospital marketing their low infection rates.

are intentionally diverse and typically include the following: physicians, nursing staff, infection control practitioners, quality assurance personnel, risk management personnel, and representatives from microbiology, surgery, central sterilization, and environmental services (Lee & Lind, 2000).

To perform its charter, an infection control committee is involved in many areas related to infection control, including planning, monitoring, evaluating, educating, and dealing with reportable information. Planning involves the development of policies and procedures as well as interventions to help prevent and reduce infections. Monitoring includes minimizing risks, identifying problem areas, and taking corrective measures to reduce risk. Evaluation is a continuous cycle of reviewing policies and procedures in place and improving protocols in all departments of the healthcare facility. Education entails training hospital staff in the latest infection control measures and communicating any policy changes to them. The committee members must keep up to date on new strains of bacteria and other changes in their respective areas of expertise. They must also track infection outbreaks and report them to the appropriate public health officials.

Nosocomial Infection Rate

One of the main functions of an infection control committee is to track **nosocomial infections**, which are infections acquired by patients during their stay in a hospital. They may be viral, bacterial, or fungal, and commonly include pneumonia (e.g., ventilator-associated pneumonia) and infections of the bloodstream, the urinary tract, the gastrointestinal system, and surgical sites (Custodio, 2016).

Nosocomial infections are a concern because they are so prevalent. The Centers for Disease Control and Prevention (CDC) reported in 2014 that 1 of 20 hospital patients develops a nosocomial infection every day (CDC, 2014). Moreover, these infections can be a death sentence for patients whose immune systems are already compromised, such as young children, older adults, or those with a preexisting condition. Again, according to the CDC, there were an estimated 722,000 HAIs in US acute care hospitals in 2011, of which roughly 75,000 hospital patients died during their hospitalizations, with costs in the tens of millions (CDC, 2015). Given the high costs—both in terms of lives and dollars—associated with nosocomial infections, it is clear why it is so important for hospital infection control committees and others to track their rate of occurrence.

To calculate nosocomial infection rates, one must first be able to distinguish between infections that result from patient care and those that the patient developed before being admitted. Then, the rate itself is calculated by using the following formula:

$$\frac{\text{All nosocomial infections occurring in a specified period} \times 100}{\text{All discharges, plus deaths for that period}}$$

Hands-on Statistics 7.1: Calculate Nosocomial Infection Rate by Hand

Let's practice calculating the nosocomial infection rate by hand, using the following example. Sea View Hospital for the month of August had 220 discharges, including deaths. Of the 220 patients discharged, 5 had a nosocomial infection. Follow these directions to calculate the nosocomial infection rate:

1. Multiply the number of nosocomial infections by 100:

$$5 \times 100 = 500$$

2. Divide the result by the number of discharges:

$$500/220 = 2.27$$

The result is that 2.27% of patients discharged in August had nosocomial infections.

In the next example you can try the same process using Excel.

Hands-on Statistics 7.2: Calculate Nosocomial Infection Rate Using Excel

To calculate the nosocomial infection rate in Excel, follow these directions:

1. Open a new, blank Excel workbook.
2. Enter the data as shown in **Figure 7.2**, but leave cell E4 blank.

E4 =(D4*100)/B4

	A	B	C	D	E
1			Sea-View Hospital		
2			August 2014		
3		Discharges Including deaths)		HAI	% of HAI's for August
4		220		5	2.272727273

Figure 7.2 Using Excel to calculate HAI percentage.

3. In cell E4, enter the following function:

=(D4*100)/B4

Click out of the cell, and 2.272727273 should appear, as in Figure 7.2.

Specific Infection Rate

In addition to nosocomial infections, other specific infections or infection types may be tracked, such as upper respiratory infections, HIV/AIDS, or bloodstream infections. The rates of these infections can be calculated for individual facilities and compared with the national average or state average.

The infection rate calculation formula is as follows:

$$\frac{\text{All infections of a specific pathogen or type occurring in a specified period} \times 100}{\text{All discharges including deaths during that period}}$$

Hands-on Statistics 7.3: Calculate Infection Rate by Hand

Let's practice calculating the infection rate by hand, using the following example. Ocean Breeze Hospital for the month of February had 340 discharges. Of those 340 discharges, 16 had an upper respiratory infection. Follow these directions:

1. Multiply the number of upper respiratory infections by 100:

$$16 \times 100 = 1{,}600$$

2. Divide the result by the number of discharges:

$$1{,}600/340 = 4.70\%$$

The result is that 4.7% of patients discharged in February had upper respiratory infections.

Hands-on Statistics 7.4: Calculate Infection Rate Using Excel

To calculate the infection rate in Excel, follow these directions:

1. Open a new, blank Excel workbook.
2. Enter the data as shown in **Figure 7.3**, but leave cell D4 blank.
3. In cell D4, enter the following function:

=(C4*100)/A4

(continues)

Click out of the cell, and 4.705882353 should appear, as in Figure 7.3.

	A	B	C	D
1	Ocean Breeze Hospital			
2	February 2014			
3	Discharges		URI's	Infection rate %
4	340		16	4.705882353

Figure 7.3 Ocean Breeze Hospital infection rate.

DID YOU KNOW? Infections were rampant among the wounded during the Civil War because doctors of the time did not know that unwashed hands and unsterilized surgical instruments passed staph and strep infections from patient to patient. These infections, which we now know as postoperative infections, were known at the time as "surgical fever."

Now you know!

Postoperative Infection Rate

Postoperative infection rate is the frequency with which infections occur in hospital patients following surgery. Postoperative infections are one type of nosocomial infection and play a role in increasing mortality and morbidity as well as increasing the cost of health care.

Postsurgical wounds may be classified according to the type and level of contamination and/or infection that occurs in them.

A *clean contaminated wound* occurs when an operative instrument enters into a colonized organ or cavity of the body, but only under elective and controlled circumstances. The most common contaminants are within our own body; for example, a procedure known as a sigmoid colectomy involves cutting into a part of the colon, which contains the bacterium *Escherichia coli*. Pulmonary resection and gynecologic procedures also involve areas of the body that naturally host contaminants. Infection rates for these procedures average 4% to 10%, which can be improved with specific preventive strategies.

A *contaminated wound* (note the absence of the word *clean* this time) is one in which gross contamination is found at the surgical site in the absence of infection. Examples include laparotomies that are performed for penetrating injuries that have intestinal spillage and any elective spillage with gross contamination.

Dirty wounds are those in which an infection is already present when a surgical procedure is performed in the area. Surgical exploration of the abdomen for acute bacterial peritonitis or intra-abdominal abscesses is a classic example of a dirty wound. Unusual pathogens are often associated with a dirty wound if it has occurred at a hospital or nursing care facility.

The formula for calculating postoperative infection rate is:

$$\frac{\text{All infections in clean surgical cases occurring during a specified period} \times 100}{\text{All surgical operations during that period}}$$

Hands-on Statistics 7.5: Calculate Postoperative Infection Rate by Hand

Let's practice calculating the postoperative infection rate by hand, using the following example. In the month of June, Randolph Memorial Hospital performed 694 surgical operations. Of these, 6 resulted in postoperative infections in clean surgical cases. Follow these directions:

1. Multiply the number of postoperative infections by 100:

$$6 \times 100 = 600$$

2. Divide the result by the number of surgical operations:

$$600/694 = 0.86\%$$

Thus, 0.86% of surgical operations performed in June resulted in postoperative infections.

Hands-on Statistics 7.6: Calculate Postoperative Infection Rate Using Excel

To calculate the postoperative infection rate in Excel, follow these directions:

1. Open a new, blank Excel workbook.
2. Enter the data as shown in **Figure 7.4**, but leave cell C5 blank.
3. In cell C5, enter the following function:

   =(B5*100)/A5

   Click out of the cell, and 0.864553314 should appear, as in Figure 7.4. Clicking the cell and formatting the number to two decimal places will round to a more reasonable value.

C5 fx =(B5*100)/A5

	A	B	C
1		Ranger Rick Hospital	
2		June 2014	
3			
4	Surgical operations	Number of PostOp Infections	Percentage of PostOp Infections
5	694	6	0.864553314

Figure 7.4 Percentage of postoperative (PostOp) infections.

DID YOU KNOW? The Surgical Care Improvement Project (SCIP) was a US quality initiative, undertaken through the partnership of many national organizations, that tracked 20 perioperative measures to improve quality of care and reduce surgical complications. Of these measures, 9 were reported to the public. Although participation in SCIP was not mandated, the Centers for Medicare and Medicaid Services reduced reimbursement by 2% to facilities that did not report SCIP performance measures.

The purpose of this initiative was to decrease the number of HAI instances. However, tying reimbursement (or withholding it in this case) due to the number of HAIs did not seem to lower the number of instances, according to published research findings in 2013 in the *New England Journal of Medicine*. In fact, the researchers stated, "We found no evidence that the 2008 CMS policy to reduce payments for central catheter-associated bloodstream infections and catheter-associated urinary tract infections had any measurable effect on infection rates in US hospitals" (Lee et al., 2012).

Now you know!

Complication Rates

Complications are medical conditions that the patient develops during hospitalization. They can occur for many reasons, such as falls, medication reactions, blood transfusion reactions, and cardiopulmonary resuscitation.

Complications are typically classified as avoidable or unavoidable. Avoidable complications result from deficient care on the part of the healthcare provider. For example, if a patient developed a pulmonary embolism during a procedure and was not given an appropriate prophylaxis before the procedure, the pulmonary embolism would be an avoidable complication resulting from an error in judgment or policy. However, if that same patient received appropriate prophylaxis and still developed the pulmonary embolism, the complication would then be considered unavoidable and accredited to the patient's disease.

Some complications are considered to be physician-related, such as unplanned return to surgery, surgery at the wrong site, and failure of surgical anastomoses.

Complication rate, or the frequency with which complications occur, can be calculated monthly or semiannually depending on how often the facility wishes to review these statistics. Complication rates are severity-adjusted to reflect the most common types of complications by procedure, being categorized as minor or major. Calculating this rate allows the facility to benchmark its performance against that of other facilities of similar size and to implement best practices to eliminate complications due to provider error and change policies and procedures to reflect changes in medical care and protocol.

Complication rates can also be calculated for specific complications, such as allergic reactions to medications and blood transfusions and complications related to coronary artery bypass grafts, spinal surgery, regional anesthesia in children, and carotid endarterectomy. Such rates can be used not only to track the prevalence of specific complications but to evaluate the effectiveness of interventions. For example, in a study on spinal surgeries, researchers found that patients who received bone-morphogenetic protein (BMP) had a much higher risk of complications than those who did not receive the BMP, as well as higher hospital charges and lengthier hospital stays. Complications associated with use of this drug, which was approved by the Food and Drug Administration in 2002 for use in surgeries of the anterior lumbar spine, specifically have occurred in anterior cervical fusion and posterior cervical process surgeries. Further research is needed to develop and refine guidelines for use and to further study the long-term risks and benefits associated with BMP. Research of this nature is profoundly beneficial to the patient and important in reducing complications.

Complication rate is computed with the following formula:

$$\frac{\text{All complications during a specified period} \times 100}{\text{All discharges in that period}}$$

Hands-on Statistics 7.7: Calculate Complication Rate by Hand

Let's practice calculating the nosocomial infection rate by hand, using the following examples. The Good Samaritan Hospital in Texas for the months of October to December had 22 complications and 2,244 discharges. Follow these directions:

1. Multiply the number of complications by 100:

$$22 \times 100 = 2{,}200$$

2. Divide the result by the number of discharges:

$$2{,}200/2{,}244 = 0.98\%$$

Thus, 0.98% discharges from October to December were associated with complications.

Now we will consider an example that involves a specific complication rate. Sea View Hospital had 18 patients who received the same type of medication, and 4 of those patients had a severe allergic reaction to the medication. Follow these directions:

1. Multiply the number of patients who had an allergic reaction by 100:

$$4 \times 100 = 400$$

2. Divide the result by the number of patients who received the medication:

$$400/18 = 22.22\%$$

Thus, 22.22% of the patients who received the medication had an allergic reaction to it.

Hands-on Statistics 7.8: Calculate Complication Rate Using Excel

To calculate the complication rate in Excel, follow these directions:

1. Open a new, blank Excel workbook.
2. Enter the data shown in **Figure 7.5**, but leave cell E5 blank.
3. In cell E5, enter the following function:

=(C5*100)/A5

Click out of the cell, and 0.980392 should appear, as in Figure 7.5. You may round the number to two decimal places by right clicking and selecting *Format Cells*, the *Number* tab, and *Number*, and then entering "2" in the *Decimal places* field.

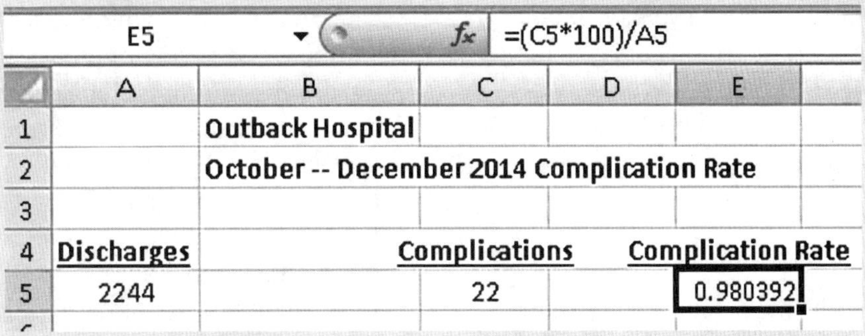

E5 | fx =(C5*100)/A5

	A	B	C	D	E
1		Outback Hospital			
2		October -- December 2014 Complication Rate			
3					
4	Discharges		Complications		Complication Rate
5	2244		22		0.980392

Figure 7.5 Complication rate computed using Excel.

Cesarean Section Rate

Cesarean delivery, also known as cesarean section, is the most frequently performed surgery in the United States, with about one-third of deliveries being by that method (Palermo, 2014). Cesarean section rates have risen drastically up to about 2006, when there was a slight decline, due in part to the Joint Commission's mandate that hospitals try to reduce the number of cesarean sections performed.

Cesarean sections are often performed for the following reasons: previous cesarean section, three or more previous vaginal births, fetal distress, breech presentation, failure to progress in the first stages of labor, placenta previa, cephalo-pelvic disproportion, and antepartum hemorrhage. Trends that have had an impact on the rate of cesarean sections in recent years include increases in the following: the number of multiple births due to in vitro fertilization, maternal age, weight gain before and during pregnancy, and the number of first-time mothers having a cesarean section, which escalates the need to have a cesarean section with subsequent children. Certainly the impact of the mandate by the Joint Commission will continue to lower the frequency of this procedure.

Cesarean section rates vary across the United States and around the world. There has been significant debate about the increasing number of cesarean sections over the last few years and their appropriateness. Cesarean sections can create unnecessary risk to the mother, such as hospital-acquired infections and lengthier stays in the hospital, making them more expensive than vaginal deliveries. Factors affecting the decision to undergo a cesarean section include the patient's type of insurance, patient preference, the physician's approach, and the hospital's policy. The World Health Organization, concerned over the increased number of cesarean sections performed around the world, recommends performing this procedure only if there is a medical indication for it; specifically, it should be performed in only 15% of deliveries. China is a country of special concern. Among the Chinese hospitals researched in 2007 and 2008, 46% of the babies born were delivered by cesarean section, making China the top-ranking nation for cesarean deliveries in the world (Hvistendahl, 2012). From 2007 to 2016, cesarean deliveries dropped to 32% due to the encouragement of doctors and nurse midwives that women with low-risk pregnancies have a vaginal delivery (Wang & Hesketh, 2017).

Maintaining statistics on cesarean section rates is important to understand not only the rise in cesarean sections but also the associated complications that can occur in the mother and fetus. A reduction in the number of cesarean sections performed would result in a culture change and decreases in payment to the facilities and utilization of services.

The formula for calculating cesarean section rate is as follows:

$$\frac{\text{All cesarean sections during a specified period} \times 100}{\text{All deliveries in that period, including cesarean sections}}$$

Hands-on Statistics 7.9: Calculate Cesarean Section Rate by Hand

Let's practice calculating the cesarean section rate by hand, using the following example. St. Mary's Hospital in the month of February had 280 deliveries, of which 6 were cesarean sections. Follow these directions:

1. Multiply the number of cesarean sections by 100:

   $$6 \times 100 = 600$$

2. Divide the result by the number of deliveries:

   $$600/280 = 2.14\%$$

Thus, 2.14% of the deliveries in February were cesarean sections.

Hands-on Statistics 7.10: Calculate Cesarean Section Rate Using Excel

To calculate the cesarean section rate in Excel, follow these directions:

1. Open a new, blank Excel workbook.
2. Enter the data shown in **Figure 7.6**, but leave cell D4 blank.
3. In cell D4, enter the following function:

   =A4*B4/C4

   To round the answer to two decimal places, right click on the cell, select *Format Cells*, the *Number* tab, and *Number*, and then enter "2" in the *Decimal places* field. The result should be 2.14, as shown in Figure 7.6.

D4 fx =A4*B4/C4

	A	B	C	D
1	C-section with Excel			
2				
3	Feburary C-sections		Deliveries	Percentage
4	6	100	280	2.14

Figure 7.6 Cesarean section percentages computed using Excel.

DID YOU KNOW? According to Greek mythology, Asclepius, known as the god of healing, was the son of the god Apollo. Asclepius's mother was the princess Coronis. Apollo assigned the duty of educating his son to the centaur Chiron, who was skilled in many different areas, including healing, a subject in which Asclepius excelled. He became so good that he could restore life to the dead! However, as ruler of the underworld, and the dead, Hades was not pleased with Asclepius, as he was concerned that if Asclepius made all men immortal, it would put him out of business. He made his case to Zeus, who killed Asclepius with a thunderbolt.

Asclepius has retained a place in medicine to this day. Many believe that the serpents around the medical caduceus symbol are a tribute to the coiled serpent that adorned his staff. Also, the names of two of his daughters, Hygeia, meaning "health," and Panacea, meaning "all-healing," have very obvious connections to health care.

Now you know!

Consultation Rates

After a patient has received a diagnosis for a condition, the patient or a healthcare professional may request a second opinion regarding the diagnosis from another physician, which is known as a **consultation**. Consultations are usually requested when the patient's diagnosis is beyond the initial physician's realm of expertise, the physician lacks training or experience, or an ethical dilemma is involved. General internists are the most likely to request consultations.

A request for the consultation must be made prior to the second physician seeing the patient and must be documented in the medical record by both the attending physician and the consulting physician. The consulting physician reviews documentation from the medical record and visits the patient. Once the consultation is completed, the physician dictates a consultation report, giving his or her findings and making suggestions for further treatment. All consultations must be documented in the patient's record, allowing the attending physician to review the results of the findings. The "three R's" to remember related to consultations are *request*, *render*, and *respond*.

Avoid confusing the terms *referral* and *consultation*. A *referral* means the patient's actual care is being referred to someone else, whereas a *consultation* means the other physician is providing only recommendations, not care.

Consultation rate, or the frequency with which a physician requests consultations, can be used in any service section of the hospital, such as pediatrics, surgery, obstetrics, and so forth. Each hospital chooses which statistics they wish to track.

The consultation rate may be calculated as follows:

$$\frac{\text{All patients receiving a consultation} \times 100}{\text{All patients discharged}}$$

Hands-on Statistics 7.11: Calculate Consultation Rate by Hand

Let's practice calculating the nosocomial infection rate by hand, using the following example. The medicine unit of Blue Sky Hospital discharged 224 patients in December, and 16 of those patients received consultations. Follow these directions:

1. Multiply the number of patients who received consultations by 100:

$$16 \times 100 = 1{,}600$$

2. Divide the result by the number of patients discharged:

$$1{,}600/224 = 7.14\%$$

The result is that 7.14% of the patients discharged in December received a consultation.

Hands-on Statistics 7.12: Calculate Consultation Rate Using Excel

To calculate the consultation rate in Excel, follow these direction:

1. Open a new, blank Excel workbook.
2. Enter the data shown in **Figure 7.7**, but leave cell B6 blank.
3. In cell B6, enter the following function:

=(B4*100)/A4

To round the answer to two decimal places, right click on the cell, select *Format Cells*, the *Number* tab, and *Number*, and then enter "2" in the *Decimal places* field. The result should be 7.14, as shown in Figure 7.7. Note that in the previous Excel example you did not use parentheses, yet this time you did. Either approach will work, based on the rules of operator precedence. Multiplication and division are handled in order, left to right. However, addition and subtraction would need to be forced to happen first by using parentheses, if required.

B6 fx =(B4*100)/A4

	A	B
1	Consultantion Rate Blue Sky Hospital	
2		
3	Discharged patients	Received consultations
4	224	16
5		
6	*Consultation rate =*	*7.14*

Figure 7.7 Calculate the consultation rate for Blue Sky Hospital using Excel.

General Occurrence Rates

Occurrence rates may be calculated for many different purposes in various areas of the hospital. The formula to calculate general occurrence rates is as follows:

$$\frac{\text{How many times an event occurred} \times 100}{\text{How many times the event could have occurred}}$$

DID YOU KNOW? When the superbug, better known as methicillin-resistant *Staphylococcus aureus* (MRSA), hit Greece in 2007, it wreaked havoc because physicians in that country had overprescribed antibiotics more than any other county. In fact, the use of antibiotics outside of hospitals was twice the median of other countries (Kresge & Gale, 2012). One of the solutions was to lower the nurse-to-patient ratio, which resulted in Greece having one of the lowest nurse-to-patient ratios in the world. Other measures included nurses washing their hands more often and infected patients being separated from others.

Now you know!

Hands-on Statistics 7.13: Calculate General Occurrence Rates by Hand

Let's practice calculating another rate by hand, using the following example. Water's Edge Hospital has 650 medical care beds plus 100 psychiatric care beds. In August, the psychiatric facility had 100 patients in house, 15 of whom received a consultation. Follow these directions:

1. Multiply the number of consultations by 100:

$$15 \times 100 = 1{,}500$$

2. Divide the result by the number of patients in house:

$$1{,}500/100 = 15\%$$

Thus, 15% of patients in house in August received psychiatric consultations.

Of course, the preceding calculation could be performed in a similar fashion with Excel, or another spreadsheet application, as we did for the previous Excel examples.

Statistical Measures and Tools

In previous chapters, you have used R as a powerful tool to analyze statistical data. In this section, you will learn about R commander (Rcmdr), a powerful **graphical user interface (GUI)** for R, including how to install it and use it to calculate various statistical measures. A GUI presents a visual metaphor for user interaction; in contrast, with a text-based interface, all commands are typed, using a mouse, touch screen, or voice input. You will also learn about text mining and visualization using R-Project and Wordle.

R Commander Graphical User Interface

R commander gives the user an interface to R that is more common (when compared with other applications) than the native R interface and gives easy access to commonly used features of R. It is, like most R tools, free and easy to install. Follow along to learn how to install R commander.

Hands-on Statistics 7.14: How to Install R Commander

1. To learn what Rcmdr has to offer and find learning resources, visit the Rcommander website (http://www.rcommander.com/).
2. To install Rcmdr, first open your installation of R.
3. From the *Packages* menu in R, select *Install Package(s)*.
4. Select *0 Cloud* as a mirror.
5. Scroll down in the list until you find *Rcmdr* and select *OK* to install. The program will now retrieve needed files via your internet connection. Allow the installer to retrieve any missing files from the CRAN mirror site.
6. At this point, R Commander is installed. Whenever you start R and want to use R Commander, just type "library(Rcmdr)" and press Enter at the prompt to load R Commander. Note the capital *R* in Rcmdr; many powerful software applications are case sensitive in this way. **Figures 7.8** and **7.9** show how to load R Commander and what the interface looks like when loaded.

```
> library(Rcmdr)
Loading required package: splines
Loading required package: car

Rcmdr Version 2.0-1

Warning messages:
1: package 'Rcmdr' was built under R version 3.0.2
2: package 'car' was built under R version 3.0.2
```

Figure 7.8 Loading R Commander from the command prompt.

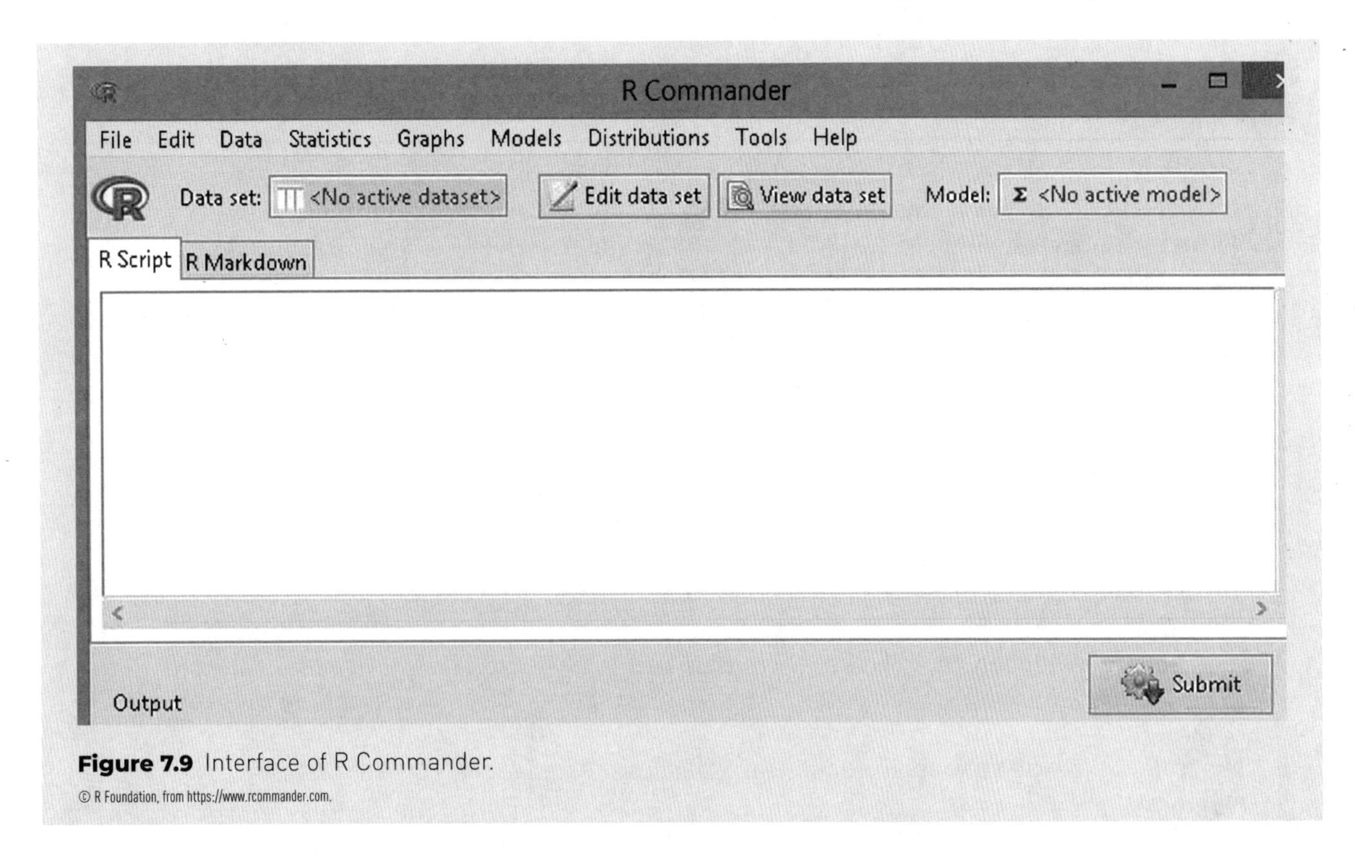

Figure 7.9 Interface of R Commander.

Now that you have successfully installed R Commander, we will explore some of the features and try out one of the built-in statistical methods that you used in a previous chapter to see how R Commander can streamline our analysis process.

Hands-on Statistics 7.15: Explore Some Features of R Commander

1. If you do not already have R Commander loaded into R, at the prompt, type "library (Rcmdr)" and press Enter to start it.
2. A feature that will be familiar to anyone who has used a word-processing or other office application is the *File/Save As* menu. In this case, it is *File/Save script as* to save a program you are writing. Also, you can open a previously saved script from the *File* menu. We will not examine the R Markdown format in this text, but it is a subset of the Markdown language that allows easier integration of R results into HTML web pages.
3. The *Data* menu allows you to import raw data for R to work on. In the next Hands-on Statistics exercise, we will see how to use this menu.
4. The *Statistics* menu displays many commonly used statistical methods so that they are easily available. We will also use one of these in the next Hands-on Statistics exercise.
5. The *Graphs* menu gives you the ability to easily visualize your data.
6. The *Models* and *Distributions* menus allow you access to some popular analysis models and significance distributions. Recall from previous chapters that we had to examine a table for significance based on degrees of freedom (df) and our level of significance. With the *Distributions* menu, you can produce your own table, among other things.
7. The last two menus allow you to install packages easily and search for help on topics.
8. To put into action some of these interesting features of R Commander, in the next Hands-on Statistics exercise, we will use a one-way analysis of variance (ANOVA), just as we did in the previous chapter, except that now we will use R Commander as our GUI.

Hands-on Statistics 7.16: Perform a One-Way Analysis of Variance Using R Commander

1. Start R Commander at the R command prompt by typing "library(Rcmdr)" and pressing Enter.
2. Next, select the following sequence: *Data* menu, *Import data*, and *From text file, Clipboard, or URL*.
3. Name the dataset "onewayanova" and for *Field separator*, select *Commas*, as shown in **Figure 7.10**.
4. Find the file "CH07_One_Way_Anova.csv" in Chapter 7 of the eBook and download it. Note these data are similar to data used in a previous chapter but in a two-column format.

   In *Messages* at the bottom, it should show *The dataset onewayanova has 39 rows and 2 columns*.
5. Select the *View Data Set* button at the top of the R Commander screen to display your raw data, which should look similar to the raw data in **Figure 7.11**.
6. Select *Statistics, Summaries, Active data set* to display summary information about your data, such as minimum, maximum, median, and mean (**Figure 7.12**). The *Grades* column is the only one with valid results, however, as the class column uses 1, 2, and 3 to note A, B, and C classes.
7. Select the following options, in sequence: *Data menu, Manage variables in active data set, convert numeric variables to factors... , Class, Use numbers*. This will convert the text column to numeric.
8. Select the following options, in sequence: *Statistics menu, Means, One Way ANOVA, OK* (**Figure 7.13**).

   Interpreting the results, Pr(>F) of .0242 is the *P* value of significance. Assuming we chose a significance level of .05, we can accept H_1 as the hypothesis because .0242 is less than .05. So we can reject the null hypothesis and accept the hypothesis that there is a difference in grades.
9. To graphically display your results, select the *Graphs* menu in R Commander, then *Plot of means*, and then switch to the R program. You will see a visual representation of the three classes (**Figure 7.14**).

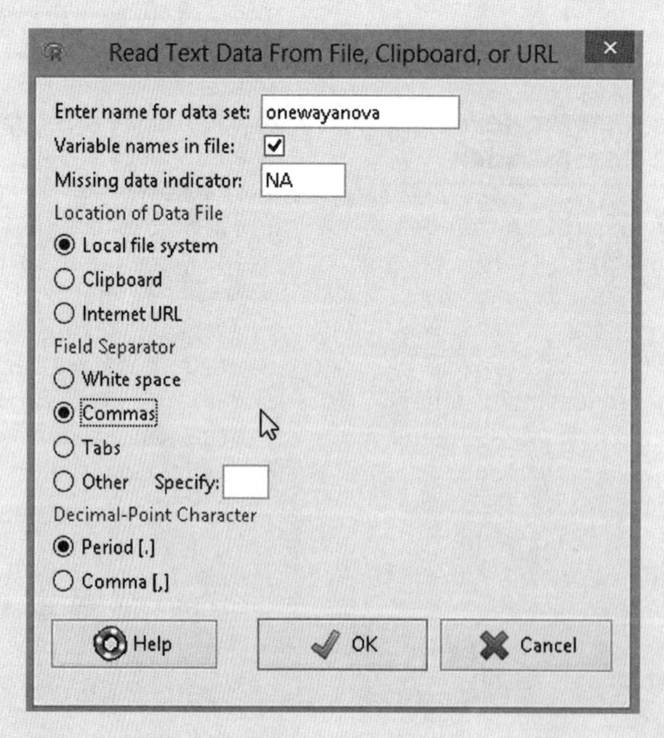

Figure 7.10 R Commander import screen.

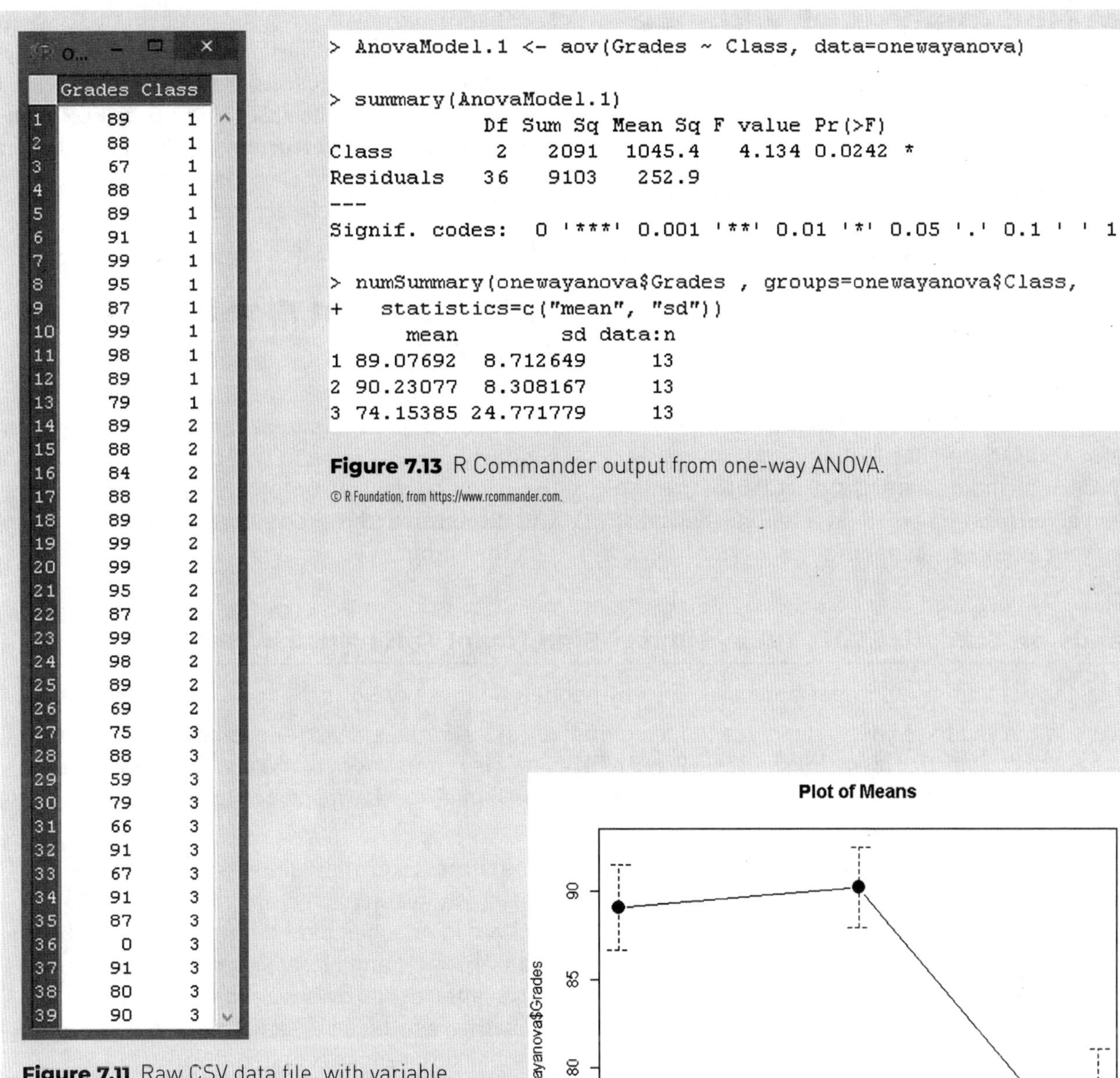

	Grades	Class
1	89	1
2	88	1
3	67	1
4	88	1
5	89	1
6	91	1
7	99	1
8	95	1
9	87	1
10	99	1
11	98	1
12	89	1
13	79	1
14	89	2
15	88	2
16	84	2
17	88	2
18	89	2
19	99	2
20	99	2
21	95	2
22	87	2
23	99	2
24	98	2
25	89	2
26	69	2
27	75	3
28	88	3
29	59	3
30	79	3
31	66	3
32	91	3
33	67	3
34	91	3
35	87	3
36	0	3
37	91	3
38	80	3
39	90	3

Figure 7.11 Raw CSV data file, with variable names, to be imported.


```
> AnovaModel.1 <- aov(Grades ~ Class, data=onewayanova)

> summary(AnovaModel.1)
            Df Sum Sq Mean Sq F value Pr(>F)
Class        2   2091  1045.4   4.134 0.0242 *
Residuals   36   9103   252.9
---
Signif. codes:  0 '***' 0.001 '**' 0.01 '*' 0.05 '.' 0.1 ' ' 1

> numSummary(onewayanova$Grades , groups=onewayanova$Class,
+   statistics=c("mean", "sd"))
      mean        sd data:n
1 89.07692  8.712649     13
2 90.23077  8.308167     13
3 74.15385 24.771779     13
```

Figure 7.13 R Commander output from one-way ANOVA.

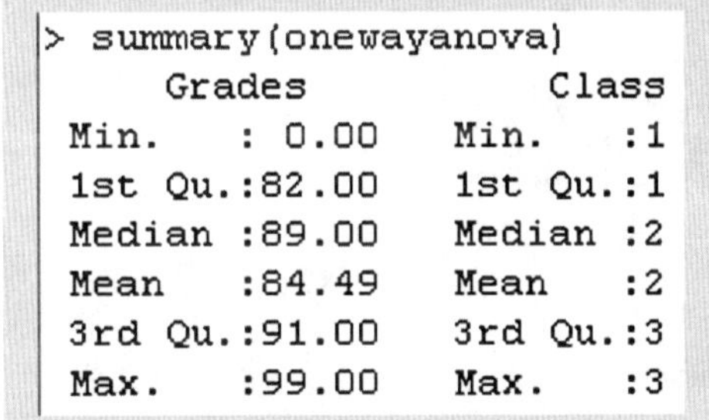

```
> summary(onewayanova)
     Grades          Class
 Min.   : 0.00   Min.   :1
 1st Qu.:82.00   1st Qu.:1
 Median :89.00   Median :2
 Mean   :84.49   Mean   :2
 3rd Qu.:91.00   3rd Qu.:3
 Max.   :99.00   Max.   :3
```

Figure 7.12 Summary of raw data.

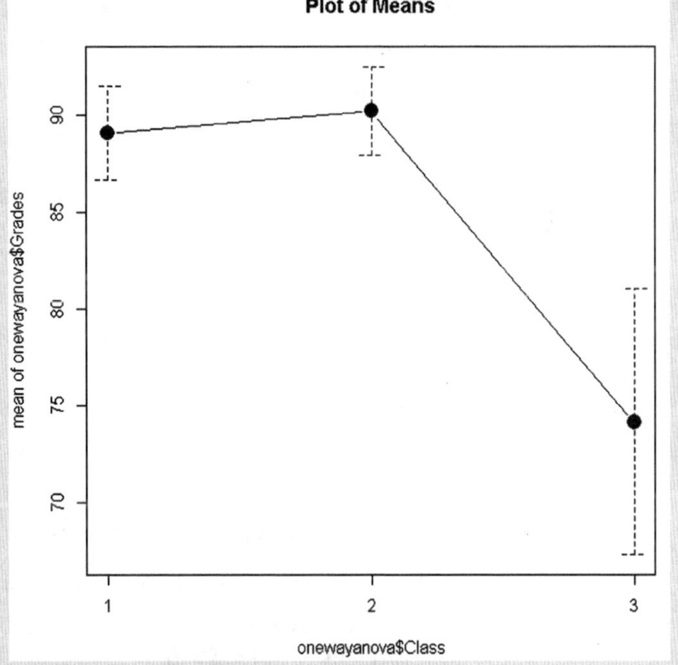

Figure 7.14 Graphic displaying means of three classes of grades.

Now that you have learned how R Commander can streamline your analysis process, in this case with an ANOVA, you will conduct further analysis to give you more information on the data. Recall that an ANOVA informs you whether there is a significant difference among sets of data you are examining, such as sets X, Y, and Z. However, what ANOVA cannot tell you is which is different. However, several other methods may be used for this determination.

Post-Hoc Analysis of Variance

After you have completed an ANOVA and found a difference—that is, you have rejected the null hypothesis—you may use **post-hoc analysis** to determine if there is a significant difference between the groups, which refers to several types of post-hoc analysis methods. Put another way, it determines which factors are significantly different (based on your specification of significance). These tests are done after an ANOVA or other tests and are done for further analysis and clarification of the data under examination.

There are several methods of post-hoc analysis for pairwise comparison after finding significance using an ANOVA, including a straight pairwise comparison with no adjustment for type I error and methods that adjust your results to minimize type I error. (Recall that in type I error you reject the null hypothesis when it is true; that is, you mistakenly claim that a treatment had little to no effect.) Such methods include Bonferroni, Holm, Fisher LSD, and the **Tukey honest significant difference (HSD)**, named after statistician John Tukey. Next you will examine only the Tukey HSD, which is good at reducing type I errors.

Tukey Honest Significant Difference

A pairwise comparison using R Commander is fairly easy and is basically the same process as that used for conducting a one-way ANOVA, except that you select the pairwise comparison option. The next example will demonstrate this process and show how to interpret the output.

Hands-on Statistics 7.17: Tukey Honest Significant Difference With R

To perform a pairwise comparison using R Commander, follow these directions:

1. Assuming you have your dataset loaded from the previous example, select the following options in sequence: *Statistics menu*, *Means*, *One Way ANOVA*, *Pairwise comparison of means*, *OK* (**Figure 7.15**).
2. Scroll up in the output window to show the results for *Simultaneous Tests for General Linear Hypotheses* (**Figure 7.16**).

Note that 1, 2, and 3 are the classes, with grades for each. To determine differences, the Tukey HSD gives a *P* value to determine whether a difference exists between each class compared. Pr(>|t|) shows us that for class 2 compared with class 1 (2 – 1), there is no significant difference (*P* = .9813, which is greater than .05). For class 3 compared with class 1 (3 – 1) and class 3 compared with class 2 (3 – 2), the *P* values are .0561 and .0369, respectively, the latter of which is less than .05.

The conclusion is that classes 1 and 2 had about the same results, whereas class 3 is different from both classes 1 and 2, with class 2 showing a little more difference (because the *P* value for class 2 was lower than for class 1). Moreover, we may conclude that if three different methods of teaching were used on the three classes, the methods used for the first two classes had about the same effect, whereas the method used for the third class was different (and in fact had less-satisfactory results given the overall student grades).

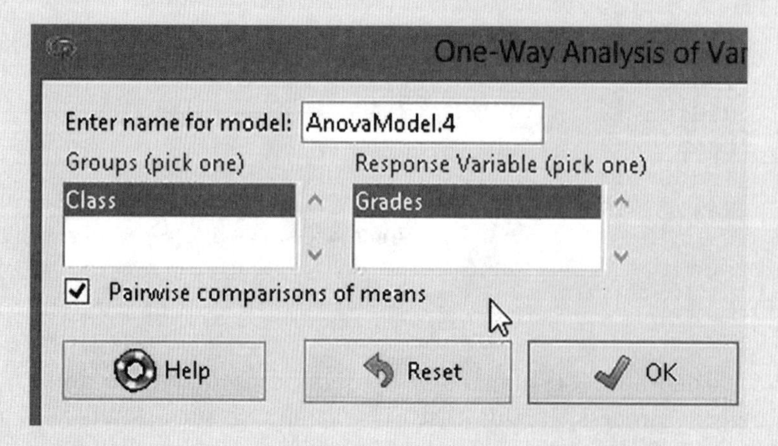

Figure 7.15 Pairwise Tukey selection screen.

Output

```
          Simultaneous Tests for General Linear Hypotheses

Multiple Comparisons of Means: Tukey Contrasts


Fit: aov(formula = Grades ~ Class, data = onewayanova)

Linear Hypotheses:
             Estimate Std. Error t value Pr(>|t|)
2 - 1 == 0     1.154      6.237   0.185   0.9813
3 - 1 == 0   -14.923      6.237  -2.393   0.0561 .
3 - 2 == 0   -16.077      6.237  -2.578   0.0369 *
---
Signif. codes:  0 '***' 0.001 '**' 0.01 '*' 0.05 '.' 0.1 ' ' 1
(Adjusted p values reported -- single-step method)
```

Figure 7.16 Output from pairwise comparison of means.

How Does Your Hospital Rate?

After a quick search of the internet for information on HAIs, Sally finds much more information than she realized was available. She discovers that there are different types of infections, different required and optional reporting standards for HAIs (some by the federal government and some on a state-by-state basis), sites and dashboards to show data on infection rates by facility, and efforts by different groups to decrease the occurrence of these costly and life-threatening infections. In short, her eyes are opened to the reality of HAIs. Study the infection rates of hospitals in your area.

Consider the following:

1. Which infections are most prevalent in your area?
2. What reporting requirements related to HAIs must hospitals in your area follow?
3. What organizations are combating HAIs in your area, and what programs are in place?

DID YOU KNOW? According to one study, 4.5 infections were reported for every 100 hospital admissions in 2002 (Klevens et al., 2007). However, this statistic does not distinguish between different types of infections. HAIs may be grouped into three broad areas, based on how the infection is acquired: (1) medical device-associated infections—such as those associated with intravenous or central lines, ventilators, and catheters—which account for about two-thirds of all HAIs; (2) surgical-site infections, and (3) gastrointestinal infections (*Clostridium difficile*-associated disease) from contaminated surfaces, contaminated hands, or improperly sterilized instruments. Since 2002, more hospitals have started reporting HAI's. Using the following website, https://arpsp.cdc.gov/profile/infections, you will notice that rates of C-diff have had a significant drop. Use the site to find updates on different HAI's.

Now you know!

HAI Reporting and Tracking

The National Healthcare Safety Network (NHSN), the nation's largest and most-used HAI reporting and tracking system, helps states and facilities identify problems and track progress in lowering the rate of HAIs. Although requirements are changing, key reporting requirements to NHSN include the following:

- Central line–associated bloodstream infections
- Surgical site infections

- Catheter-associated urinary tract infections
- *C. difficile* laboratory-identified (LabID) event
- MRSA bacteremia LabID event
- Healthcare worker flu vaccination
- Surgical Care Improvement Project

The process that takes place for reporting is as follows:

1. The hospital infection control committee reviews all collected information on infections for the period under consideration. If it is determined that the infections are valid and qualify for reporting, they are tabulated. The local hospital has veto power to adjudicate events that do not meet their criteria.
2. Data are sent to the CDC.
3. The CDC collects the data and reports them to the Department of Health and Human Services (HHS) and to the Secretary of Health and Human Services.

Both the CDC and HHS use these data to help all concerned parties make informed decisions to minimize instances of HAI. Trends can be identified and problem areas can receive targeted help, all to the benefit of patients.

These data are available at Medicare's Hospital Compare website (http://www.medicare.gov/hospitalcompare). By entering your zip code, you can find hospital rankings for many different areas, including, most importantly, instances of HAIs.

Text Mining and Visualization Using R-Project and Wordle

To continue with data mining explorations, next you will examine **text mining**, which is another method to reveal hidden information in unorganized text or numeric data. Song (2013) explained that while "data mining predominately relies on statistical methods to uncover trends in structured data, text-mining techniques seek to make sense of information that is unstructured, such as a doctor's scribbles on a patient's chart." Text mining, especially in health care, is a growing field and is especially useful in that the amount of nonsorted and textual data is increasing at an alarming rate. Raja, Mitchell, Day, and Hardin (2008) noted that "the documents that comprise the health record vary in complexity, length and use of technical vocabulary" which "makes knowledge discovery complex." This challenge supports the value of using text mining to make sense of textual data.

The three general steps of text data mining are (1) preprocessing of the data, (2) text mining, and (3) postprocessing, when the results are summarized and reported, possibly as graphics. In the next Hands-on Statistics exercise, you will work through these steps. You will recall from a previous chapter that you manually created Wordle visualizations to graphically display numeric data. Now you will harness the power of R to text mine data and use the resultant data to create visualizations from information extracted with R-Project.

Hands-on Statistics 7.18: Reading a Simple Local File Into R-Project for Text Mining

In this example you will take a small text file and text mine it for information to use later in a word cloud. Make sure you have first installed R-Project and R-Studio. Follow these directions:

1. There is an excerpt from the Virginia Department of Health website on HAIs named "hai.txt" that will be used for this example. Download the file found in Chapter 7 of the eBook to your desktop. The file is plain text and has a hard return (Enter) at the end of the document to avoid read errors.
2. Check the working directory of R and change it to where you want. To find the current directory type:

getwd() to check

Make sure the desktop is the directory from an R-Studio prompt to show the working directory. It may return something similar to the following:

"c:/users/*younameorlogin*/documents"

If it is not the desktop (so you can read the text file you just downloaded), then set the working directory to the desktop by typing the following from the R-Studio console prompt:

setwd("c:/users/*yournameorlogin*/desktop")

Be sure to replace *yournameorlogin* with the proper login or name. This step will set the current working directory to the desktop, where your simple text file is. Use your up arrow key to toggle through previously used commands to re-run getwd(), and check that you changed the directory to your desktop.

3. Next we will set a pointer to where the text data file is so R can use it for processing:

HAIFILE="hai.txt"

This could be a complete path as well, such as:

/users/*yourfilename*/hai.txt

Now read the file into a variable by entering the following:

hai_data=readLines(HAIFILE)

Check the length of the file to make sure it was read in properly (**Figure 7.17**):

length(hai_data)

```
> HAIFILE="hai.txt"
> hai_data=readLines(HAIFILE)
> length(hai_data)
[1] 29
>
```

Figure 7.17 Length reported for HAI.txt file.

If you open the HAI file in a program like Notepad, you will find 29 lines, as reported by the length function. The next task was to convert these documents into a **text corpus** (a structured dataset of textual information that is put into a format better for data analysis). Complete this step by importing the text mining library:

1. Select the *Tools* menu, then *Install Packages . . .*, and in the center text box named *Packages*, type "tm" and click the *Install* button to install the tm package and any related dependency libraries.
2. Load the library by typing the following:

   library(tm)

3. Create a vector variable holding the data by typing the following (**Figure 7.18**):

   mydoc.vec <- VectorSource(hai_data)

```
> mydoc.vec <- VectorSource(hai_data)
```

Figure 7.18 Creating a vector in R-Project.

4. Next, you will create our corpus and clean it up with some preprocessing (referred to as transformations in the tm package user guide), such as converting all text to lowercase and removing punctuation. Available options can be viewed from the tm package by typing "getTransformations()" as shown in the screen capture in **Figure 7.19**:

```
> getTransformations()
[1] "as.PlainTextDocument" "removeNumbers"
[3] "removePunctuation"    "removeWords"
[5] "stemDocument"         "stripWhitespace"
```

Figure 7.19 Viewing options in the tm package.

Further information can be found in the tm documentation; however for now you will perform our preprocessing document cleanup by typing the following:

myworkingdoc.corpus <- Corpus(mydoc.vec)

myworkingdoc.corpus <- tm_map (myworkingdoc.corpus, tolower)

myworkingdoc.corpus <- tm_map (myworkingdoc.corpus, removePunctuation)

myworkingdoc.corpus <- tm_map (myworkingdoc.corpus, removeNumbers)

myworkingdoc.corpus <- tm_map (myworkingdoc.corpus, removeWords,stopwords("english"))

Note that the last line should be typed as a single phrase.

5. Now that the document is cleaned up, you will see some ways to analyze it. To show the 29 lines, which are referred to as documents since you can load in an entire directory of documents if wanted, type the following:

   inspect(myworkingdoc.corpus)

(*continues*)

For each line, as possible (some are blank), you will see key words that were discovered after the document preprocessing before step 5. Now that you know the number of each line (or document if you loaded in many documents), you could use inspect on only one line, or document. For example, to examine the last line (29) only, type the following:

```
inspect(myworkdingdoc[29])
```

This shows you exactly the same information as shown earlier on line 29.

6. Now use the TermDocumentMatrix function to create a matrix of word counts:

```
TDM <- TermDocumentMatrix
(myworkingdoc.corpus)
TDM
```

The first line creates the matrix, and the second shows how many words and how many documents (or lines in our case, which is what documents represent).

7. Refine the findings even more:

```
findFreqTerms(TDM, lowfreq=2)
```

This function shows terms that show up at least twice in the document. The next function finds associations between words associated with *often* at a .25% level or higher:

```
findAssocs(TDM, 'often', .25)
```

Lastly, remove any sparsely (not often) used words:

```
TDM2 <- removeSparseTerms
(TDM, sparse=.95)
```

Your findings of frequent terms used in the document, and terms associated with the word *often* can be mined by typing the following:

```
findFreqTerms(TDM, lowfreq=2)
findAssocs(TDM, 'often', .25)
```

Now that we have text mined some information from the HAI.txt document, we will use R-Project's ability to integrate a word cloud data visualization of the frequently used words, based on how often they are used.

Hands-on Statistics 7.19: Create a Word Cloud Using R-Project

Word clouds are graphical visualizations of data. You made word clouds with Wordle in a previous chapter, so this will extend that topic to creating word clouds using the wordcloud library available for R-Project. This will allow R-Project to take your data and display the results as a visualization, which may help you convey findings to your intended audience in a more meaningful way. Follow along to see this process work:

1. Start R then, import the word cloud library by selecting Packages, 0 Cloud, scroll down to wordcloud, then "yes" to prompts. You will need internet access for this.
2. Next at the command prompt of R, type the following:

```
library(wordcloud)
library(tm)
```

3. Make sure your current working directory is your desktop (or where you saved your hai.txt file to) by typing the following:

```
getwd()
setwd("c:/users/freddy/desktop/")
```

The line *getwd()* allows you to see where your working directory is, and *setwd* sets the working directory.

4. In place of *freddy*, type the name you logged in under. Then type the following:

```
wordcloud (myworkingdoc.corpus,colors
=brewer.pal(8,"Dark2"),random.
order=FALSE)
```

Note the output in **Figure 7.20**.

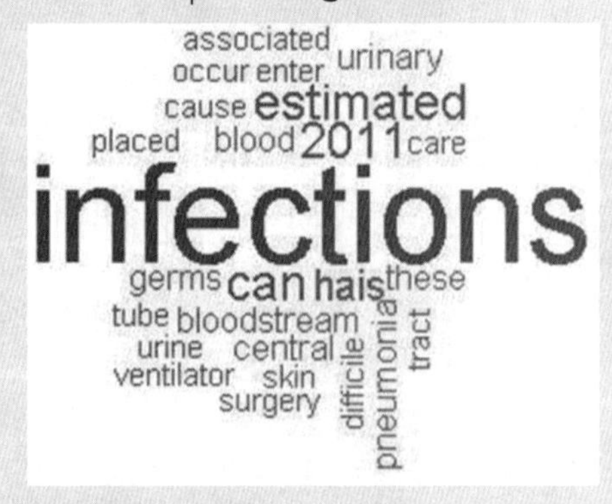

Figure 7.20 Word cloud created with R-Project wordcloud library.

At this point you could create a screen capture of the output in R-Studio and insert it into either a presentation or word processing document, depending on your needs. You could also control the number of colors displayed and other features in the final graphic by examining the documentation freely available on optional settings in the wordcloud package; however, just using the basic settings yields very interesting results to visually display your data.

How Does Your Hospital Rate?

The surgery is a success, and Sally is most impressed with the doctor and staff who cared for her. The medical staff were meticulous about hand washing and hygiene, and the operating room had a camera facing the entry door to enforce hand washing by all who entered the surgical area.

Consider the following:

1. How does your hospital rate on hand hygiene?

Global Perspective

Even though there have been major national projects to get medical professionals to agree on HAI reporting standards, not all states in the United States have adopted state-mandated reporting laws (other than what the federal government via the CDC requires). 37 of 52 states and territories have laws requiring HAI data submission as of 2013, which has increased since 2011 (Herzig, Reagan, Pogorzelska-Maziarz, Srinath, & Stone, 2015).

Certainly, though, the trend is moving toward additional legislative action on HAI reporting, and there has been a drastic increase of state-mandated reporting since 2004 nationally.

The state of California in 2011 reported 10,938 cases of hospital onset and 14,924 cases of hospital-associated *C. difficile* infections, better known as C. diff. The data reported for the year 2011 were calculated based on inpatient service days, which were 14,253,455. The majority of deaths involve patients who are 65 years or older.

In England, hospitals are required to report all C. diff cases to the government. By reporting these cases, they have been able to reduce C. diff infections by 50% since 2008. England as well as other European countries require healthcare facilities to have antibiotic control programs to aid in reducing C. diff (Eisler, 2012). In 2011, England and Wales reported 2,053 deaths lower than the 2,704 reported in 2010 (Office for National Statistics, 2012). The United States does not have such a national protocol in place at this time but does require facilities participating in Medicare or Medicaid to report C. diff infection data since 2013.

The website healthmap.org is a great way to track current health outbreaks around the world. The interactive map on the site allows you to track the latest outbreaks of a disease for states or countries. The information is valuable in calculating healthcare statistics and tracking outbreaks of deadly diseases. Using your internet browser, visit https://www.healthmap.org/ and explore some of the current global data.

Chapter Summary

In this chapter you have covered key information about HAIs, specifics about the types of infections, and federal and state reporting requirements. Hospital infection control committees were discussed, as were methods to calculate HAIs. Many different formulas to calculate different types of infection rates were explained. Lastly, the chapter showed methods to perform text data mining and how to display the results graphically with data visualization using a word cloud.

Apply Your Knowledge

1. Visit the following web page from the South Carolina Department of Health and Environmental Control: https://www.scdhec.gov/health-professionals/healthcare-associated-infections-hai/hida-public-reports. Click on *Interactive Comparison Tool*, and then compare the data for one or more years. Explain what trends you found. Have HAIs increased or decreased? Contrast surgical site versus CLABSI infection data.

2. Visit the following web page from CesareanRates.org (https://www.cesareanrates.org/state-dashboards). Choose six states, and graph the statistics relating to cesarean sections. Write an analysis of your findings.
3. Windy Hospital for the months of January through March reported the following information for surgical service:
   - Surgical operations: 1684
   - Patients operated on: 1668
   - Postoperative infections: 24
   - Postoperative deaths: 2

   Calculate the postoperative infection rate and postoperative death rate.
4. Sunshine Hospital has reported the following data for the months of April through June:
   - Cardiology had 464 discharges and 5 complications.
   - Orthopedics had 568 discharges and 8 complications.
   - Medicine had 610 discharges and 14 complications.
   - Surgery had 526 discharges and 10 complications.

   Based on the information given, calculate the complication rate for each medical service.
5. Oceanside Hospital reported 246 discharges from the medicine service for the month of June. Of those discharged patients, 24 patients were consulted on. Calculate the consultation rate.
6. Nurses, and in fact all hospital staff, are required to have knowledge of infection control methods. Depending on the role of the staff person, there are different requirements. Pick a hospital and determine the following:
   a. Minimum set of requirements all staff must know
   b. Proper hand-washing technique approved by the facility (If you cannot find this information for your facility, report on the technique approved by the CDC.)
7. Prepare a compelling and attention-catching PowerPoint presentation showing how and why hand washing lowers infection rates. Be as dramatic and impressive as you can be!
8. Using a document related to emerging infectious diseases (globally) of your choice, create a word cloud after performing a text mining operation to "clean up" the document and find frequent terms or associations. Create a PowerPoint presentation with your graphic, a few bullets of information, and some interesting graphics to help review the key diseases.
9. Similar to question 8, using another healthcare–related document of your choice (perhaps the rising cost of health care), create a word cloud after performing a text mining operation to "clean up" the document and find frequent terms or associations. Prepare a short narrative to explain your reasoning and findings.
10. Create an Excel spreadsheet with one tab for each rate of calculation you have learned about in this chapter. Make sure the tabs are labeled, such as "Infection Rate," "C-section," etc.

References

Centers for Disease Control and Prevention. (2014). The burden. Retrieved September 20, 2014, from https://www.ncbi.nlm.nih.gov/pubmed/1757688

Centers for Disease Control and Prevention. (2015). Healthcare-associated infections. Retrieved July 7, 2015, from http://www.cdc.gov/HAI/surveillance/

Custodio, H. (2016). Hospital acquired infections. *Medscape*. Retrieved November 14, 2017, from http://emedicine.medscape.com/article/967022-overview

Eisler, P. (2012). Far more could be done to stop the deadly bacteria C. diff. *USA Today*. Retrieved March 14, 2013, from http://usatoday30.usatoday.com/news/health/story/2012-08-16/deadly-bacteria-hospital-infections/57079514/1

Herzig, C. T. A., Reagan, J., Pogorzelska-Maziarz, M., Srinath, D., & Stone, P. (2015). State-mandated reporting of health care-associated infections in the United States: Trends over time. *American Journal of Medical Quality, 30*(5), 417–424. Retrieved May 20, 2015, from https://www.ncbi.nlm.nih.gov/pmc/articles/PMC4272669/#__ffn_sectitle

Hvistendahl, M. (2012). Cesarean nation: The cautionary tale of how China came to have the world's highest C-section rate. *Slate*. Retrieved June 5, 2014, from https://slate.com/human-interest/2012/01/cesarean-nation-why-do-nearly-half-of-chinese-women-deliver-babies-via-c-section.html

Klevens, R. M., Edwards, J. R., Richards, C. L., Horan, T. C., Gaynes, R. P., Pollack, D. A., & Cardo, D. M. (2007). Estimating health care-associated infections and deaths in US Hospitals, 2002. *Public Health Reports, 122*, 160–166.

Kresge, N., & Gale, J. (2012, February 9). Greek doctors battle hospital superbug as crisis depletes budget. *Bloomberg*. Retrieved October 20, 2013, from http://www.bloomberg.com/news/2012-02-09/greek-doctors-battle-hospital-superbug.html

Lee, F., & Lind, N. (2000, June 1). The infection control committee. *Infection Control Today*. Retrieved November 9, 2014, from https://www.infectioncontroltoday.com/general-hais/infection-control-committee

Lee, G., Kleinman, K., Soumerai, S, Tse, A., Cole, D., Fridkin, S., . . . Jha, A. (2012). Effect of nonpayment for preventable infections in US hospitals. *New England Journal of Medicine, 367*, 1428–1437.

Manring, M. M., Hawk, A., Calhoun, J. H., & Andersen, R. C. (2009). Treatment of war wounds: A historical review. *Clinical Orthopedics & Related Research, 467*(8), 2168–2191.

Office for National Statistics. (2012). Deaths involving *Clostridium difficile*: England and Wales: 2011. Retrieved August 22, 2012, from https://www.ons.gov.uk/peoplepopulationandcommunity/birthsdeathsandmarriages/deaths/bulletins/deathsinvolvingclostridiumdifficileenglandandwales/2012-08-22

Palermo, E. (2014, November 5). C-section rates continue to decline in the US. *Livescience.com*. Retrieved from http://www.livescience.com/48616-cesarean-section-rates-decline-us.html

Raja, U., Mitchell, T., Day, T., & Hardin, J. M. (2008). Text mining in healthcare. Applications and opportunities. *Journal of Healthcare Information Management, 22*(3), 52–56.

Song, M. (2013, April 1). Opinion: Text mining in the clinic. *The Scientist*. Retrieved from https://www.the-scientist.com/opinion/opinion-text-mining-in-the-clinic-39531

Wang, E., & Hesketh, T. (2017). Large reductions in cesarean delivery rates in China: A qualitative study on delivery decision-making in the era of the two-child policy. *BMC Pregnancy and Childbirth, 17*, 105. Retrieved from https://www.ncbi.nlm.nih.gov/pmc/articles/PMC5716234/

Web Links

Rcommander: http://www.rcommander.com/

Healthcare-Associated Infections: http://www.cdc.gov/hai/

Reduced Incidence of Postoperative Infections: http://www.medscape.com/viewarticle/736766

Making Health Care Safer: http://www.cdc.gov/VitalSigns/Hai/StoppingCdifficile/

Complications in Surgical Patients: http://archsurg.jamanetwork.com/article.aspx?articleid=212419

Deaths Involving *Clostridium difficile*: England and Wales: 2011: http://www.ons.gov.uk/ons/rel/subnational-health2/deaths-involving-clostridium-difficile/2011/stb-deaths-involving-clostridium-difficile-2011.html

Agent Used in Spinal Surgery Linked to Higher Complications Rate, Greater Inpatient Charges: http://esciencenews.com/articles/2009/06/30/agent.used.spinal.surgery.linked.higher.complications.rate.greater.inpatient.charges

Mum Depicts Every Real Pregnancy Problem Ever in Hilarious Illustrations: http://www.maternityandinfant.ie/pregnancy-mum/rising-number-csections/

Method of Delivery and Pregnancy Outcomes in Asia: The WHO Global Survey on Maternal and Perinatal Health 2007–08: http://www.who.int/reproductivehealth/topics/best_practices/GS_in_Asia.pdf

Caesarean Nation: http://www.slate.com/articles/double_x/doublex/2012/01/cesarean_nation_why_do_nearly_half_of_chinese_women_deliver_babies_via_c_section_.html

A National Survey of US Internists' Experiences With Ethical Dilemmas and Ethics Consultation: http://www.medscape.com/viewarticle/471368_5

Table of Iatrogenic Deaths in the United States: http://www.ourcivilisation.com/medicine/usamed/deaths.htm

Monitoring Hospital-Acquired Infections to Promote Patient Safety—United States, 1990–1999: http://www.cdc.gov/mmwr/preview/mmwrhtml/mm4908a1.htm

National Healthcare Safety Network (NHSN) Report, Data Summary for 2009, Device-Associated Module: http://www.cdc.gov/nhsn/PDFs/dataStat/2010NHSNReport.pdf

Hospital C-Section Rates Vary Widely: http://www.medicalnewstoday.com/articles/257231.php

Greek Doctors Battle Hospital Superbug as Crisis Depletes Budget: http://www.bloomberg.com/news/2012-02-09/greek-doctors-battle-hospital-superbug.html

International Nosocomial Infection Control Consortium Findings of Device-Associated Infections Rate in an Intensive Care Unit of a Lebanese University Hospital: http://www.ncbi.nlm.nih.gov/pmc/articles/PMC3326952/

Reduction in *Clostridium difficile* Rates After Mandatory Hospital Public Reporting: Findings From a Longitudinal Cohort Study in Canada: http://www.plosmedicine.org/article/info%3Adoi%2F10.1371%2Fjournal.pmed.1001268

QuickStats: Rates of *Clostridium difficile* Infection Among Hospitalized Patients Aged ≥65 Years, by Age Group—National Hospital Discharge Survey, United States, 1996–2009: http://www.cdc.gov/mmwr/preview/mmwrhtml/mm6034a7.htm

CHAPTER 8

Health Information Management Statistics

There are truths which are not for all men, nor for all times.

—Voltaire

CHAPTER OUTLINE

Introduction
Functions of Health Information Management
Requirements to Work in Health Information Management
Labor Cost and Compensation
- Transcription Cost and New Technology
- Other Costs Associated With Health Information Management
- Productivity
- Healthcare Facility Staffing
- Measuring Utilization

Types of Financial Reports
- Readmission Rate Reports
- Discharge Reports

Two Types of Budgets: Operational and Capital
- Operational Budget
- Capital Budget

Data Mining With Naive Bayes and R-Project
Global Perspective
Chapter Summary
Apply Your Knowledge
References
Web Links

LEARNING OUTCOMES

After completing this chapter, you should be able to do the following:

1. Define and discuss the requirements to become employed in a health information management (HIM) department.
2. Discuss the importance of calculating the HIM budget.
3. Define and be able to calculate labor cost and compensation.
4. Define and be able to calculate transcription costs for in-house or outsourcing contexts.
5. Define and be able to calculate other costs associated with HIM.
6. Define and be able to calculate productivity standards.
7. Define and be able to calculate staffing levels.
8. Identify and be able to calculate statistics from financial reports used in the HIM department.
9. Explain the difference between operational and capital budgets, and demonstrate how to calculate payback and return on investment.

10. Demonstrate how to output text and graphics to files using R-Project.
11. Demonstrate naive Bayes data mining techniques.

KEY TERMS

Capital budget
Case mix index
Case mix report
Discharge report
Electronic data management systems (EDMS)
Enterprise report management (ERM)
Full-time equivalent (FTE)
Naive Bayes classification (NBC)
Operational budget
Patient encounter
Payback period
Personal time, fatigue, and delay factor (PF&D)
Productivity
Readmission rate report
Release of information
Return on investment
Transcription

How Does Your Hospital Rate?

The health information management department at Lakeview Hospital is understaffed in the coding department by four coders: two inpatient coders, one outpatient coder, and one emergency department coder. Sandra is advertising the coding positions available. She has received several applications for the positions, but only four have the professional credentials required for the job. Two of these applicants have just become credentialed and have no coding experience. The other applicants have experience coding but are not credentialed. She decides to give them all a coding test. Once all the applicants have completed testing, she reviews the results and finds that all of them have not only passed but excelled. Thus, the test has not helped her determine the best candidate.

Consider the following:

1. Which applicants would you hire and why?
2. How would you make a decision about these candidates?
3. What other criteria would you consider?

Introduction

In this chapter, we discuss the many staff positions and statistical calculations associated with the health information management (HIM) department. We will explore financial reports used for tracking medical charts that have not been coded for reimbursement purposes, HIM cost analysis functions, employee compensation, and other functions important to the HIM department. Lastly, we will examine prediction and classification with linear regression analysis as well as the related naive Bayes data mining algorithm.

Functions of Health Information Management

A professional in the area of HIM collects, analyzes, and protects medical information related to hospital patient care. The HIM department is responsible for maintaining a budget that relates to staffing, labor cost, and productivity and for determining the need to hire new employees within the HIM department. Each department in the hospital maintains its own budget. Other uses of these statistics are related to employee compensation, absentee rates, and many other standards that pertain to HIM and the employees who work within the HIM department. Employees in HIM work closely with patients, doctors, and insurance providers. HIM directors report to the chief financial officer and, in small facilities, to the chief executive officer. The HIM staff also has relations outside the facility with vendors who provide services in transcription, coding, accounting, and billing.

Requirements to Work in Health Information Management

Although jobs in an HIM department are typically on-site, options are increasing for workers to telecommute. According to US Department of Labor

Statistics, HIM is one of the nation's fastest-growing health occupations. Salaries of up to $80,000 or more a year are possible, and some facilities offer sign-on bonuses of up to $10,000 and compensation to cover relocation expenses. Of course, formal education, certifications, and geographical area are key determinants of pay. Most employers require applicants for jobs in HIM to be registered health information technicians (RHITs), registered health information administrators (RHIAs), or certified coding specialists (CCSs). Desired qualities in HIM professionals include analytical, interpersonal, and technical skills, along with an orientation to details. Employment for health information technicians is expected to increase by 22% between 2012 and 2022.

DID YOU KNOW? If you have ever heard the saying, "Eat, drink, and be merry for tomorrow we shall die," you can thank the real "father of medicine," Imhotep, who was the ancient Egyptian who first said it, and not the Greek Hippocrates who lived 2,000 years later. His philosophy on life was to follow the mantra of his saying.

The Greeks borrowed Imhotep's philosophy, poems, and medical knowledge and rewrote their own history to attribute his work to Hippocrates. Sir William Osler, a famous Canadian physician of the 1800s who developed the phrase "life in the fast lane," notes that Imhotep was "the first figure of a physician to stand out clearly from the mists of antiquity. Imhotep diagnosed and treated over 200 diseases, including: 15 diseases of the abdomen, 11 of the bladder, 10 of the rectum, 29 of the eyes, and 18 of the skin, hair, nails, and tongue" (African Heritage Foundation, 2015). He also practiced dentistry, pharmacology, and surgery. As Osler said, "The good physician treats the disease; the great physician treats the patient who has the disease" (Cadigan & Nickson, n.d.).

Now you know!

Labor Cost and Compensation

Labor cost can be an enormous cost for HIM departments, depending on the size of the hospital. HIM departments have always been considered a non–revenue-producing department; that is, they do not provide a service that is directly related to patient care. Labor cost can consume 70% of the profits of a healthcare facility.

Calculating labor cost helps the HIM department plan their fiscal year budget. In addition to salary, such calculations must include salary increases, bonuses, paid time off, sick days, health insurance, mandatory Social Security contributions, 401K contributions, and other financial incentives that employees may receive in the next fiscal year. In fact, these benefits often constitute 30% of an employee's total compensation. Professionals must learn to budget generously for labor costs to avoid falling short, as employees may quit or transfer to another department, and, until those positions can be filled, other employees may need to work additional hours.

Hands-on Statistics 8.1: Calculate Labor Cost and Compensation by Hand

To practice calculating labor cost, consider the following examples.

Scenario 1: Lucy Childs works 40 hours per week in the HIM department at Masters Hospital at a rate of $16/hour. Additionally, she receives benefits that are valued at 35% of her base salary. Follow these directions:

1. Multiply the number of hours worked per week by the number of weeks in a year:

   40 hours per week ×
   52 weeks in a year = 2,080 hours in a year

2. Multiply the number of hours worked in a year by the hourly rate:

   2,080 hours per year × $16/hour =
   $33,280 (base salary)

3. Multiply the base salary by the benefits percentage:

   $33,280 × 0.35 = $11,648 (benefits based on percentage of base salary)

4. Add the base salary to the value of benefits:

   $11,648 + $33,280 = $44,928
   (total compensation for a year)

Thus, total compensation for Lucy would be $44,928, which includes benefits.

Scenario 2: Owen works 20 hours a week in the HIM department at Masters Hospital at a rate of $12/hour with no benefits. Follow these directions:

1. Multiply the number of hours worked per week by the number of weeks in a year:

   20 hours per week × 52 weeks in a year = 1,040 hours in a year

2. Multiply the number of hours worked in a year by the hourly rate:

   1,040 hours per year × $12/hour = $12,480

Thus, total compensation for Owen would be $12,480.

Hands-on Statistics 8.2: Calculate Labor Cost and Compensation Using Excel

To calculate the labor cost in Excel, follow these directions:

1. Open a new, blank Excel workbook.
2. Enter the data shown in **Figure 8.1**, but leave cell D7 blank.
3. In cell D7, enter the following function:

   =C3*C4*C5*C6+(C3*C4*C5)

   Click out of the cell, and 44928 should appear, as in Figure 8.1.

D7 fx =C3*C4*C5*C6+(C3*C4*C5)

	A	B	C	D	E
1		Lucy Childs labor cost			
2					
3		Hours a week	40		
4		Work weeks	52		
5		Hourly pay rate	16		
6		35% benefits	0.35		
7				44928	

Figure 8.1 Employee compensation using Excel.

Transcription Cost and New Technology

Some functions in the HIM department are contracted out to various companies, eliminating the need for additional employees and reducing cost of labor and benefits to the department. Some of the functions that can be contracted out are release of information, **transcription**, and coding. As new technology emerges, such as voice recognition, digital dictation, and web-based transcription, there will be a decreasing need to have transcription. However, until all types of healthcare facilities have implemented this technology, the supervisor of transcription will need to understand how to determine labor cost for a transcriptionist.

Hands-on Statistics 8.3: Calculate Transcription Cost by Hand

To practice calculating transcription cost, consider the following example. The HIM department at Riverview Hospital has three full-time transcriptionists who can each produce 1,100 lines per day. Maddie earns $14.00 per hour, for an annual salary of $29,120 ($14/hour × 2,080 hours per year). George earns $13.50 per hour, for an annual salary of $28,080 ($13.50/hour × 2,080 hours per year). Christy earns $14.50 per hour, for an annual salary of $30,160 ($14.50/hour × 2,080 hours per year). Now you need to determine the employee's annual productivity, as well as the unit cost.

To calculate annual productivity, follow these directions:

1. Multiply the number of lines completed per person per day by the number of transcriptionists:

   1,100 × 3 = 3,300

2. Multiply the number of work days in a week by the number of weeks in a year:

   5 × 52 = 260

3. Multiply the number of lines completed per day by all three transcriptionists by the number of workdays in a year:

   3,300 × 260 = 858,000

Thus, the three transcriptionists together complete about 858,000 lines per year.

To calculate unit cost, add the salaries of the three transcriptionists:

$29,120 + $28,080 + $30,160 = $87,360

This formula can be applied not only to transcription, but to any position that requires an employee to meet productivity goals, such as coders.

DID YOU KNOW? **Personal Time, Fatigue, and Delay (PF&D)** allowance is a time allowance required by the US Department of Labor, Wage and Hour Division. They recognize that normal working fatigue prevents employees from producing at their highest rate throughout the workday. Examples of PF&D factors include cleanup, restocking or refilling of machines, and worker fatigue, to name a few. All can reduce the hourly amount a worker can produce.

Employers must take this nonproductive time into consideration when determining theoretical piece rates for workers by including in their time study estimates a PF&D factor. This factor will be, in many cases, 10 minutes per hour; however, a range from 9 minutes and up (or > 15% of the work hour) is acceptable to the Department of Labor. These allowances will include breaks and rest periods during a workday. Employers not applying PF&D allowances for qualified individuals may incur back-wage liability. Note that the PF&D factor for disabled individuals will be increased beyond the allowance for a non-disabled worker.

As an example, an employer who has a PF&D allowance of 10 minutes per hour would use a factor of 0.1667 to the standard hourly time performed by the non-disabled worker. A 10-minute PF&D (16.67% of a work hour) factor is equal to 1.2, computed as follows:

$$1.2 = 60 \text{ minutes}/(60 \text{ minutes} - 10\text{-minute break})$$

Even though medical transcriptionists are measured at an hourly production rate, this PF&D allowance would not be factored into their time/piece rate computation.

Now you know!

Hands-on Statistics 8.4: Calculate Transcription Cost Using Excel

To calculate labor costs, annual productivity, and unit costs, follow these directions:

1. Open a new, blank Excel workbook.
2. Enter the data shown in **Figure 8.2**, but leave cells E4 to E6 and D8 to D11 blank, because you will enter formulas in these cells.
3. In column C, highlight cells C4, C5, and C6, right click in the highlighted area, and select *Format Cells*, the *Number* tab, and *Currency*, which will force two decimal places of precision for the hourly pay.
4. In cell E4, enter the following formula:

   =C4*D4

   Click out of the cell, and 29,120.00 should appear, as in Figure 8.2.

	A	B	C	D	E
1		**Riverview Hospital HIM Department**			
2					
3			Pay per hour	Hours	
4		Employee 1	$14.00	2080	**$29,120.00**
5		Employee 2	$13.50	2080	**$28,080.00**
6		Employee 3	$14.50	2080	**$30,160.00**
7					
8		Annual productivity (lines)		**3300**	
9		Work days in a year		**260**	
10		Lines for all transcriptionist:		**858000**	
11		Unit Cost		**$87,360.00**	

Figure 8.2 Riverview Hospital HIM transcription costs.

5. Click on cell E4 to highlight it, and then hover the cursor over the lower right corner of the cell until it becomes a black plus sign. Click and drag the cursor downward to copy the formula down through E6. You can then highlight then entire E column by clicking on it, and set it to *Currency* format as you did in Step 3.
6. In cell D8, enter the following formula:

=1100*3

7. In cell D9, enter the following formula:

=5*52

8. In cell D10, enter the following formula:

=D8*D9

9. In cell D11, enter the following formula:

=E4+E5+E6

If you would like, you may add underlining and boldface font to make your output look exactly as shown in the example.

Other Costs Associated With Health Information Management

Another cost that should be considered is a charge for **release of information**. Health information departments can charge a reasonable fee for releasing patient information under the guidelines of the Health Insurance Portability and Accountability Act (HIPAA), also known as the Privacy Rule. Deciding on a reasonable fee requires some time studies, which may involve the following:

- Reviewing how long it takes to request and log the information
- Locating patient information
- Retrieving the information
- Reviewing for prior disclosures
- Billing for the release of information and completing the release of information log

Consideration must also be given for nonlabor expenses. These expenses include postage, supplies, equipment, and contracts, as well as salaries. Other costs associated with paper health records include those for providing secure storage for health information and for providing precise access to the stored information. For example, many large facilities have off-site information warehouses, which require both storage and access fees to retrieve information. Continuing education of the HIM staff is also a hidden cost.

HIM departments can cut costs by using storage media such as digital tape, CDs, DVDs, solid-state, and cloud storage. All of these media are environmentally friendly, can reduce waste and storage costs, and can save on paper, toner, and service charges for copiers.

Increased use of electronic medical records (EMRs) in place of paper medical records will result in employees spending less time locating and retrieving stored medical records, which will result in lower costs. However, many facilities still use paper records, so such costs still must be accounted for.

Moreover, a move to an EMR system entails costs of its own. One cost is that associated with scanning paper records into electronic formats, which requires an **electronic data management system (EDMS)**. Such a system allows the facility to scan, move information from transcription into the EMR, add digital signatures, and increase the speed for information access. For facilities that have a large amount of pre-existing paper files, outside groups are often contracted to convert paper documents into electronic format.

EDMSs actually are a group of technologies that work together to manage the storage, retrieval, indexing, and disposition of records, along with information assets of the organization. Foremost, the EDMS includes document management, forms processing (including e-forms), imaging, enterprise report management, and workflow management.

The **enterprise report management (ERM)** system incorporates payroll, cost accounting, managed care, and detailed patient information in one database. This information can be retrieved from any area in the healthcare facility by authorized personnel. With this type of quick access, an enterprise report manager allows managers to view variance analysis, productivity, and clinical utilization along with many other types of reports. These reports can be scheduled daily, weekly, or monthly and can be output as Excel files or Adobe PDFs. Incorporating this type of technology allows the HIM department to keep track of when physicians are signing charts, when and how employees are using the system, and many other statistics. New technology will always be an influential factor in helping improve the quality of a progressive facility.

With patients experiencing increasing access to their physicians via e-mail, insurance companies have begun reimbursing physicians for their time spent e-mailing patients. You should note that any e-mails or text messages are considered legal components of a health record. Certainly the security of such systems will be questioned and improved as usage increases; however, as demands of computer security are constantly changing, these improvements in security likewise must be ongoing.

Hands-on Statistics 8.5: Calculate Release of Information Cost Using Excel

To calculate release of information cost in Excel, follow these directions:

1. Open a new, blank Excel workbook.
2. Enter the data shown in **Figure 8.3**, but leave cells B7 to B16 blank. You will enter formulas into these cells.
3. In cell B7, enter the following formula:

   =ROUND(100/C4,2)

   In this formula, you are dividing the cost of supplies ($100) by the average requests per month (420), which is in cell C4, and rounding the result using binary (base 2) math instead of decimal (base 10) math, to minimize rounding errors. You may find more information on this type of rounding on the internet by entering the search phrase "off by one error."
4. In cell B8, enter the following formula:

   =160/C4

	A	B	C	D
1		**Release of Information Cost**		
2		**Sunset Harbor Hospital**		
3				
4	Average requests per month:		420	
5				
6	Non-Labor Cost	Cost Per Request		
7	Supplies: $100	$0.24	all charges per month	
8	Equipment: $160	$0.38		
9	Postage: $335	$0.80		
10	Copy Service Contract: $150	$0.36		
11	Wages: $15 per hr:			
12	**Wage=15 X 2080 (52 weeks x 40 hrs) work hrs per year*			
13	*for salary = 31,200 / by 12 months =*			
14	*$2,600 per month*	$2,600.00		
15	*formula =(15*2080)/12*			
16	**Total->**	**$7.97**	equals cost per request	

Figure 8.3 Release of information cost.

5. In cell B9, enter the following formula:

=335/C4

6. In cell B10, enter the following formula:

=150/C4

7. Select cells B7 to B16, and right click to format them as *Currency* and to two decimal places of precision.
8. In cell B14, enter the following formula:

=(15*2080)/12

9. In cell B16, enter the following formula:

=B14/C4+B7+B8+B9+B10

So, in reality, the total cost for processing one release of information request is $7.97.

Productivity

Productivity measurements allow the HIM department to categorize the hospital's labor as either a product or service. Coders must be able to maintain a coding productivity standard for emergency department records, inpatient records, and outpatient surgeries. For example, a coder must be able to code 25 emergency department records in 1 hour and 15 outpatient surgeries in 1 hour. However, many factors influence how much the coder can realistically produce in 1 hour, including how long the coder has been coding that day before the hour in question, prior experience in coding that particular type of record, and other functions the coder is required to perform in the HIM department (e.g., perhaps it is a small facility or there are staffing issues). It stands to reason that if a coder has experience in coding outpatient surgeries, he or she would have no problem meeting productivity standards. However, if the coder has no experience, allowances would have to be made and the coder given an extended time frame to meet the productivity standards.

With regard to quality of data input, most HIM departments have a coding accuracy of 97% to 98%. Usually the coding supervisor pulls a random sample of charts and verifies them for accuracy. The frequency of this sampling, though, depends on quality control measures established by the facility. Based on the supervisor's findings, if the coder needs improvement in certain areas, he or she will be given the necessary training to bring performance up to standards.

Healthcare Facility Staffing

As you can imagine, predicting staffing needs is difficult, especially if the healthcare facility is large. One organization, Mercy Health Systems, has a 31-hospital networking system that services hospitals in Missouri, Kansas, Oklahoma, and Arkansas (Kyruus, 2015). This organization implemented enterprise-wide, web-based scheduling software to improve overall performance. This software has taken the guesswork out of trying to schedule staff and determine overtime needs, which has saved them thousands of dollars, a critical goal in today's ever-competitive environment.

Patient encounter is one way that healthcare organizations determine the need for **full-time equivalent (FTE)** workers for the month and part-time workers, as well. A patient encounter is considered any contact between the patient and a healthcare provider, which can include physician interactions, laboratory work, x-ray services, and any other services that could be provided to the patient.

Hands-on Statistics 8.6: Calculate Staffing Hours Needed by Hand

To calculate staffing levels, consider the following example. The HIM department at Sunnyside Hospital needs to establish the number of inpatient coders needed. The number of inpatient records to be coded per hour is 8, the number of hours worked per day is 7.5, allowing for a 30-minute lunch break. This means the coder would need to code 60 records per day. The daily discharges from the facility are 85. Based on this information, the calculation would be as follows:

85/60 = 1.4 FTE

(continues)

Now that we have calculated the FTE, you can see that because you are going to hire only a full-time person, you can round the result to show that you need only one full-time employee, based on FTE analysis. If you were to hire a part-time person as well, you would take 0.4 (representing the amount of time that exceeds the FTE's capacity) and multiply that by 40 hours to get the number of hours the part-time employee would need to work. The second employee would need to work 16 hours a week (0.4 × 40 = 16).

Hands-on Statistics 8.7: Calculate Staffing Minutes Needed by Hand

You can also calculate staffing time by minutes. Follow these steps:

1. Divide the length of time it takes to code a record by the number of hours worked to find out how many records can be coded in a given number of hours.
2. Multiply the total discharges for a week by the number of minutes it would take to code a record.

Hands-on Statistics 8.8: Calculate Staffing Minutes Needed Using Excel

To practice calculating staffing minutes needed using Excel, consider the following example. At Happy Valley Hospital, Karen needs to hire coders, but she does not know how many are needed. On reviewing her coders' performance, she finds that it takes a coder 30 minutes to code an inpatient chart and that the coders have a total of 34 hours a week of productivity. Total discharges for the week are 475. Follow these directions:

1. Open a new, blank Excel workbook.
2. Enter the data shown in **Figure 8.4**, but leave cells B7 to B9 and C9 blank.
3. In cell B7, enter the following formula:

   =B3*B5

   This formula calculates the total minutes required to code all discharges each week by multiplying the minutes required to code each chart by the number of discharges for the week.
4. In cell B8, enter the following formula:

   =B7/60

   This formula converts the minutes required to code all discharges each week to hours, by dividing the minutes by 60.
5. In cell B9, enter the following formula:

   =B8/B4

   This formula calculates how many full-time employees are needed to code all of the discharges each week, based on the total hours required for coding and on the productivity hours per week.

	A	B	C
1	**Happy Valley Hospital**		
2			
3	Time to code an inpatient chart:	30	
4	Productivity hours per week	34	
5	Discharges for the week	475	
6			
7	Minutes required to code	14250	
8	Hours required (min. / 60)	237.5	
9	Employees needed	6.985	***=-->7 employees needed***

Figure 8.4 Excel calculating staffing needs by minutes.

6. In cell C9, enter the following text:

'= --> 7 employees needed

Note and be sure to include the single quotation mark preceding the equal sign in this text. Without it, Excel will treat this text as a formula, which we do not want in this case. Preceding the equal sign with a single quotation mark tells Excel to ignore the formula option (in this cell only) and changes the cell to text only, displaying exactly what you type.

Measuring Utilization

To operate efficiently, hospitals must measure not only the cost of medical services provided to patients but also how resources are utilized in providing these services. This type of reporting will permit hospital administrators to quantify whether there is an increase or decrease in services provided and assist in determining the need for additional staffing or equipment. Indicators can point out issues related to accessibility, comprehensiveness, productivity, and continuity of care.

Utilization of health care is a multidimensional concept. It may be examined from multiple viewpoints, including those of the physician and of the patient. The physician's viewpoint is based on the number of hospitalizations and the number of patients seen and therefore tends to be more quantitative and objective. The patient's perspective, communicated via reports on services received, is based on such factors as the availability of a nurse and the quality of services and meals, and therefore tends to be more qualitative and subjective.

Types of Financial Reports

There are many types of financial reports that are important to HIM and other departments within the hospital facility. For example, a **case mix index** records the Medicare Severity Diagnosis-Related Group (MS-DRG), title of the MS-DRG, number of patients, number of patient days, average length of stay (ALOS), and the total charges. The **case mix report** records the intensity of resources and clinical severity for specific groups and compares these data against those of other groups. A good way to differentiate the two is to remember that a case mix index presents specific data, whereas a case mix report is a summary of findings.

These financial and statistical reports are used to guide decisions made by upper management. By linking medical and financial data, hospital administrators have a better overall view of the facility, which allows them to manage healthcare costs in a more efficient and effective manner. With the increasing cost of medical care and a 1% decrease in reimbursement from Medicare under the Affordable Care Act, hospitals must make alterations in their financial and medical operations to stay competitive and avoid cost overruns.

Clearly, the case mix index is very valuable to a facility. You can calculate it using the following formula:

$$\frac{\text{Sum of the weights of MS-DRGs for all patients discharged during a specified period}}{\text{All patients discharged during that period}}$$

Hands-on Statistics 8.9: Calculate Case Mix Index

Let's practice calculating the case mix index using the following example. Dr. Milligan has 8,146 discharges in the first quarter of 2016. The total case mix is 18,946.4663. Follow these directions:

Divide the total case mix by the number of patient discharges:

$$18{,}946.4663/8{,}146 = 2.32586132$$

Thus, the case mix index in this scenario is 2.32586132.

This formula would be used for each type of payer, such as Medicare, Medicaid, and other third-party payers. The case mix index indicates the facility's utilization of health services in providing care to patients.

Readmission Rate Reports

The **readmission rate report** summarizes inpatient services and tracks medical readmissions. Readmission times can vary among hospitals but typically are 7, 15, or 30 days. For the Centers for Medicare and Medicaid Services (CMS), the standard benchmark for readmission is 30 days. Note that any patient who is transferred to another facility is not counted as a readmission. Effective October 1, 2012, section 3025 of the

Affordable Care Act added section 1886 to the Social Security Act, which requires CMS to reduce payments to the inpatient prospective payment system (IPPS) for hospitals with excessive readmission rates. The following four adjustments were made to the IPPS final rule:

1. Identification of which hospitals are subject to the Hospital Readmissions Reduction Program
2. Provision of a methodology for calculating a healthcare facility's readmission payment adjustor factor
3. Designation of the portion of the IPPS payment that is used to calculate readmission payment adjustor amount
4. Explanation of how hospitals can review readmission information and submit corrections before the readmission rates are made public

Discharge Reports

The **discharge report** shows the services provided by the hospital by department, such as cardiology or surgery. The report also shows the total number of patients, total number of days, and ALOS. It analyzes patient data by age group, operations, emergency admissions, and consultations. Consultations are tracked by the number of patients and the number of referred patients. The formula for calculating consultation rate is as follows:

$$\frac{\text{All patients receiving consultations during a specified period} \times 100}{\text{All discharges and deaths for that period}}$$

From this report, you can calculate gross death rates, net death rates, gross autopsy rates, and net autopsy rates. To calculate the patient days, you need the total number of patients and patient days for the period under review.

Hands-on Statistics 8.10: Calculate Average Length of Stay From a Discharge Report

To calculate ALOS, use the following example. Sunnyside Hospital for the first quarter had 896 patients and 9,225 patient days. Follow these directions:

Divide the total patient days by the number of patients:

$$9{,}225/896 = 10.29$$

Thus, the ALOS is 10.29 days.

Two Types of Budgets: Operational and Capital

A budget is instrumental in helping healthcare organizations achieve targets for revenue and spending. Hospital budgets, which are typically millions of dollars, are planned a year in advance, at the beginning of the preceding fiscal year. In this way, planning is done a year before actual funds are spent. A fiscal year can be any 12-month period the facility chooses, such as January 1 to December 31. The HIM department is no different from other units in the hospital and submits a budget report for such items as supplies, travel, education requirements, benefits, wages, contracts for maintenance, and other expenses associated with the department. Two types of budgets should be considered: operational and capital.

Operational Budget

The **operational budget** compares the amount that was budgeted for with the actual amount spent for each department. The amount budgeted for may be different from the amount each department spent. This difference can be either a favorable variance (you spent less than budgeted) or an unfavorable variance (you overspent). Reports are computed to determine these conditions. Consider **Table 8.1** to see the variance in money spent versus money budgeted.

Looking at Table 8.1, you will notice that wages are budgeted for $900,000 and the HIM department has used only $62,000. So, the variances associated with the categories of wages and travel are actually favorable; more was budgeted than was needed. However, subscriptions had a cost overrun of $700, yielding an unfavorable variance.

Table 8.1 Variance in Money Spent Versus Money Budgeted

Item	Budget	Actual	Variance
Wages	$900,000	$62,000	$838,000
Travel	$10,000	$3,000	$7,000
Subscriptions	$1,500	$2,200	$700

Hands-on Statistics 8.11: Calculate Favorable Variances

Follow these directions to calculate the percentage of a favorable (in this case) variance:

1. Subtract the actual amount spent from the budgeted amount to determine the amount of variance:

$$\$900,000 - \$62,000 = \$838,000$$

2. Determine the percentage of variance by multiplying the variance by 100 and dividing the result by the budgeted amount:

$$\frac{838,000 \times 100}{900,000} = 93\%$$

The 93% means the HIM department is under budget by this amount, which would increase profits for the healthcare facility. Of course, it also means someone did not correctly estimate the wage needs for the year.

For unfavorable variances, such as with the subscriptions in the previous example, you would find the result as follows:

1. Subtract the actual from the budgeted amount to determine the variance:

$$\$2,200 - \$1,500 = \$700$$

2. Next, determine the percentage:

$$\frac{700 \times 100}{1,500} = 46.66\%, \text{ which rounds to } 47\%$$

In this example, the department was 47% over budget for subscriptions. An unfavorable variance means the actual cost of an item was greater than the expected cost and will decrease profits.

Another way to calculate a variance is as follows:

Variance/Actual cost = Percentage of variance

Consider that the amount budgeted for a fiscal year for contract services is $800, but the actual cost was $950. To determine the variance, follow these directions:

1. Subtract the actual from the budgeted amount:

$$\$950 - \$800 = \$150$$

2. Next, determine the percentage:

$$\frac{150 \times 100}{950} = 15.79\% \text{ variance}$$

Capital Budget

The second budget we will discuss is the **capital budget**. The capital budget is used for items that typically have a cost of over $500, such as an x-ray machine, da Vinci robotic arm, or large computer system for facility-wide use. However, each facility sets the dollar amount above which approval is needed to purchase the equipment. In some facilities, that amount is $1,000. Capital budget items that should be usable for more than a year, such as with a generator or a large-volume printer, are typically depreciated over a 5-year or longer lifespan. Note that each department has an allotted amount of money it spends on equipment. Lastly, any special projects must have prior approval before they can be added in the capital budget.

The director of the department that is requesting to purchase an item may require supervisors to compute what is called a payback period. The **payback period** is how long it will take to recover the cost of the equipment. The formula for calculating the payback period is as follows:

$$\frac{\text{Cost of the item}}{\text{Yearly incremental cash flow}}$$

You may also use an accounting method called return on investment. **Return on investment** allows a healthcare organization to assess the effectiveness of its investments and is also a performance measure. It is the extent and timing of gains from investments compared against the magnitude and timing of the investment cost. The formula for return on investment is as follows:

$$\frac{\text{Average yearly incremental cash flow}}{\text{Cost of the item}}$$

Two important factors to consider related to this measure are that (1) it is used for asset purchase decisions, such as machinery and computer systems, and (2) funding and approval for projects and programs depend on positive return on investment. A drawback of return on investment is that expected returns and cost will appear as they are predicted; it is not a reflection of actual performance and cannot be used to gauge variance, particularly for impacts beyond the payback period.

Hands-on Statistics 8.12: Calculate the Payback Period by Hand

To calculate the payback period, use the following example. The HIM department at Sunnyside Hospital is in need of a new scanner, which costs $5,000. The supervisor, Leah, has been asked to calculate the payback period because this is an expensive item. Leah has estimated that a savings of $800 a year is feasible if the new unit is purchased. Remember that the longer the payback period is, the less desirable the investment would be. Payback periods should be as short as possible to make the most of the investment.

Keep in mind, however, two concerns regarding the use of the payback period: (1) benefits that occur after the payback period are ignored and do not measure profitability, and (2) the payback period ignores the time value of money.

The calculation for the payback period would be as follows:

$$\frac{\$5{,}000 \text{ (cost of the scanner)}}{\$800 \text{ per year savings}} = 6.25 \text{ years}$$

Therefore, it would take 6 years and 3 months to pay back the investment on the new scanner. Do you think this would be a good payback period and investment? Why or why not?

Hands-on Statistics 8.13: Calculate Payback Period and Return on Investment Using Excel

Now we will calculate payback period and return on investment using Excel. Follow these directions:

1. Open a new, blank Excel workbook.
2. Enter the data shown in **Figure 8.5**, but leave cells E5 and E6 empty.
3. In cell E5 enter the following formula:

   =C3/C4

4. In cell E6 enter the following formula:

   =C4*100/C3

What you have just learned is mathematically simple, yet offers important information to guide purchasing decisions.

	A	B	C	D	E	F	G
1		**Sunnyside Hospital**					
2							
3		Scanner cost	5000				
4		Savings per year	800				
5				**Pay-back=>**	***6.25***	Years pay-back	
6				**ROI =>**	***16***	%	

Figure 8.5 Computing return on investment.

Hands-on Statistics 8.14: Output Text and Graphics to Files Using R-Project

Now we will turn to R-Project. You have already been making graphics with R and using features such as a screen capture (print screen) to save the graphics. Now we will have R create a graphics file for us. These saved files could then be edited or simply inserted into a Word document or presentation. It takes only a few simple commands to redirect the output from the screen (R console) to a file. This process works for text output from R (e.g., the mean of a set of data) or graphic output (e.g., a bar chart). Follow these directions:

1. First, create some data from which to generate output. Start R-Studio (or R-Project) and create a vector (list) of numbers named *a*, as shown:

   ```
   a <- c(1,2,3,4,5,1,4,7)
   ```

2. Find the mean of *a* by entering the following:

   ```
   mean(a)
   ```

3. To set R to output both to a text file (not an image or graphic) and to the console, enter the following:

   ```
   sink("myfile.txt", append=FALSE, split=TRUE)
   ```

   Note: Do not add to an existing file named "myfile.txt," but create a new one and display output in console and to the text file.
4. Enter the following:

   ```
   mean(a)
   ```

   The output will be directed to the console and a file named "myfile.txt," which will be in the present working directory of R unless you specify otherwise. You will remember that the functions getwd() and setwd("c:/the/complete/pathyouwant") are the proper functions to display the current working directory and set a new one to a location of your choice. Of course, rerunning getwd() to make sure you set it properly is never a bad idea.
5. Open the text file you just created with a text editor or word processor, and you will see your mean in the file.
6. Stop outputting your text to a file and output only to the console by typing the following:

   ```
   sink()
   ```

   Note that when you open a connection or file stream to the text file, it will be locked and in use (i.e., you will not be able to delete or move it) until you close the stream with sink() or close R-Project.
7. To plot only the data points to the console, enter the following:

   ```
   plot(a)
   ```

8. To output to a jpg file, use the following (Joint Photographic Experts Group):

   ```
   jpeg("c:/myplot.jpg")
   ```

   Note: You may use *png*, *bmp*, *pdf*, or another term instead of *jpeg*.

   ```
   plot(a)
   dev.off()
   ```

   This last step turns off the output to a jpeg file.
   Note that just like sink(), you must turn off writing to an image file.

Next we will examine how to apply another data mining concept, naive Bayes.

Data Mining With Naive Bayes and R-Project

Every day there is more information available to healthcare professionals. This information can be used to summarize past data or to predict future events. Two well-known data mining algorithms used for this classification are backpropagation neural network (BNN) and **naive Bayes classification (NBC)**. We will examine only NBC in this text. NBC has been used for document classification, spam filtering, recommender systems, and prediction of heart disease, to name but a few areas.

Named after Thomas Bayes, a mathematician from the early 1700s, NBC is a probable classification algorithm, or predictive classifier. "Naive" refers to an assumption that there is independence between all the observations or variables. For example, if you were rolling two dice, the value of one die rolled has no effect on the other die.

The NBC algorithm classifies or predicts future datasets based on previous (posterior) classification of datasets. It is similar to regression analysis and can be more accurate given less data to analyze than a linear regression, which we will discuss in the next chapter.

NBC uses an *a priori* (Latin for "before") function to classify the data and then to calculate the means of the classifications and standard deviation. This training phase then gives the algorithm enough information to make predictions about the likelihood for future similar datasets.

As a simple example of how to use an NBC, the following example shows how the built-in Iris dataset (also known as Fisher's or Anderson's iris dataset) works with R. This dataset has 150 instances, 50 from each of three types of iris, with a measurement of the length and the width of the petals for each of the 50 instances. Given that there are slight differences between the species, the mathematician Ronald Fisher in 1936 developed a model that can differentiate each species based on the variation between the measurements. In the next example, we will summarize past data in the iris dataset.

Hands-on Statistics 8.15: Using Naive Bayes on the Iris Dataset

1. Load R-Studio.
2. Select *Tools* and *Install Packages*; then type in "klaR" and click *Install*.
3. Also install *caret*. Note we used this package for linear regression as well since this library is quite powerful.
4. At the R-Studio console prompt, load the libraries by typing the following:

```
library("caret")
library("klaR")
```

5. The iris flower dataset is built in, as well as one for chemistry. Load the iris dataset into two variables such that there are data for *x* and labels for the different flowers in *y*:

```
x = iris[, -5]
```

There are 150 observations, five variables, and three types of iris:

```
y = iris$Species
```

To see the observations of data on the three types of iris, type the following (*x* is for data, and *y* is for labels):

```
x
y
```

6. Now train the model on posterior data by entering the following:

```
model=train(x,y,'nb',trControl=trainControl('none'))
```

We created a model with our *x* and *y* data, chose *nb* for naive Bayes, and no resampling.

7. Next view the summary data so far:

   model

   Cohen's kappa is a statistic used to measure inter-rater reliability on categorical items. As such it is greater than .8, so we can assume the results are confident. You will recall from previous chapters that confidence is an important factor to consider in evaluating statistical results.
8. Now create the model named *finalModel*, as shown in **Figure 8.6**, and use class as the qualifier:

   predict(model$finalModel,x)$class
9. Lastly, to finalize output, we will build a confusion matrix to see how the classifier works:

   table(predict(model$finalModel,x)$class,y)
10. To see a scatterplot matrix of the original data, type the following:

    splom(x)

    Note: Groupings are to the right of the screen, where one group is set off (setosa), whereas there is some overlap between the other two types of iris, as shown in **Figure 8.7**.

As you can see from the output, there is only slight confusion of virginica and versicolor, with setosa accurately labeled at 50, and the other two very close at 47 each. Because there were 50 of each type in the original set, the model identified the three iris types well, considering no labels were used and this was all based on differences in the sepal and petal measurements.

How does this example relate to health care? In fact, techniques such as NBC are in use right now by doctors and other medical professionals. For example, Pattekari and Parveen described in 2012 a web-based prediction system for heart disease in which patients answer online questions, which are compared with a trained dataset in the system to detect (or prescreen for) the likelihood of heart disease. Certainly it is no replacement for a doctor, but if you consider the increasing number of people who require medical services, an automated system to prescreen patients allows medical professionals to work smarter instead of harder.

```
> table(predict(model$finalModel,x)$class,y)
            y
             setosa versicolor virginica
  setosa         50          0         0
  versicolor      0         47         3
  virginica       0          3        47
> |
```

Figure 8.6 Output of types of iris.

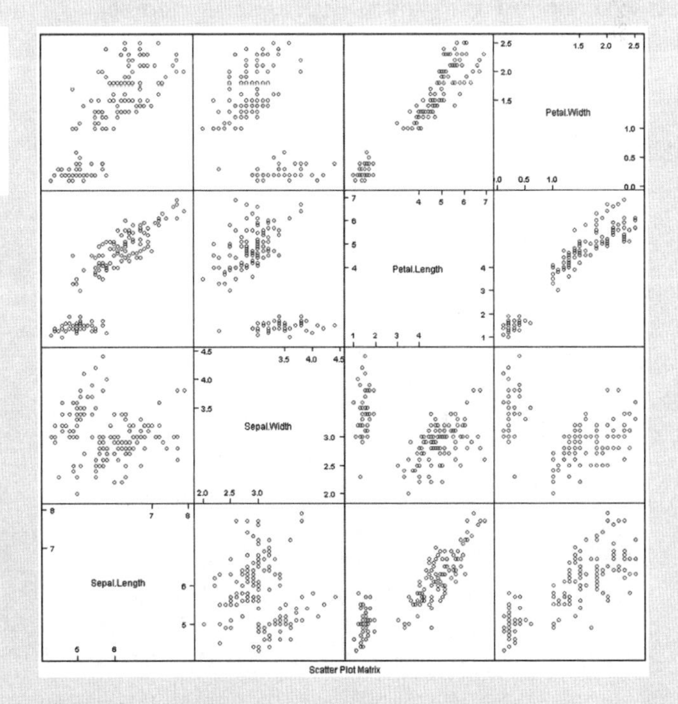

Figure 8.7 Scatter plot matrix of iris datasets.

How Does Your Hospital Rate?

Sandra has finally made her decision. She decides to hire the applicants with the credentials. Having RHIT credentials gives them credibility and shows their commitment to stand out and make themselves marketable in the HIM field. These newly credentialed RHITs will have the opportunity to gain experience and grow within the HIM department. HIM directors or coding supervisors face a difficult decision when presented with existing employees or qualified applicants (in this case with experience working in the field) who are not credentialed.

Global Perspective

Many diseases are prevalent in the United States and around the world. Tropical and subtropical areas of the world are high-risk areas for dengue fever. Dengue fever is a deadly infection, spread by mosquitoes, that is difficult to treat. This disease can sometimes take an even worse form known as severe dengue or dengue hemorrhagic fever. The World Health Organization (2019) estimates that about half of the world's population is at risk for dengue fever.

It is estimated that 390 million infections occur each year around the world, with a small number of those cases turning into the hemorrhagic version (WHO, 2019). Both versions start off with similar symptoms. In dengue hemorrhagic fever, however, tiny spots of blood develop on the skin and even larger spots under the skin and around bruised areas. In later stages, patients appear to be in a shock-like state.

Only 9 countries had experienced severe dengue prior to 1970, but this disease is now considered endemic (common within a defined area) in more than 100 countries (WHO, 2019). Asia is the most heavily affected region, bearing over 70% of the disease burden. The Americas, including the United States, are also affected, but reported incidence has been on the decline in this region in recent years—by 73% from 2016 to 2017 (WHO, 2019). With no known treatment, early detection of the illness and prevention strategies are paramount to control dengue and severe dengue fever.

Chapter Summary

In this chapter, we covered HIM functions, requirements to work in a HIM department, budgets, financial reports, productivity, and nonlabor costs. We also reviewed calculating labor costs, transcription costs, other costs, staffing level needs, case mix index, ALOS, and payback period. Data mining and predictive classification were examined with the NBC algorithm and R-Project software. Finally, we explored the ability of R-Project to export graphics to files with the sink() function.

Apply Your Knowledge

1. There are 685 requests per month, and the cost of postage is $498. Calculate the cost per request for this month by hand.
2. There are 289 requests per month and the cost of supplies is $48. Calculate the total cost of supplies for the month.
3. Coder A coded 428 inpatient records and worked 160 hours for the month of February. What is the average number of records coded per hour?
4. Coder B coded 588 outpatient surgery records, and 500 were coded accurately. Calculate the percentage of accurately completed charts.
5. Coder C accurately completed coding of 395 records and worked a total of 160 hours. Calculate the completed medical records per hour.
6. Lucy earns $12.23 per hour, Rosie earns $15.75 per hour, Buffy earns $13.45 per hour, and Henrietta earns $14.68 per hour. The employees have been given a 4.2% increase. What is each employee's new salary?
7. The HIM department has budgeted $600 for travel, and the actual amount totaled $1,100 for the year. Calculate the variance.
8. How should your facility handle the question of hiring a credentialed but inexperienced applicant versus a noncredentialed but experienced applicant for a coding position? Prepare a report for your manager with several reasons for and against hiring each individual. Reference what other facilities are doing when faced with this decision.

9. Find an example, not already listed in the text, of how naive Bayes data mining is being used in health care.
10. What is an average number of records a coder can enter per hour? Cite where you found your figure.
11. List at least one good website for information on exporting images and graphics from R-Project.
12. Use the sink() function in R-Project to create three different graphics files. You may use any function covered so far in the text.
13. Use the NBC method on a dataset of your choice. Prepare a short summary, with graphics of your findings. Note, this may take some time.

References

African Heritage Foundation. (2015). Inhotep: The true father of medicine. Retrieved March 20, 2016, from http://www.afrikanheritage.com/imhotep-the-true-father-of-medicine/

Brimmer, K. (2013) Cut labor cost with smarter scheduling, October 2013 print issue: Healthcare Finance News; Retrieved April 6, 2015, from http://www.healthcarefinancenews.com/news/research-ties-staffing-sustainablity

Cadigan, M., & Nickson, C. (n.d.). Sir William Osler quotes or Oslerisms. *Life in the Fastlane*. Retrieved from http://lifeinthefastlane.com/resources/oslerisms/

Kyruus. (2015). Mercy Health selects Kyruus to enhance enterprise-wide patient access. Retrieved September 2, 2013, from https://www.kyruus.com/news/mercy-health-selects-kyruus-to-enhance-enterprise-wide-patient-access

MedlinePlus, retrieved from McDonnell, Steve. Total Labor Cost vs. Annual Labor Cost; Business & Entrepreneurship: Retrieved December 5, 2013, from: http://yourbusiness.azcentral.com/total-labor-cost-vs-annual-labor-cost-27092.html

Pattekari, S. A., & Parveen, A. (2012). Prediction system for heart disease using naive Bayes. *International Journal of Advanced Computer and Mathematical Sciences, (3)*3, 290–294. Retrieved April 9, 2014, from http://bipublication.com/files/IJCMS-V3I3-2012-2.pdf

World Health Organization. (2019, November 4). Dengue and severe dengue. Retrieved December 30, 2019, from https://www.who.int/en/news-room/fact-sheets/detail/dengue-and-severe-dengue

Web Links

Lattice Graphs: http://www.statmethods.net/advgraphs/trellis.html

How to Become a Medical Records or Health Information Technician: http://www.bls.gov/ooh/Healthcare/Medical-records-and-health-information-technicians.htm#tab-4

Demands for Health Information Technology Graduates on the Rise: http://www.ajc.com/news/business/demand-for-health-information-technology-graduates/nQgKn/

Outbreaks: Displaying Outbreaks, Cases, and Deaths From Viral and Bacterial Diseases Which Have the Potential to Indicate Biological Terrorism: http://outbreaks.globalincidentmap.com/eventdetail.php?ID=14871 and http://outbreaks.globalincidentmap.com/home.php

Aedes aegypti: http://en.wikipedia.org/wiki/Aedes_aegypti

Methods of Reporting Hospital Financial Information (abstract): http://www.ncbi.nlm.nih.gov/pubmed/6380277

Financial Management of Hospitals (abstract): http://www.ncbi.nlm.nih.gov/pubmed/6375357

Hospitals Readmissions Reduction Program (HRRP): http://www.cms.gov/Medicare/Medicare-Fee-for-Service-Payment/AcuteInpatientPPS/Readmissions-Reduction-Program.html

Readmission Rates: http://www.mayoclinic.org/quality/readmission-rates.html

Return on Investment (ROI): http://www.readyratios.com/reference/profitability/return_on_investment_roi.html

Payback Period: http://www.investopedia.com/terms/p/paybackperiod.asp

CHAPTER 9

Research Methodology and Ethics

Research is formalized curiosity. It is poking and prying with a purpose.

—Zora Neale Hurston

CHAPTER OUTLINE

Introduction
Types of Research
Research Process
  Step 1: Identify the Problem
  Step 2: Research the Problem
  Step 3: Develop Research Questions
  Step 4: Determine the Type of Data Needed, Sample Size, and Methods of Analysis
  Step 5: Collect Data
  Step 6: Analyze the Data
  Step 7: Draw a Conclusion
  Step 8: Report the Results and Implications for Further Research
Research Ethics and the Abuse of Human Subjects
  Late 1700s: Edward Jenner and the First Smallpox Vaccine
  1850s: J. Marion Sims, the Father of Gynecology
  1900: Walter Reed and Yellow Fever Transmission
  1932 to 1972: Tuskegee Study of Untreated Syphilis
  1964: The Declaration of Helsinki
  1979: The Belmont Report
The Institutional Review Board
  Review Process
  Exemption and Types of Review
  Informed Consent
  Membership
Data Dictionary
Statistical Measures and Tools
  Common Nonparametric Statistics
  Regression Analysis: Simple Linear Regression
Global Perspective
Chapter Summary
Apply Your Knowledge
References
Web Links

LEARNING OUTCOMES

After completing this chapter, you should be able to do the following:

1. Describe two basic types of research.
2. List and describe the steps of the research process.
3. Describe the history of abuse of human subjects and the development of protective measures.

4. Explain the purpose of an institutional review board, and describe the review process.
5. Describe a data dictionary.
6. Differentiate parametric and nonparametric statistical measures.
7. Perform a chi-square test of independence.
8. Perform a Mann-Whitney *U* test.
9. Perform a linear regression analysis.
10. Create a timeline with online tools.

KEY TERMS

Applied research
Assent
Belmont Report
Beneficence
Chi-square test
Consent
Declaration of Helsinki
Drug trial
Ethics
Exploratory research
Informed consent
Institutional review board (IRB)
Justice
Linear regression
Mann-Whitney *U* test
Respect for persons
Scientific notation

How Does Your Hospital Rate?

Joan recently learned that her glioblastoma—a common and deadly type of malignant brain tumor—has returned. After talking with her neuro-oncologist, Dr. Juan, she decides to apply to join a clinical trial of a new, experimental medication that is currently being performed at Johns Hopkins University in Baltimore, Maryland. Dr. Juan contacts the hospital to make an appointment for Joan to see whether she meets the qualifications to enter the trial.

Consider the following:

1. If you were Joan, what questions would you have about participating in the trial?
2. Under what conditions would you agree to participate?
3. What protections would you expect to be in place regarding how the trial is carried out?

Introduction

With the increasing use of technology in the healthcare environment, health information management (HIM) professionals are in higher demand than ever. HIM professionals understand the importance of collecting accurate and timely data and integrating new technology that continues to influence and increase the role of HIM professionals. HIM professionals collect, organize, and analyze healthcare data. Their primary responsibilities are to provide quality data to support the delivery of excellent health care while equalizing the right to privacy and providing access for legitimate users. In traditional and nontraditional roles, HIM professionals' responsibilities have shifted to include evidence-based quality improvement initiatives, participation in clinical research projects, and risk management.

HIM professionals bring many attributes to the research process, including the following:

- Expertise in research methodology
- Knowledge of information systems and application of information technology for solving clinical problems
- Knowledge of workflow
- Knowledge of theory and curriculum development

Besides the methods used to conduct research, HIM professionals also must be aware of the ethics of research. Research conducted in an unethical manner calls into question not only the reliability of the findings produced by the study but also the character and reputation of the researcher or researchers. Even more important is the protection of human subjects.

The goal of this chapter is to equip you with both the methodology and the ethical principles required to conduct effective and responsible research. This chapter first identifies the two major types of research, applied and exploratory, and then outlines the steps in the research process. Next, ethical issues involved in research are addressed, including the historical abuse of human subjects. Institutional review boards and their role in regulating research to prevent such abuses are also covered. The chapter then introduces common nonparametric statistical concepts, including the chi-square test of independence and the Mann-Whitney *U* test, along with regression analysis and the use of linear regression in R-Project.

Types of Research

Research may be divided into two broad types: applied and exploratory.

Applied research—also known as action, intervention, or collaborative research—is less theoretical in nature and aimed more at improving a process or practice, changing a situation, or solving a problem. For instance, one team of researchers conducted a study with the goal of helping improve patient experiences with hospital mealtime (Dickinson, Welch, Ager, & Costar, 2005). Such research is important because improving nutrition for patients improves not only patient healing but overall hospital satisfaction.

Exploratory research, on the other hand, focuses on learning about new subjects or those for whom there is little existing information and on developing knowledge of a topic rather than improving a process. For example, if you want to find out whether the crisis center of your hospital is serving the needs of the local population, you would conduct an exploratory study. However, if you want to minimize emergency department wait times at the same hospital, you would conduct an applied study.

A great example of an actual exploratory research study is an article titled "An Exploratory Study of Mental Health and HIV Risk Behavior Among Drug-Using Rural Women in Jail" (Staton-Tindall et al., 2015). The study investigated whether there was a relationship between depression, anxiety, posttraumatic stress disorder, and certain types of drug use and risky sexual activity in the population under examination. The researchers found that mental health significantly correlated with severity of certain types of drug use, as well as engagement in risky sexual activity. The study concluded with suggestions for further research into the problem areas, but no firm solutions to the problem were given, as they would have been in an applied research study.

Research Process

Now that you know more about the types of research, we will consider an eight-step framework for conducting research.

Step 1: Identify the Problem

Begin your research by specifically identifying your area of study and stating the problem you wish to examine, in as much detail as you can. For example, you might seek to determine the best interventions for lowering the blood pressure levels of all employees of a company who have hypertension.

Step 2: Research the Problem

The next step would be to research the existing literature for studies on this topic, to determine what is already known about it. For instance, in your research you will likely identify the available interventions for lowering blood pressure, such as meditation, exercise, and diet, and learn about their effectiveness in other contexts. This information will help you to decide how to set up your own study.

Step 3: Develop Research Questions

This step involves formulating a question or questions that your research is intended to answer. For instance, the research question in this case might be, "Does a weekly exercise program of 30 minutes of walking on a treadmill 5 days per week at moderate intensity lower blood pressure in the staff?" The null hypothesis (H_0) is that there is no significant decrease in blood pressure in the group after the interventions were implemented. The alternative hypothesis (H_1) would be that there was a significant decrease in overall blood pressure levels due to the intervention.

Step 4: Determine the Type of Data Needed, Sample Size, and Methods of Analysis

Next you will determine the sample size (or the entire population), methods of data collection, and methods of data analysis. In terms of the blood pressure study, you would probably want to note how many staff members approximately you would like to have participate, the process for actually measuring blood pressure, when the measurements are to be taken in relation to the beginning or end of the shift, how often they will be taken, the average blood pressure, the significant difference of blood pressure, outlier data, which subjects are on blood pressure medication, and whether there were national statistics on blood pressure of medical staff.

Step 5: Collect Data

In this step you actually collect the data. For instance, you and any colleagues who are assisting you would perform blood pressure assessment on subjects at the determined times with sphygmomanometers (blood pressure gauges) and stethoscopes.

Step 6: Analyze the Data

After you have collected the data, it is time to use the statistical measures you chose to reveal your findings. You should consider outlier data, sample size, significance level (if needed), distribution of the data (normal or not), and sample type (random or not). For example, you could collect blood pressure data on all members in the study, or you might just take a sample from a random group that would represent all of them.

Step 7: Draw a Conclusion

Now that you have statistical results, you will summarize what you have discovered. You should be able to make some conclusions about the findings, as well. Did you reject the null hypothesis? Were there any issues with data collection or outliers?

Step 8: Report the Results and Implications for Further Research

Lastly you will report the findings. In this step, you must determine the best way to present your data—numerically, graphically, or both. You must also determine any implications for further research. For example, if the study found that walking on a treadmill for 30 minutes per day lowered blood pressure, a question that might warrant further research is, "Would walking for an hour lower it even more?"

Research Ethics and the Abuse of Human Subjects

So far, we have considered methods you can employ to conduct research effectively. However, research must also be conducted ethically. **Ethics** is a field of study that involves matters of right and wrong. Three tenets of ethical medical research in the United States are respect for persons, beneficence, and justice. **Respect for persons** is a concept that stresses that human beings have intrinsic and unconditional moral worth and should be treated as if there is nothing of greater value than they are. So the value of research findings would take a backseat to the protection of the participant. With **beneficence**, researchers have the welfare of the research participant as a goal of any study. Think *benefits* when you are trying to understand this term. You could consider that the benefits from the study should outweigh the potential risks associated with participation. As an example, given that the research intervention has some possible chance of harm to the subjects, the chance that it can offer strong insight into a cure for the disease is more important to the subjects. **Justice** can be defined as the fair treatment and selection of participants. They should not be selected because they are convenient or easy to influence. Also unless key to the study, care should be used in the makeup of participants with regard to race, sex, and special population status (e.g., prisoner, minor, poor, mentally handicapped).

One key concern in research ethics is how human subjects are treated. Are they respected as persons? Is their welfare considered and protected? Are they treated justly? Fortunately, organizations have taken measures internationally to protect subjects. In 1978 the **Belmont Report** outlined in detail what should be considered when conducting research involving human subjects to ensure they are treated safely and fairly. For example, if a certain experimental drug therapy has a very low chance of curing a disease but carries a high risk of serious side effects, regulations protecting human subjects might prohibit the study.

Sadly, throughout history, researchers have not always treated human subjects ethically. In some cases, they have even abused them. The following discussion highlights some of these arguably questionable cases.

Late 1700s: Edward Jenner and the First Smallpox Vaccine

In the late 1700s, smallpox, known as "the scourge of mankind," killed people on a massive scale. English doctor Edward Jenner had found that dairymaids had a lower instance of catching this deadly disease. Jenner postulated that their exposure to the less deadly cowpox virus had given them some immunity. To find out if this was true, Jenner excised samples from a dairymaid's cowpox lesion and injected it into an unknowing healthy 8-year-old boy. The subject developed a fever and showed symptoms of the cowpox disease but did not die. Two months after his recovery, Jenner injected the boy with smallpox to see whether he would catch smallpox. As it turned out, the boy had developed immunity by being exposed to a similar disease. Based on this finding, Jenner created the world's first smallpox vaccine, which paved the way for other vaccinations, which saved millions of lives globally. However, in looking back with the ethical considerations we now have, it is clear that he put nonconsenting human subjects in mortal danger.

1850s: J. Marion Sims, the Father of Gynecology

Dr. J. Marion Sims, known as the "father of gynecology," was an innovator of new techniques in gynecology and changed the field of women's reproductive health. His accomplishments, however, came at the expense of his human subjects. In the mid-1850s, Sims built a makeshift hospital on his South Carolina farm and initiated a series of experimental gynecological operations on numerous female African slaves, both his own and those brought to him for treatment. One particular medical problem he was treating was a vesicovaginal fistula, which is a tear between the vagina and bladder.

DID YOU KNOW? A cadaver provides an excellent learning tool for physicians by allowing them to study a real body without the risk of harming a living person. Even this practice, however, has its ethical concerns.

In England prior to the mid-1800s, dissection of human cadavers was banned except on the bodies of executed murderers. This restriction resulted in a shortage of eligible cadavers and provided a revenue stream for some enterprising individuals who dug up the bodies of the recently deceased and sold them to physicians. Two men, Hare and Burke, skipped the digging and smothered more than a dozen lodgers in an Edinburgh boardinghouse from 1827 to 1828 to keep their inventory of cadavers stocked.

Ethical issues surrounding the use of cadavers continue today. The future of cadaver use may be in the form of what James Park refers to as a living cadaver or a body in which the brain is dead but other systems still function. With appropriate permission from family members or a statement from the deceased in their advance directive for medical care, researchers would be allowed to perform dissection on these bodies (Park, 2014). Certainly this concept of a living cadaver and the rights of such individuals will be a future debate in medical ethics.

Now you know!

1900: Walter Reed and Yellow Fever Transmission

In 1900, Walter Reed, after whom the Walter Reed Hospital is named, was researching how yellow fever is transmitted. Although some at the time correctly suspected that mosquitoes are the primary carriers of this disease, the fact had yet to be proven. In his research, Reed injected 22 Spanish immigrants with yellow fever. These impoverished research subjects were offered $200 to participate and $500 if they contracted yellow fever. To this particular group of subjects, the money was too good to pass up, even at the very real risk of death. By today's ethical standards, the money offered was coercive due to the amount.

How Does Your Hospital Rate?

Joan travels to Duke University for her first appointment with the medical research team. She learns that she must meet the following criteria to participate in the trial: she must have only one tumor, and the tumor must be recurrent, surgically accessible, no smaller than 1 centimeter, no larger than 5 centimeters, and located at least 1 centimeter away from any ventricles. After further testing and careful review of the results, the medical research team finds that Joan meets the qualifications for the clinical trial. The treatment that Joan will be given is a genetically engineered poliovirus (PVS-RIPO) that is being tested as an anticancer treatment.

Consider the following:

1. If you were on the team of researchers conducting the trial, what information about Joan would be important to obtain?
2. What information would be critical to give to Joan about the trial?
3. What measures would you take to help comfort Joan and make her feel safe?

DID YOU KNOW? Dr. Lorenzo Frink was an early experimenter in pain management studies in eastern North Carolina. Smith and Smith (2014) note that Dr. Frink conducted experiments on slaves on his plantation. As punishment, they were chained in his attic and underwent pain studies and experiments with lobotomies. Dr. Frink's research was being referenced even as late as the late 1990s by medical students. When the original home of Dr. Frink was sold, chains and manacles were found in the attic.

Now you know!

1932 to 1972: Tuskegee Study of Untreated Syphilis

The infamous 40-year study conducted on the campus of the Tuskegee Institute in Macon County, Alabama, beginning in 1932 illustrates the importance of informed consent in medical experimentation. The study was conducted by the US Public Health Service (USPHS) and explored the natural history of syphilis and its treatment in African Americans. The experimental group consisted of 399 syphilitic men, and the control group consisted of 201 uninfected men. To keep subjects in the study, the USPHS offered them free medical exams and free meals. Subjects were not given curative treatments but were simply observed to determine the course of the disease. They were not informed of the presence of the disease, and after one subject learned he had the disease in 1947, he was denied treatment. The USPHS tried to end the experiment on several occasions but was unsuccessful, in part due to insistence by the Centers for Disease Control and Prevention in the late 1960s that the study should continue.

Brandt (1978) notes that even the underlying purpose of the study, based on social Darwinism and racism, was unethical, as it was intended to discourage sexual relationships between black men and white women: "The Negro, doctors explained, possessed an excessive sexual desire which threatened the very foundations of white society." The racism and ethical breaches in this case contributed to one of the darkest chapters in US history. The study was formally ended 1972. In 1973, attorney Fred Gray filed a class action suit on behalf of the men, women, and children who were affected by this unethical research and succeeded in winning more than $9 million dollars in damages for the plaintiffs. In 1997, President Clinton issued a formal apology to the subjects of the experiments.

1964: The Declaration of Helsinki

In response to this long history of human subject abuse, the World Medical Association met in Helsinki, Finland, and developed the **Declaration of Helsinki**. This document established groundbreaking ethical standards with regard to human research. First adopted in June 1964 and subsequently revised numerous times, the declaration was the first major global effort to regulate human subject research. A key part of the document states that, "In the field of research a fundamental distinction must be recognized between clinical research in which the aim is essentially therapeutic for the patient and the clinical research the essential object of which is purely scientific and without therapeutic value to the person subjected to the research" (World Medical Association, 2018).

1979: The Belmont Report

Just as the Declaration of Helsinki established international ethical standards for human subject protection, the Belmont Report established national ones for the United States. The purpose of the Belmont Report was to summarize the basic ethical principles identified by the National Commission for the Protection of Human Subjects of Biomedical and Behavioral Subjects, an organization created when the National Research Act was signed into law in 1974.

One result of the Belmont Report was the establishment of **institutional review boards (IRBs)**, whose responsibility it is to determine whether proposed research would be harmful to subjects.

Another result was the development of the concept of beneficence. Beneficence, as discussed previously, is the value of having the welfare of the research participants as a goal of any research studies or clinical trials. Coercion of any kind violates the principle of beneficence because it causes subjects to act against their own best interests. For example, prisoners might fear punishment by guards if they were to refuse treatment. Similarly, citizens of a country ruled by an autocratic or dictatorial government and prisoners of war may feel they have no choice but to participate in medical experiments. Even the position of authority that a physician holds in a society, such as that of Nazi Germany, can contribute to the subject feeling coerced to participate.

Another key concept related to human subject protection, established by the case of *Schloendorff vs. Society of New York Hospital*, is that of informed consent. As a result of this case, hospitals are required to inform patients of all risks and benefits associated with a study and to request and obtain the patient's consent to participate in the experiment. Informed consent is one of the major tenets of the approved IRB process a researcher should include, if appropriate, to the research methodology.

The Institutional Review Board

Before you conduct research in an institutional setting, you must obtain approval from an IRB at the institution by providing written details of how subjects will be selected and treated and how resulting data will be collected and disseminated. Any organization that receives federal grant money for medical research or that conducts research on human subjects is required by law to have an IRB. The purpose of the IRB is to provide ethical and regulatory oversight of research subjects, particularly of vulnerable populations such as children, inmates, pregnant women, human fetuses, and individuals with compromised mental capabilities.

Review Process

The National Research Act Public Law 99-158, the Health Research Extension Act of 1985, and the National Commission for the Protection of Human Subjects of Biomedical and Behavioral Research have guidelines for how to conduct research when human subjects are involved. These regulations mandate that researchers receiving federal grant funds make sure the research is reviewed and approved by the institution's IRB. In fact, it is a good idea to obtain IRB approval even for research that is not federally funded. Considerations that should be made for subjects include but are not limited to the following:

- The welfare and rights of subjects are protected.
- Informed consent is obtained where appropriate.
- There is no unreasonable physical, mental, or emotional risk to participants.
- The benefits of the research outweigh the risks to the subjects.
- Appropriate IRB certification has been granted.

Consider the following steps as a general guideline for the IRB process:

1. Researchers should have a certification on file with the institution, such as the Collaborative Institutional Training Initiative (CITI) or the National Institutes of Health (NIH). In the case of off-site requests to conduct research, this requirement may be waived, although it is advisable to have certification as a prerequisite in all cases.
2. The researcher(s) submit a request to conduct research (which may vary by organization) to the designated IRB contact. Generally this will also involve submitting the research goals, methods, informed consent forms (if needed), and confidentiality information. At this point the lead reviewer examines the research proposal and determines whether the research is exempt from review, requires only expedited review, or requires full board review.
3. Researchers are informed of a need for more information, approval, or denial.

Exemption and Types of Review

Research presented for IRB approval may be found to be exempt from review or to require expedited or full IRB board review. A designated person on the IRB board makes an initial screening of the research proposal to determine the type of review required. At many institutions, any research involving human subjects requires IRB approval. Federal regulations document 45 CFR 46.102(f) states that a human subject is "a living individual about whom an investigator conducting research obtains (1) data through intervention or interaction with the individual, or (2) identifiable private information."

If the IRB initial screening determines that the study involves no human subjects, then the study is exempt. The study may also be found to be exempt on the following bases:

- The study presents no greater risks of harm than those that a subject might encounter in their normal daily life or performance of work.
- The proposed study involves tests of educational effectiveness, such as aptitude tests and tests of instructional effectiveness.
- The proposed study involves data collection procedures, such as interviews or surveys, that protect participants' identities.
- The research involves the use of existing data, which protects the identity of subjects.

If human subjects are involved and the study is not otherwise found to be exempt, it is examined to see whether an expedited process (considering that risk to subjects is minimal or more) or a full

IRB review (if significant concerns or risks are evident) is required. An expedited review might be conducted by only one or a few members of the IRB and not the full board, whereas a full board review would be conducted by all members. At the end of the review, the board may ask for clarification, ask for changes, or deny the research based on their findings.

The following factors would increase risk to subjects and thus would receive special consideration by the board:

- The research involves the administration of drugs or other substances.
- The research involves pregnant women, fetuses, etc.
- The research may present subjects with life-threatening conditions.
- The research involves intrusive procedures (e.g., the use of scopes, biopsies).
- There is clear evidence from prior research of potential dangers to the subjects.
- The research might have a legal risk for the subject or affect the participants' privacy (e.g., substance abuse, sexual behavior/ preference).

Informed Consent

The participation by human subjects in any research project must be consensual. Therefore, the study must include a clearly established way for potential subjects to opt out if they do not want to participate, with no repercussions from the researchers. The researcher is strongly advised to obtain **informed consent** from subjects, which includes two elements: (1) the potential subject has enough information to make an informed decision, and (2) there is no coercion or pressure to participate directed toward the potential subject, and participation is completely voluntary.

The use of an informed consent form is strongly recommended unless the use of the form may adversely affect the study or the subjects fall under a special population status, in which case the party responsible for the subjects would provide assent to the research. Consent may be obtained without a physical form, or it may be waived as unnecessary, depending on the level of risk posed by the research. If approval from the subjects was obtained by a different method, this should be noted in the research proposal.

Consent is given by subjects who have reached the legal age for consent, which is 18 years old in the United States. **Assent** is an agreement to participate given by someone other than the subject on behalf of the subject, in cases in which the subject is not able to give legal consent. Children, for example, cannot participate in a research study unless their guardian or parent provides assent. Note that the relationship of the researcher to the subject (e.g., if the subjects are family members) does not negate the need for consent.

Membership

According to federal guidelines, an IRB should be composed of at least five members, one of whom is the IRB chair. The members should have no conflicting interest with the research, should be made up of men and women, should include people from different professions, should include at least one person with a background in science, should include one person not affiliated with the institution, and can "invite individuals with competencies in special areas" as needed (US Department of Health and Human Services, 2019).

Next you will learn about a data dictionary, which is important to ensure consistent and reliable data collection, and review some statistical measures that deal with data that may not fit a normal distribution of data.

Data Dictionary

Known by other names, including data definition matrix, content specifications, and metadata repository, a data dictionary, by general definition, is a description of a database(s) or format for cells in a spreadsheet or database. In business settings, the layout of a data dictionary is quite detailed and designed so that data are as accurate and error-free as possible.

A data dictionary might be a document or might be a component software application in a large commercial system. Regardless of the size, though, a data dictionary describes the data held in a field, its required status, or its length or format, if the field is linked to other data files. As an example with R-Project, the dataMeta library allows for the creation of a data dictionary for an R dataset; in addition to definitions of the variables and descriptions of the variables (e.g., "cell *INS. Provider* holds insurance information"), it also allows tracking of when the last change was made to the document and who made it so that change management is available. On a smaller scale, there may be no formal document, but the concept of a field that holds data in a proper format is the same (e.g., a date field where the date must be entered in a specific format, such as day/month/year).

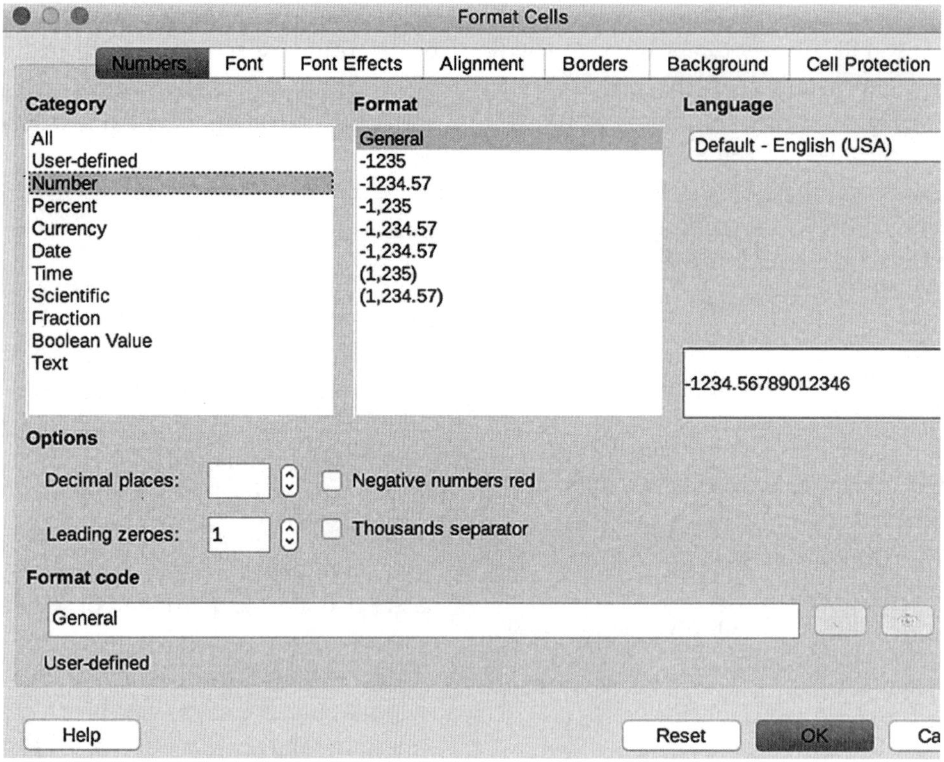

Figure 9.1 Cell properties in a spreadsheet.

As an example, if you were creating a patient entry database, you might require a first and last name. A middle initial might be only one character in length, whereas the first and last names could be up to 30 characters. A name field might use text, whereas an age field would need to use a whole number (not fractional). Lastly, certain fields, such as name and age, may well be required fields, not allowing you to move on until a value is entered. Name and contract information from each patient visit would be tied back (related) to a master patient index database, so that a check-in and check-out database might have address and name information tied to a patient list database. Since the doctors in a practice would be limited to only a few, the field to select which doctor patients are seeing could be a drop-down list so that there was no way to mistype the doctor's name. You can see how just that one aspect would reduce errors in typing the doctor's name, since the names would have been pre-typed in another linked database.

Spreadsheets also utilize a data dictionary via *cell properties*, allowing users to format cells to specific properties, very similar to how it would be done in a database. With either Excel or Open Office, you right-click on the cell and choose items such as text or numeric, a Boolean field option (yes or no, male or female), number of digits of precision, and date/time, etc. Note in **Figure 9.1** a screen capture from Open Office showing options under *cell properties* by right clicking on the cell.

Statistical Measures and Tools

Common Nonparametric Statistics

Parametric statistical methods, discussed in previous chapters, are based on a normal distribution of data. For example, consider that you have all of the patient data from two large groups, one given a treatment and the other a placebo. Because the groups were randomly selected from the population, the results from the sample may be generalized to the entire population.

For situations in which a normal distribution cannot be assumed, however, nonparametric methods should be used. Known as distribution-free tests (i.e., not based on a normal distribution), nonparametric tests are based on fewer assumptions about the data in question and can be used on rank and ordinal data. So, although parametric methods have greater power and provide good results with fewer data, nonparametric methods are required in many cases.

Two examples of nonparametric tests are the **chi-square test** and the **Mann-Whitney *U* test**, which are discussed in the following sections.

Chi-Square Test of Independence

There are two chi-square tests. The chi-square goodness-of-fit test is a one-variable test used to determine how close a categorical variable aligns with a theoretical model and, thus, whether the sample data of a study are a fair representation of the population under study. The chi-square test of independence, in contrast, is a two-variable test used to determine whether there is an association between the two variables. This chapter focuses on the chi-square test of independence.

The chi-square test of independence examines categorical data, such as yes, no, or neutral, rather than quantitative values, such as weight, height, or test scores. For example, in a survey, respondents might be classified according to gender and voting preferences on a topic, such as whether they agree, are neutral, or disagree. The chi-square test of independence could be used to determine whether their gender is related to their choice.

This test is best used with a random sample and a population at least 10 times as large as the sample, but it may be used with a smaller sample-to-population ratio, if the data at hand do not permit it. Stating the population and sample size in the results helps validate the findings.

Hands-on Statistics 9.1: Chi-Square Test of Independence Using Excel

1. Open a new, blank Excel workbook.
2. Enter the data shown in **Figure 9.2**, but leave cells C6, D6, and E4 to E6 blank, because you will enter formulas in these cells.
3. In cell C6, enter the following formula:

   =SUM(C4:C5)

   Copy and paste the formula from cell C6 to cells D6 and E6.
4. In cell E4, enter the following formula:

   =SUM(C4:D4)

   Copy and paste the formula from cell E4 to cell E5.
5. Enter the additional data shown **Figure 9.3**, but leave cells C10, C11, D10, and D11 blank, because you will enter formulas in these cells.
6. Enter the following formulas in the cells indicated:

   C10: = (C6*E4)/E6

   C11: = (C6*E5)/E6

   D10: = (D6*E4)/E6

   D11: = (D6*E5)/E6

   These formulas will produce the expected values by multiplying column and row totals and dividing by grand total.

E4 | fx =SUM(C4:D4)

	A	B	C	D	E	F
1						
2		Actual values	Student coffee consumption >=2 cups a day			
3			Graduate	Undergraduate	Row total	
4		Female	49	39	88	
5		Male	65	95	160	
6		Column total	114	134	248	

Figure 9.2 Chi-square with using Excel: Part 1.

(continues)

	A	B	C	D	E
1					
2		Actual values	Student coffee consumption >=2 cups a day		
3			Graduate	Undergraduate	Row total
4		Female	49	39	88
5		Male	65	95	160
6		Column total	114	134	248
7					
8					
9		Expected values	Graduate	Undergraduate	
10		Female	40.4516129	47.5483871	
11		Male	73.5483871	86.4516129	

Figure 9.3 Chi-square using Excel: Part 2.

7. Next, enter the additional data in cell C13 shown in **Figure 9.4**, and enter the following formula in cell D13:

   =CHITEST(C4:D5,C10:D11)

   This formula will return the *P* value for the data under examination.

   Considering that the alpha is .05, you next would examine the returned *P* value.

The hypothesis ($P > .05$) is that there is a relationship between gender and coffee consumption, meaning these variables are dependent. However, this hypothesis is not supported in this example, because $P = .0228$, which is less than the alpha.

Thus, the null hypothesis ($P < .05$) is true; there is no relationship between gender and coffee consumption, meaning these variables are independent. Thus, we conclude that there is no difference between male and female graduate and undergraduate students' coffee consumption.

D13 fx =CHITEST(C4:D5,C10:D11)

	A	B	C	D	E
1					
2		Actual values	Student coffee consumption >=2 cups a day		
3			Graduate	Undergraduate	Row total
4		Female	49	39	88
5		Male	65	95	160
6		Column total	114	134	248
7					
8					
9		Expected values	Graduate	Undergraduate	
10		Female	40.4516129	47.5483871	
11		Male	73.5483871	86.4516129	
12					
13			P < .05 accept null P =	0.022819897	

Figure 9.4 Chi-square example showing CHITEST formula: Part 3.

Mann-Whitney U Test

The Mann-Whitney *U* test (also known as the Mann-Whitney-Wilcoxon, Wilcoxon rank-sum, or Wilcoxon-Mann-Whitney test) is a nonparametric test that may be used in place of a *t* test in studies in which there are two samples, but with one population having larger values than the other. This test is more appropriate than the Student *t* test in studies in which a normal probability distribution may not be assumed, the distribution is unknown, or the sample size is very small (from less than 30 to 50). You could, in fact, try both tests if the only difference is size and compare the results.

Hands-on Statistics 9.2: Mann-Whitney *U* Test Using R-Project

To practice using the Mann-Whitney *U* test in R-Project, imagine you have data from a blood analysis done to determine whether a certain treatment affects levels between two groups of informed and consenting subjects in a research experiment. Follow these directions to analyze the data:

1. Start R-Project, and at the command prompt, create two vectors of data (g_1 and g_2) by entering the following:

```
g_1 = c(12, 9, 17, 19, 10, 11, 15)
g_2 = c(17, 23, 19, 16, 19, 21, 21, 17, 16, 22)
```

2. To perform the Mann-Whitney *U* test on these sets of data, enter the following:

```
wilcox.test(g_1,g_2)
```

The results of the test will appear, as shown in **Figure 9.5**.

Note in Figure 9.5 that the *P* value is .009265, which means the hypothesis $P > .05$ is false. This finding is statistically significant because the probability that it occurred by chance is only .009, or a bit less than 1%. Moreover, as indicated in the results of the test in Figure 9.5, the alternative hypothesis is true, so you can reject the null and conclude the treatment does have some effect. By default, R-Project uses a continuity correction factor in the final calculation. If the samples are larger in number of values, it could be turned off. If smaller, the result is continuity-corrected for the interval but not for the estimate. The example shows a small number of values, but this feature could still be turned off by running the test with an optional argument, such as the following:

```
wilcox.test(g_1, g_2, correct = FALSE)
```

Generally, though, it is best to use the correction factor.

```
> wilcox.test(g_1,g_2)

        Wilcoxon rank sum test with continuity correction

data:  g_1 and g_2
W = 8, p-value = 0.009265
alternative hypothesis: true location shift is not equal to 0

Warning message:
In wilcox.test.default(g_1, g_2) : cannot compute exact p-value with ties
```

Figure 9.5 Output of Mann-Whitney *U* test with R-Project.

Regression Analysis: Simple Linear Regression

Regression analysis is used to make predictions of future data based on historical data. The initial use of regression analysis dates back to 1805, when Legendre and Gauss used the least squares method for astronomical observations. Francis Galton coined the word *regression* in the 19th century to communicate a biological phenomenon.

In **linear regression**, the relationship between historical data for one independent variable and one dependent variable is used to predict future values for the dependent variable. Regression analysis can assist in explaining how values of a dependent variable can change when an independent variable is changed. Regression analysis may be used to predict future inventory needs, future sales, future commodity or stock prices, or productivity gains. In the next chapter, you will learn about a more complex version called multiple regression.

Hands-on Statistics 9.3: Linear Regression Using R-Project

To practice using linear regression in R-Project, consider the following example. A company can produce a number of widgets on an assembly line, given a certain amount of time spent by the workers. You would like to know whether there is a significant relationship between time spent by the workers and the number of widgets produced. Examine the findings in this test at the .05 significance level. Therefore, $P < .05$ will mean we accept the hypothesis that there is a relationship between time spent and widgets produced, whereas $P > .05$ will mean we accept the null hypothesis that there is no relationship between the dependent and independent variables.

1. Start R-Project or R-Studio, depending on which interface you like better.
2. Check your working directory by typing "getwd()" and pressing Enter.
3. Change your working directory to one where you can access the data file we will create in the next step, or make sure you can save to the directory listed in step 2.
   To switch to the Windows desktop, for a user named Sam that we are logged in as, type "setwd("c:/users/sam/desktop")" and press Enter. Next you should retype "getwd()" and check that the directory is correct.
4. Create the following data in Notepad or another text editor and save to either your desktop or your working directory.
5. After creating the file and saving to the desktop or your working directory, type "list.files()" and press Enter to see your files, as shown in **Figure 9.6**.
6. Import the file into R-Project by entering the following:

```
mydata=read.table("data.txt",header=TRUE)
```

Note that "header" means the first line is a title only. Also, our data already had spaces included; however, when using comma-separated values, add the following:

```
sep=","
```

```
Minutes Units
23 1
29 2
49 3
64 4
74 4
87 5
96 6
97 6
109 7
119 8
149 9
145 9
154 10
166 10
```

Figure 9.6 Raw data in text file.

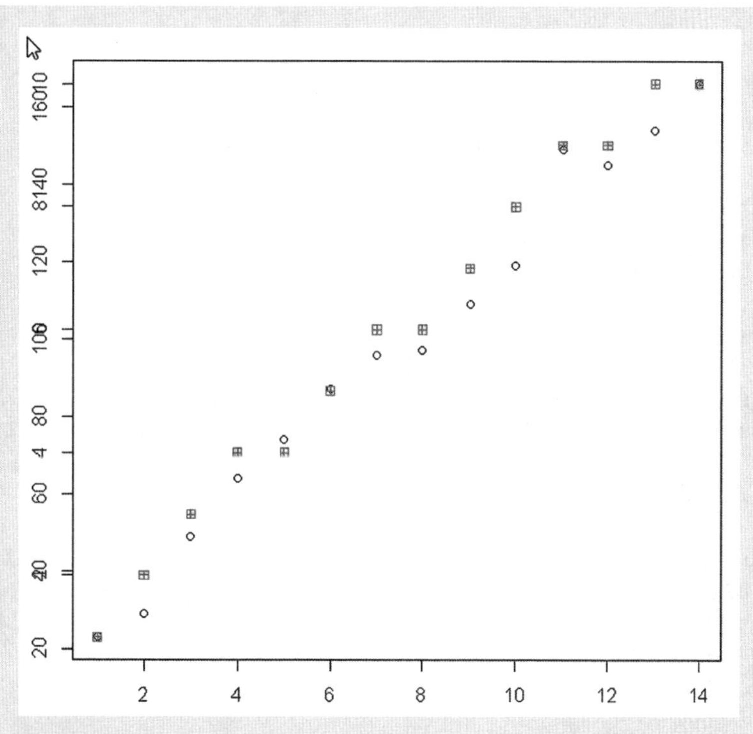

Figure 9.7 Plotting both on a scatterplot.

7. Check that it was read by typing "mydata" and pressing Enter.
8. Now, you can visually display a scatterplot in a few different ways (**Figure 9.7**):
   a. To show both sets of data in four panes, enter the following:

```
pairs(mydata)
```

   b. Or, to combine both in one graphic, follow these steps:
      i. Enter the following to plot the first column:

```
plot(mydata$Minutes)
```

      ii. Add the next plot to the existing plot instead of creating a different one by entering the following:

```
par(new=TRUE)
```

      iii. Plot the unit data as green squares by entering the following:

```
plot(mydata$Units, pch=12,col="green")
```

      iv. In the URL's listed, examine the following URL for: www.endmemo.com/program/R/pchsymbols.php for information on different types of symbols.
9. Next, create the regression analysis by completing these steps:
   a. To create a linear model of the data, enter:

```
mymodel=lm(mydata$Minutes~mydata$Units)
```

   b. To display the regression analysis, as shown in **Figure 9.8**, enter:

```
summary(mymodel)
```

   c. The ~ symbol is a *tilde*, located on the upper left of the keyboard. The variables are *mydata$Minutes*, which is the *Y* or dependent variable, and *mydata$Units*, which is the *X*, or independent, or predictor, variable.

(continues)

```
> summary(mymodel)

Call:
lm(formula = mydata$Minutes ~ mydata$Units)

Residuals:
    Min      1Q  Median      3Q     Max
-9.2318 -3.3415 -0.7143  4.7769  7.8033

Coefficients:
             Estimate Std. Error t value Pr(>|t|)
(Intercept)     4.162      3.355    1.24    0.239
mydata$Units   15.509      0.505   30.71 8.92e-13 ***
---
Signif. codes:  0 '***' 0.001 '**' 0.01 '*' 0.05 '.' 0.1 ' ' 1

Residual standard error: 5.392 on 12 degrees of freedom
Multiple R-squared:  0.9874,    Adjusted R-squared:  0.9864
F-statistic: 943.2 on 1 and 12 DF,  p-value: 8.916e-13
```

Figure 9.8 Output from regression analysis.

Note the *P* value is 8.916e-13, which is scientific notation for *P* = .0000000000008916, a number much lower than .05, so you can accept the hypothesis that there is a relationship.

With scientific notation, if the exponent, in this case −13, is negative, you move the decimal point to the left 13 places, padding with zeros. If positive, move to the right. If zero, leave it alone.

DID YOU KNOW? Large numbers are often displayed in **scientific notation**. This format is used to display a number by replacing part with E+n, where E (exponent) multiplies the preceding number by 10 to the *n*th power. For instance, two-decimal scientific format would display the number 12345678901 as 1.23E+10, which is 1.23 times 10 to the 10th power. To convert the number from scientific notation back to normal, move the decimal point 10 spaces to the right: 12300000000. You could also key in 1.23 on a calculator and use the Exp (exponent function) with 10 for an approximate result. Microsoft Excel and some other applications often can display only 15 digits of precision. If a number is larger, the number may be displayed in scientific notation.

Now you know!

Hands-on Statistics 9.4: Built-in Datasets in R-Project

R-Project has many built-in datasets useful for testing and experimentation. To see all of the datasets available, follow these directions:

1. Type "data(package = .packages(all.available = TRUE))" and press Enter.
2. Look in the boot dataset and find the dataset "*breslow, Smoking Deaths Among Doctors*."
3. Load the package by typing "library(boot)" and pressing Enter.
4. Display the data in the specific dataset by typing "breslow" and pressing Enter. Of course, you would reference the data as in previous examples. Some datasets are available, though, without the need to import a library, such as the dataset on the geyser Old Faithful.

Hands-on Statistics 9.5: Predict Future Values Using R-Project

To predict some data with R-Project, follow these steps:

1. Type "faithful" and press Enter at the R-Project prompt. Note there are 272 records in the Old Faithful geyser dataset. If you had a text file, as in the previous example, and would like to try this exercise with different data, the file would be similar to the following data:

Eruptions	Waiting
3.600	79
1.800	54
3.333	74

   You will use this dataset to predict some values that are not present; note the names of the column heads, because you will need to use them in a moment.
2. Create a linear regression model by using the lm function, as you did previously. To make the "faithful" dataset available so you can reference variables in it directly, type the following and then press Enter:

   attach(faithful)

   modelfaithful = lm(eruptions ~ waiting)

   Use either <- or = as an assignment operator, and use = as a named-parameter specifier for functions.
3. To predict a range for how long the eruption will last, assuming an 80-minute wait between eruptions, type "mydata = data.frame(waiting=46)" and press Enter.
4. Type "predict(modelfaithful, mydata, interval="predict")" and press Enter.
5. Take the "faithful" dataset out of memory by typing the following:

   detach(faithful)

Figure 9.9 shows an upper and lower range for the number of eruptions after a 46-minute wait with a default confidence level of 95%.

Note that there are differing opinions on whether to use "attach()" or not. If you did not use it, you would simply type the name of the dataset followed by "$variable," as in "faithful$eruptions." In any case, remember to detach the dataset when done to save memory.

```
> attach(faithful)
> modelfaithful  =  lm(eruptions ~ waiting)
> mydata=data.frame(waiting=46)
> predict(modelfaithful, mydata, interval="predict")
       fit       lwr      upr
1 1.60487 0.6195259 2.590213

> > detach(faithful)
```

Figure 9.9 Range of eruptions after a 46-minute wait.

DID YOU KNOW? Ted Kaczynski, also known as the "Unabomber," was considered a child prodigy who was accepted into Harvard University at age 16 and later earned a PhD in mathematics at the University of Michigan. Between 1959 and 1962 while a student at Harvard, Kacyzynski and 21 others were the unknowing subjects of an experiment by professor Henry Murray. The subjects knew they were being studied as they defended a discussion topic but did not know the other person was a professional debating expert charged to create extremely stressful situations so that Murray could research the subjects' reactions. The stress factors included abusive attacks and assaults to their egos. Students were not debriefed afterwards as to the real intentions of the study, and most experts feel that Kacyzynski's extreme dislike of authority and technological progress was strengthened by this traumatic event.

Now you know!

How Does Your Hospital Rate?

Joan begins her treatment with PVS-RIPO and is thrilled to learn that it has already begun to reduce the size of her tumor.

Consider the following:

1. What types of clinical trials are being conducted in research hospitals in your state or region?
2. What protections are in place to ensure the safety of the subjects?
3. What has been the outcome of the trials?

Global Perspective

Although the United States has a strict policy on what kinds of medical experiments can be carried out on US soil, many drugs used by Americans are validated for safety and effect by experiments conducted overseas. Since 2008, the NIH has reported 58,788 experimental **drug trials** involving humans in 173 countries, a 2,000% increase since 1990, according to an article in *Vanity Fair* (Barlett & Steele, 2011). Duff notes that there were "6,485 clinical trials overseas in 2008 on drugs intended for American use, 23 times more than the 271 in 1990. . . . Nobody keeps track of all those trials" (Wilson, 2010). Karen Riley, a spokeswoman for the Food and Drug Administration (FDA) notes that "some clinical trials are done overseas without being submitted to the drug review agency. But, once a company files a New Drug Application . . . Phase 2 and Phase 3 trials for that purpose must be submitted and older trials may be reviewed" (Wilson, 2010).

Often these trials are conducted in countries with large numbers of poor and illiterate people who may not be aware of the risks involved nor feel that they can refuse to participate. In some countries, an elevated view of authority figures, such as doctors, increases the coercion some subjects experience to participate.

While the United States hosts the highest number of clinical trials of any single nation, roughly half of clinical trials occur without US involvement (ClinicalTrials.gov, 2020). Indeed, the FDA (2019) notes that research conducted outside of the United States after 1986 must have data that "constitute valid scientific evidence (§860.7) and that the rights, safety, and welfare of human subjects have not been violated" and must meet the standards of the Declaration of Helsinki. Whether such laws are enforced, however, is questionable.

Chapter Summary

In this chapter you have been introduced to two general types of research and a framework to conduct your research. Information was also presented to support the need for oversight during research experiments. There is no doubt that medical research is important to the overall health of the world's population, but it must be tempered with safeguards that protect human subjects from unethical procedures and experiments. In the United States, research studies involving human subjects are under the strict guidelines of the IRB. Nonparametric statistics were examined, specifically the chi-square test of independence and the Mann-Whitney *U* test, as well as regression analysis—all important methods common in the field of research. Lastly, drug trials were discussed in the overall context of safe medical research.

Apply Your Knowledge

1. The National Institutes of Health offers a free certification in Protecting Human Research Subjects. Visit http://phrp.nihtraining.com/users/login.php and complete the free certification course. It can be finished in a few hours, costs nothing, and will be a great item on your resume to add to your professional credentials.
2. Research the University of Iowa "Monster Study," which took place in 1939. Write two to three paragraphs on the research that was being performed. What were the main ethical issues involved?
3. Research the 1940s US experiments with syphilis in Guatemala. Write two to three paragraphs on the outcome of the research. What were the main ethical issues involved?
4. Referring to the webpage provided by the US Department of Health and Human Services on the Belmont Report (https://www.hhs.gov/ohrp/regulations-and-policy/belmont-report/index.html), write two to three paragraphs on the boundaries between practice and research.
5. Referring to the NIH's ClinicalTrials.gov webpage on gliobastoma research (https://clinicaltrials.gov/ct2/results?term=glioblastoma+at+Johns+Hopkins+Maryland), review the results of the clinical trial that Joan was a part of. Did they have favorable results with their patients?
6. CITI is an organization that, like the NIH, offers institutional review board (IRB) certification and is as recognized as the National Institutes of Health (NIH) certification. There are two tracks you can take: "Social and Behavioral Sciences" or "Biomedical." Referring to the CITI website (https://www.citiprogram.org/), learn about this certification and consider obtaining it.
7. Using a timeline application, such as http://timeglider.com/, create a timeline for the events leading up to the Belmont Report. Include at least 10 data points. Feel free to use reasons not mentioned in the chapter, if applicable.
8. Create a timeline of events related to questionable medical ethics in the Unites States only, using the following website as a starting point: https://www.niehs.nih.gov/research/resources/bioethics/timeline/index.cfm. List 10 to 15 key events from the distant past (pre-1800 to today) that you feel are important.
9. Create a timeline of events related to questionable medical ethics involving African Americans in the United States or one featuring events that occurred from the start of the Cold War in 1945 to its ending in 1991.
10. Using the "Old Faithful" dataset, make predictions for 60 minutes and for 75 minutes.
11. Using another dataset built into R-Project, complete a simple regression analysis.
12. Write a brief analysis (one to two paragraphs) of a research study that is being conducted at a hospital in your state or region.
13. Create a data dictionary. If you were creating a patient data dictionary for a free clinic, what fields would you include? What would the data types be? Which would be optional and which required? What descriptions would each field have? What fields would be linked to a different database? How many databases would ultimately be needed?

References

Barlett, D., & Steele, J. (2011). Deadly medicine. *Vanity Fair*. Retrieved November 15, 2013, from http://www.vanityfair.com/politics/features/2011/01/deadly-medicine-201101

Brandt, A. (1978). Racism and research: The case of the Tuskegee syphilis study. *Hastings Center Magazine,* December.

ClinicalTrials.gov. (2020). Trends, charts, and maps. US National Institute of Medicine. Retrieved February 4, 2020, from https://clinicaltrials.gov/ct2/resources/trends#LocationsOfRegisteredStudies

Dickinson, A., Welch, C., Ager, L., & Costar, A. (2005). Hospital mealtimes: Action research for change? *Proceedings of the Nutrition Society*, *64*(3), 269–275.

Park, J. (2014). The living cadaver: Medical uses of brain-dead bodies. Retrieved August 20, 2014, from https://s3.amazonaws.com/aws-website-jamesleonardpark—freelibrary-3puxk/CY-LCADA.html

Smith, A., & Smith, W. (2014). *The Dead Don't Like to be Forgotten: Tales of the South Brunswick Islands*. Morrisville, NC: Lulu Press.

Staton-Tindall, M., Harp, K., Minieri, A., Oser, C., Webster, J., Havens, J., & Leukefeld, C. (2015). An exploratory study of mental health and HIV risk behavior among drug-using rural women in jail. *Psychiatric Rehabilitation Journal, 38*(1), 45–54. doi:10.1037/prj0000107

US Department of Health and Human Services. (2019). Code of Federal Regulations. Title 21—Food and drugs: Part 56—Institutional review boards; subpart B—organization and personnel. Retrieved July 21, 2019, from http://www.accessdata.fda.gov/scripts/cdrh/cfdocs/cfcfr/cfrsearch.cfm?fr=56.107

US Food and Drug Administration. (2019). PMA clinical studies. Retrieved October 15, 2019, from https://www.fda.gov/medical-devices/premarket-approval-pma/pma-clinical-studies

Wilson, D. (2010, December 2). 6,485 Overseas clinical trials and counting. *The New York Times*. Retrieved from http://prescriptions.blogs.nytimes.com/2010/12/02/6485-overseas-clinical-trials-and-counting/

World Medical Association. (2018). WMA Declaration of Helsinki ethical principles for medical research involving human subjects. Retrieved June 9, 2018, from https://www.wma.net/policies-post/wma-declaration-of-helsinki-ethical-principles-for-medical-research-involving-human-subjects/

Web Links

Statistic Brain Research Institute: http://www.statisticbrain.com/

The Belmont Report: http://videocast.nih.gov/pdf/ohrp_belmont_report.pdf

SynDaver: Beyond Human: http://syndaver.com/

Wilcoxon Rank Sum and Signed Rank Tests: http://astrostatistics.psu.edu/su07/R/html/stats/html/wilcox.test.html

Eight Shocking Facts About Sterilization in US History: http://mic.com/articles/53723/8-shocking-facts-about-sterilization-in-u-s-history

Hippocratic Oath: http://en.wikipedia.org/wiki/Hippocratic_Oath

US Slave (blog): http://usslave.blogspot.com/2011/05/dr-j-marion-sims-medical-experiments-on.html

Ethics and Experimentation on Human Subjects in Mid-Nineteenth-Century France: The Story of the 1859 Syphilis Experiments: http://www.ncbi.nlm.nih.gov/pubmed/12955963

Top 10 Evil Human Experiments: http://listverse.com/2008/03/14/top-10-evil-human-experiments/

Alliance for Human Research Protection: http://www.ahrp.org/history/chronology.php

The Establishment of Institutional Review Boards in the US Background History: http://www.iupui.edu/~histwhs/G504.dir/irbhist.html

NIEHS Institutional Review Board: http://www.niehs.nih.gov/about/boards/irb/index.cfm

The Revision of the Declaration of Helsinki: Past, Present, and Future: http://www.ncbi.nlm.nih.gov/pmc/articles/PMC1884510/

45 CFR 46: http://www.hhs.gov/ohrp/humansubjects/guidance/45cfr46.html

USC Libraries: Research Guides: http://libguides.usc.edu/content.php?pid=83009&sid=818072

Clinical Trials (NHS): http://www.nhs.uk/Conditions/Clinical-trials/Pages/Definition.aspx

ClinicalTrials.gov (NIH): https://clinicaltrials.gov/ct2/results?term=glioblastoma+at+Johns+Hopkins+Maryland

Revolutions: Cheat Sheet for Prediction and Classification Models in R: http://blog.revolutionanalytics.com/2012/08/cheat-sheet-for-prediction-and-classification-models-in-r.html

dataMeta: Making and Appending a Data Dictionary to an R Dataset: https://cran.r-project.org/web/packages/dataMeta/vignettes/dataMeta_Vignette.html

CHAPTER 10

Data Collection and Reporting Methods

Believe you can and you're halfway there.

—Theodore Roosevelt

CHAPTER OUTLINE

Introduction
Descriptive Research and Information Collection
Sampling
- Nonprobability Sampling
- Probability Sampling
- Random Numbers and Random Sampling

Quality Question Design for Data Collection
- Bloom's Taxonomy and Question Design
- Guidelines for Question Writing
- Types of Questions

Types of Studies
- Longitudinal Study
- Case Study
- Documentary Study

Data Collection Methods
- Survey
- Interview
- Questionnaire
- Observation
- Appraisal
- Survey Tools
- Pivot Tables
- Multiple Regression Analysis

Global Perspective
Chapter Summary
Apply Your Knowledge
References
Web Links

LEARNING OUTCOMES

After completing this chapter, you should be able to do the following:

1. Identify the different approaches to sampling and determine appropriate sample sizes.
2. Demonstrate random sampling using Excel and R-Project.
3. Describe different types of data collection questions and explain how to design a quality question.
4. Analyze Likert results with R-Project.
5. Describe different types of research studies and the purposes of each.
6. Identify common data collection methods and survey tools.
7. Demonstrate how to use SurveyMonkey to create a survey.
8. Demonstrate how to use PivotTable in Excel.
9. Demonstrate multiple regression analysis using Excel.

KEY TERMS

Appraisal
Case study
Cluster sampling
Cohort study
Convenience sampling
Descriptive research
Documentary study
Forced Likert scale
Generalizability
Health care appraisal (HCA)
Information collection
Instrument
Interview
Likert scale
Longitudinal study
Multistage sampling
Network sampling
Nonprobability sampling
Observation
Panel study
Pivot table
Probability sampling
Questionnaire
Rating scale
Sample frame
Sampling bias
Selective sample
Simple random sampling
Snowball sample
Social exchange theory
Stratified sampling
Survey
Systematic random sampling
Taxonomy
Trend study
Visual analogue scale (VAS)
Volunteer sample

How Does Your Hospital Rate?

Lori is in charge of marketing for a hospital and must create a dashboard on the hospital website and an advertisement for a highway billboard comparing critical hospital information that will persuade potential customers to choose her hospital for their health care needs. She also plans to post this information on some social media sites. Lori has developed a three-step plan to accomplish this assignment: (1) determine the data needed, (2) collect the data, and (3) summarize and distribute the data to those who will post the data.

Data she would like to publicize include the following: patient satisfaction scores, emergency department (ED) wait times, cleanliness ratings, infection rates, and comparisons of key data with nearby hospitals and with national averages. Of these, Lori can easily obtain ED wait times, infection rates, and cleanliness ratings from existing local sources. Patient satisfaction data, however, she must collect herself. She also needs some national data for her project.

Consider the following:

1. What other measures would be useful for Lori to collect data on in her effort to promote her hospital?
2. What are some different ways Lori could collect data regarding patient satisfaction with her hospital?
3. What websites should she consult to obtain national data related to the measures she has outlined?

Introduction

The collection of data is critical for any large organization that is interested in being successful. Healthcare organizations, in particular, rely on data such as patient satisfaction surveys, wait times, readmission rates, and financial costs and savings to evaluate their performance and plan for continued success. In fact, patient health itself, which is the primary concern of health care, is directly impacted by data quality and analysis. Therefore, knowledge of this topic is essential to you as a healthcare professional.

This chapter will introduce you to different types of data collection as well as methods and procedures to collect information for various audiences. Study design and process considerations will be examined to help you accurately collect and analyze your findings. Because the concepts presented in this chapter build on those presented in previous chapters, feel free to use the methods you learned about in previous chapters to summarize, analyze, and display your data to the user.

Descriptive Research and Information Collection

As we learned in the previous chapter, an ethical approach to research is essential to the protection of human subjects. This ethical approach must extend to the collection and reporting of data related to a study. Moreover, research that involves data collection requires different ethical standards than does mere information collection. To better understand the differences between these two concepts, we will consider their definitions.

Information collection is the gathering of data for nonresearch purposes, such as for records, reporting (perhaps to an advisory board), or quality improvement. Such data could include patient wait times in the ED, the number of sexually transmitted infections (STIs) reported to a state health board for the month, or the quality of food in the facility cafe. Even data gathered for these purposes, however, must not be linked to individuals' identities, but must be anonymous.

Descriptive research, in contrast, involves gathering data to answer a research question, such as whether different age groups experience vastly different satisfaction rates or whether STI rates correlate with ethnicity.

If you were to give a standardized test to select health information technology (HIT) students across the United States on a specific day (all students in a certain year of study), and you examined the mean, median, and standard deviation, you would be conducting descriptive research. However, if you merely summarized scores from the students to review the quality of student learning, you would just be collecting information. The descriptive study, then, is one that involves deliberate data collection, data preparation, and, finally, data summarizing and formal reporting. Batten (1986) notes that descriptive research "is a description and interpretation of what is;" descriptive research explains and summarizes the data you are examining.

Sampling

Previous chapters have discussed the sampling concepts of population and sample as terms to describe the entire population in question and the smaller subset (the sample) used to make generalizations about the population as a whole. Another important sampling concept is **sample frame**, which is all of the people in your population you could sample. The key here is *could* sample, meaning you will probably not sample all of them in the frame, just a smaller subset.

For example, imagine you own a car dealership. All of the people who come to the lot and look at the cars and trucks are the population. Some 3,000 of these agree with the salesperson to give their home contact information. This group, then, is the sample frame. To learn how people like the new automotive lineup on your lot, you decide to call every 10th person on your list of 3,000 to conduct a survey. This group of 300 selected from the sample frame is the sample.

Nonprobability Sampling

Nonprobability sampling and **probability sampling** are two general sampling approaches you should be familiar with. Nonprobability sampling is less rigorous in terms of validity and uses data samples not based on random selection.

For example, if you stood on a busy New York City street corner and asked everyone who walked by you to answer a short three-question survey, that would really not be a random selection of the population, as far as your selection of the participants as a researcher. The people who chose to answer would be random, but your sampling design had no influence on this.

Convenience sampling, or haphazard sampling, is based on participants who are easy to reach, such as members of an audience or passersby on the street. The problem with this method is that there is no way to tell whether the group is a good example of the population at large.

A **volunteer sample** is composed of people who choose to be a part of the sample based on their interest, such as listeners of an internet radio station who participate in an online poll. In this case, your sample is chosen by the interested listeners themselves, and you have no real ability to select participants as a researcher.

If you have ever lived where it snows or have watched the Weather Channel, you will quickly grasp the idea of a **snowball sample**. As rolling a snowball in the snow causes it to grow larger and larger, the nonprobability technique of snowball, or chain-referral, sampling works to increase the size of the population under examination. With this technique, you gather information on your population as best you can, but you enlist the help of the people you have contacted to refer other members of the group you are examining. Think of how many times you have received an offer to get a discount or gift if you "refer a friend" and you will have the general idea.

Similar to the snowball method, **network sampling** uses social or other network connections to help find participants. Examples of these groups include people who always drink tea together at 4 pm, people in a certain workplace setting, a support group, or farmers who use a certain brand of tractors, etc. For example, if you were to approach people who shoot rifles at least twice a week at a firing range and ask them questions about hearing loss, then you would have a network sample.

Probability Sampling

Purposive, judgmental, subjective, or selective are all names for sampling that is the opposite of convenience sampling. The term **selective sample** describes a sample in which you select the subjects you will investigate. The goal in selecting a sample is **generalizability**, which is the ability of the sample to accurately represent the entire population you wish to examine and thus to yield the same findings as if you could study the entire population.

The opposite of generalizability, and a condition you should avoid, is **sampling bias**, which happens when the people who are selected or respond are not a representative sample of the entire population. For example, if you surveyed people leaving a grocery store about the freshness of the meat department, it could well be that the only people to answer the survey were people wanting to be helpful, who might be more inclined to give a positive response. This scenario would bias the results and not help in improving the meat department.

The main advantage offered by probability sampling methods is that these methods offer a high degree of reliability that the sample in question is a good approximation of the entire population. This reliability helps to support the validity of your findings, or at least ensure that your findings were based on a valid sample. Key probability sampling methods include the following:

- Simple random sampling
- Systematic random sampling
- Stratified sampling
- Cluster sampling
- Multistage sampling

Simple random sampling methods use a sample, chosen by various random means, that represents the entire population under examination. As mentioned in previous chapters, a capital N represents the entire population and a lowercase n represents the sample of that entire population.

One way to obtain random samples from a population is the *lottery method*. Given that all participants are numbered $1 - N$, you could put tickets with these numbers in a hat and have someone draw one or more without looking. The people whose numbers are drawn—such as 11, 5, 7, and 45—would be the sample.

A variation of the lottery method is **systematic random sampling**, in which you have a numbered list with all of the members of your population and you randomly select numbers from it, such as every third person, or whatever value you choose.

With **stratified sampling**, you establish groups, or strata, of the population, based on a predetermined characteristic, and then select a random sample from each group. For example, to obtain data about people in a state, you might break the areas into *coastal*, *middle state*, and *western* regions. Each region then would be an area from which you would randomly select a certain number of participants.

Another useful sampling method that involves groups is **cluster sampling**. With this method you create groups, or clusters, that contain members of the population under study and then randomly select some of the clusters as your sample to survey. For example, if you had 100 people, you might have clusters of every 10 people and then randomly choose some of the groups of 10 to be your sample.

Lastly, you may determine that using several of the sampling methods listed would be the best way to identify your sample, which is known as **multistage sampling**. Again you are trying to eliminate bias and select a valid random sample with any of the methods.

How Does Your Hospital Rate?

In collecting hospital data for her marketing effort, Lori contacts the health information management department at her hospital for data on ED wait times and infection rates (broken out according to areas such as surgical, central line infections, etc.). For cleanliness data, she visits Medicare.gov to compare hospitals. (To follow along, visit the following website: http://www.medicare.gov/hospitalcompare/search.html.) After entering her zip code in the *Find a hospital* search box, she selects her hospital and then chooses the *Survey of patients' experiences* tab. The results for her hospital's survey on the question *Patients who reported that their room and bathroom were "Always" clean* show that her facility is several percentage points higher than the state and national averages. There are other tabs, as well, that have valuable patient satisfaction data. All of these facts she knows could be used in her marketing materials.

Lori also needs information on the expansion of parking and other facility renovations in recent years. Because this information is nonstandard and would not be on a hospital quality comparison site, she decides to conduct a survey.

Consider the following:

1. What type of survey would you recommend that Lori set up?
2. What are some different ways that she could administer the survey?
3. How might she go about analyzing the results of the survey?

Random Numbers and Random Sampling

Random sampling requires use of a method of generating random numbers. There are two types of random numbers: pseudorandom numbers and real random numbers. Pseudorandom numbers are not truly random and are used in computer programming to test programs that require random numbers. They are good for testing because they are the same random pattern each time and thus are not really random in the pattern of numbers they generate. However, with programming, a consistent pattern of numbers that does not change, when testing a program, is important. After testing is completed, real random numbers are then used. For most statistical purposes, real random numbers are appropriate.

Hands-on Statistics 10.1: Create a Random Sample Using Excel

Excel has two random number functions, RAND() and RANDBETWEEN(*X*,*Y*). The first generates a fractional random number between 0 and 1, and the second generates a whole integer value in the range you specify by *X* and *Y*. Seeding the random number generator is handled by Excel for you, so you will be generating real random numbers, and they will be changed every time you update the sheet (e.g., make a change), so be sure you note this behavior. In the following example, you will create a fictional group of possible participants (P1–P10) and then use Excel to randomly select some of them to become your sample frame of the entire population. Follow these directions:

1. Open a new, blank Excel workbook.
2. In cell A3, enter "P1" for participant 1. Hover the mouse pointer over the lower right corner of the cell until the + icon appears. Then click and hold the left mouse button, drag the pointer down to cell A12, and release the mouse button. Note that Excel will autofill cells A4 to A12 with numbers P2 to P10, as shown in **Figure 10.1**.
3. In cell F3, enter the following formula:

   =RANDBETWEEN(1,3)

   This will allow Excel to generate a random number from 1 to 3. Once entered, use the same technique as in step 2 to autofill cells F4 to F12 with a random number for each member of the population (**Figure 10.2**).
4. Next, copy cells F3 to F12 by selecting the entire area and either selecting copy from the toolbar or entering CTRL-C on the keyboard. Then highlight B3 to B12 and select *Paste Values*. Make sure you paste only the values, though, and do not copy the formulas, because doing so will not solve the problem of sorting the results.

(*continues*)

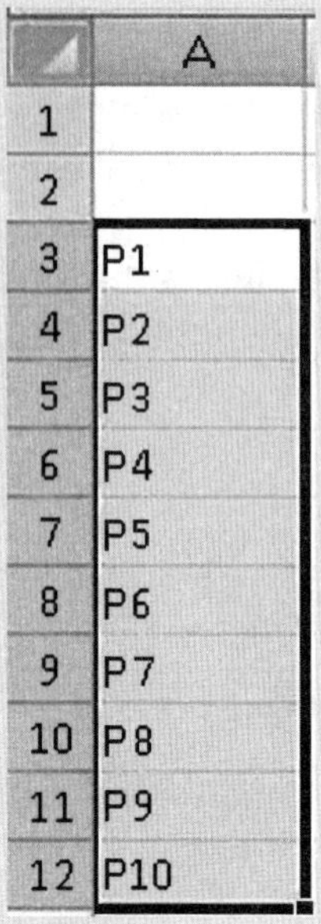

	A
1	
2	
3	P1
4	P2
5	P3
6	P4
7	P5
8	P6
9	P7
10	P8
11	P9
12	P10

Figure 10.1 Population members in Excel.

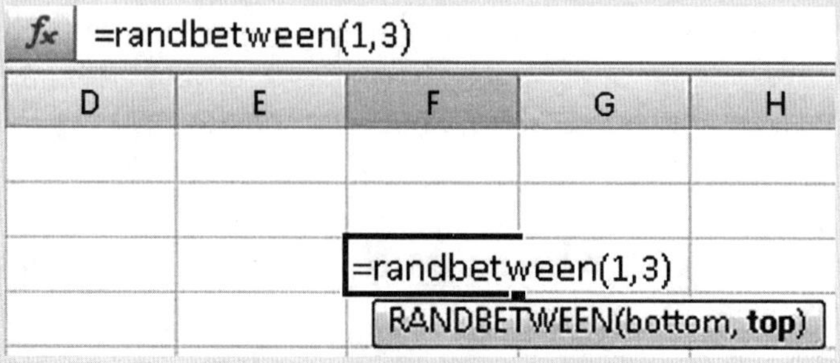

fx =randbetween(1,3)

D	E	F	G	H
		=randbetween(1,3)		
		RANDBETWEEN(bottom, **top**)		

Figure 10.2 Random function in Excel.

5. Select columns A and B and center the contents of these columns, so they are easier to read.
6. Highlight cells A3 to B12 and select the *Data* menu, *Sort*, and then *Sort Ascending* on column B smallest to largest. At this point, you have your sample frame (**Figure 10.3**), which would be the number you chose as the one to designate the participant (e.g., 1, 2, or 3).

Note in Figure 10.3 that if you had determined that *"all number twos"* would be the people selected, then P8 and P9 would be the ones you would contact. Of course, larger numbers of participants would be used, as the example is merely to demonstrate how the process works.

The aforementioned example with Excel used only real random numbers. To get more control over how random numbers are generated, you will need to use a more advanced software application, such as R-Project.

3	P1	1
4	P2	1
5	P4	1
6	P5	1
7	P7	1
8	P10	1
9	P8	2
10	P9	2
11	P3	3
12	P6	3

Figure 10.3 Participants of the sample frame.

Hands-on Statistics 10.2: Generate Pseudorandom Numbers Using R-Project

1. Start R-Project, and then from the interactive prompt (although a script would work as well), type the following function:

   set.seed(100)

   This function sets a random number generator "seed" to establish consistent pseudorandom numbers. If you did not use this command, you would generate real random numbers. However, by using this same seed value, in this case 100, you will generate the exact same sequence of random numbers each time you run R-Project.
2. Type "runif" and press Enter, and you will see the syntax to generate random numbers from a range of values. Key in the following to generate a fractional random number between 1 and 2,000 (**Figure 10.4**):

   runif(1, 1, 2000)

```
> set.seed(100)
> runif(1,1,2000)
[1] 616.2245
>
```

Figure 10.4 Setting a random number seed and generating a pseudorandom number.

3. Next, exit R-Project and do not save the workspace. Restart R-Project, and rerun the exact same *runif* command:

   runif(1, 1, 2000)

   Note you received a different random number (a real one in this case).
4. Now, set the seed back to the same value you used before by typing the following:

   set.seed(100)

5. Rerun *runif* by pressing the up arrow key a few times.

   runif(1, 1, 2000)

   You will see the exact same pseudorandom number you generated from the first R-Project session (**Figure 10.5**).

```
> runif(1,1,2000)
[1] 1516.63
> set.seed(100)
> runif(1,1,2000)
[1] 616.2245
>
```

Figure 10.5 Generating a real random number, and then resetting the seed to generate a pseudorandom number.

Note that you may generate a different value than what is shown in the figures, but the first session value will still be the same as your last generated value from this Hands-on Statistics exercise.

The previous example demonstrates how to generate fractional random numbers and how to work with pseudorandom and real random numbers. The next Hands-on Statistics exercise demonstrates the sample function in R-Project, which can return random whole number (integer) values or select random items from a list.

Hands-on Statistics 10.3: Sample Function in R-Project

1. Start R-Project. From the R command prompt, type the following function:

   sample(1:20,1)

   This function will generate a random integer number from 1 to 20.
2. To generate 10 random numbers that do not repeat from a range of 1 to 50, type the following:

   sample(1:50,10, replace=F)

3. To randomly select from a list of words, first create the list of words by typing the following function (you may choose any words you like to use in place of "cat," "dog," and "house"):

   a<-c("cat","dog","house")

4. Then, to randomly select a word from the list, type the following:

   sample(a,1)

   Of course, you may include as many words as you would like in the list and select as

(continues)

many words as you would like from the list. You could also generate pseudorandom responses by preceding the code with a seed value, as in the previous example.

Next, we will examine how to write questions that provide insight to your research question(s).

Quality Question Design for Data Collection

In addition to being able to generate random numbers to facilitate random sampling, you must be able to construct effective questions for collecting data. Noted early 20th-century educational reformer, psychologist, and philosopher John Dewey said, "A problem well-stated is a problem half-solved." This statement certainly applies to designing questions for data collection. If you do not state your questions well, you are not likely to answer them well. So, strive to design quality questions.

One consideration in question design is the level of information you hope to obtain with the question. Bloom's taxonomy is an effective tool for determining what intellectual level your questions should target. Another consideration is the type of question to write. Both of these considerations are discussed in the following sections.

Bloom's Taxonomy and Question Design

In 1956, researcher Benjamin Bloom headed a group of educational psychologists who developed a classification, or **taxonomy**, for levels of intellectual behavior important in learning as well as for teaching and assessing what was taught. The resulting model, known as Bloom's taxonomy, contains six levels of activity, which from the simplest to the most complex are as follows: remembering, understanding, applying, analyzing, evaluating, and creating. **Figure 10.6** illustrates this taxonomy, with the levels increasing in difficulty from the bottom of the pyramid toward the top of the pyramid.

When developing research questions, consider using this framework to help you write questions that address the appropriate intellectual level for the people you will be questioning and the information you will be gathering. For example, you might use an easy question level (Bloom's lowest level, remembering) if you are asking someone to recite statistics you told him or her regarding incidence rates of healthcare–acquired infections or to answer "yes" or "no": "How many infections are acquired by patients in Northside Hospital each year?" or "Have you ever had a loved one die while in the hospital?" In contrast, you would use the highest level of Bloom's taxonomy (creating) if you were asking people to design a complex system: "How would you go about designing an ideal health information system for the Hyperion Health Care system of hospitals?"

Guidelines for Question Writing

Besides Bloom's taxonomy level, you should also consider the wording, clarity, and tone of your questions. Batten and others offer the following guidelines for writing effective questions:

1. Word the questions clearly.
2. Address only one specific issue in each question.
3. Ask only questions the respondent can answer in a valid fashion.
4. Write questions relevant to your research question or questions.
5. Keep questions as short and to the point as possible.

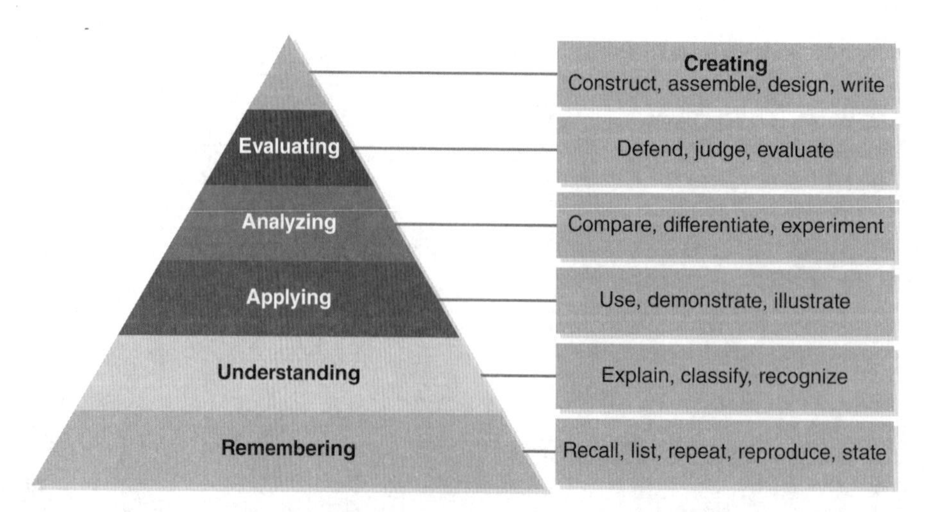

Figure 10.6 Bloom's revised taxonomy.

6. Avoid negative, biased, or leading questions.
7. Avoid slang, jargon, and technical terms respondents are not likely to understand.
8. Write response methods that you can analyze successfully.
9. When possible, sequence questions from general to specific.

Types of Questions

Next you should consider types or formats of questions, including the following. As a rule of thumb, revisit your research questions and the types of data you are trying to find to determine the best question, scale, format, and so forth for your instrument design.

Multiple Choice Questions

Multiple choice questions ask the respondent to choose from a selection of answer options, usually four or five. Typically only one option is correct, but the question may ask respondents to select more than one (e.g., "Select all that apply."). An example of a multiple choice question is as follows:

> To the best of your memory, how many times did you self-administer a dose of medication using your patient-controlled analgesia device during your hospital stay?
>
> a. None
> b. 1 to 5 times
> c. 6 to 10 times
> d. More than 10 times

Rating Scales

Rating scales enable the respondent to select a value from a continuum of possible answers, such as "least" to "most likely" or "no pain" to "worst pain." You have probably seen these in surveys in which you are asked to rate your satisfaction on a scale from 1 to 5, where 1 is "not satisfied" and 5 is "very satisfied." Some of these give you a neutral, middle option of something like, "no opinion," whereas others do not. One such scale is the **Likert scale**, along with its variation, the **forced Likert scale**. Named after Rensis Likert, this scale typically gives respondents between four to seven options.

For example, a five-item Likert scale may resemble the following:

> Make a selection based on how you would respond to the following statement: "The nursing staff responded to my expressed needs in a timely manner."

Strongly agree	Somewhat agree	Neither agree nor disagree	Somewhat disagree	Strongly disagree

And here is an example of a four-item forced Likert scale:

> Make a selection based on your response to the following statement: "My surgeon thoroughly explained all aspects of my surgery to me before my procedure."

Strongly agree	Somewhat agree	Somewhat disagree	Strongly disagree

Of course there are many examples of such scales, including "good, neutral, bad," or "never, infrequently, sometimes, frequently, or always."

Another key scale with which you should be familiar is the **visual analogue scale (VAS)**. A popular example of this type of scale is the pain scale (**Figure 10.7**). In a VAS, levels on the scale are visually depicted for the respondent. Some researchers believe it to be more accurate than the Likert scale. Probably its greatest advantage, however, is that it may be used with respondents who are unable to read or understand written scales, such as small children, people with mental disabilities, or people with dementia. The respondent can choose the graphic that best represents his or her response.

Short Answer Questions

Short answer questions pose open-ended questions that allow the respondent to answer in greater detail and depth. They can elicit good information, but their responses can be hard to quantify. There are no predetermined answers for these. Analysis requires either summary reading by researchers or the use of text analysis tools to determine patterns and key word matches. An example of this type of question is, "How was your stay at the hospital?"

Demographic Questions

Demographic questions garner information that allows you to compare responses of individuals within a population who share a common characteristic, such as age, gender, occupation, or income level. These questions help you compartmentalize the findings from your primary research question or questions. For example, you may be trying to determine how age correlates with soda consumption,

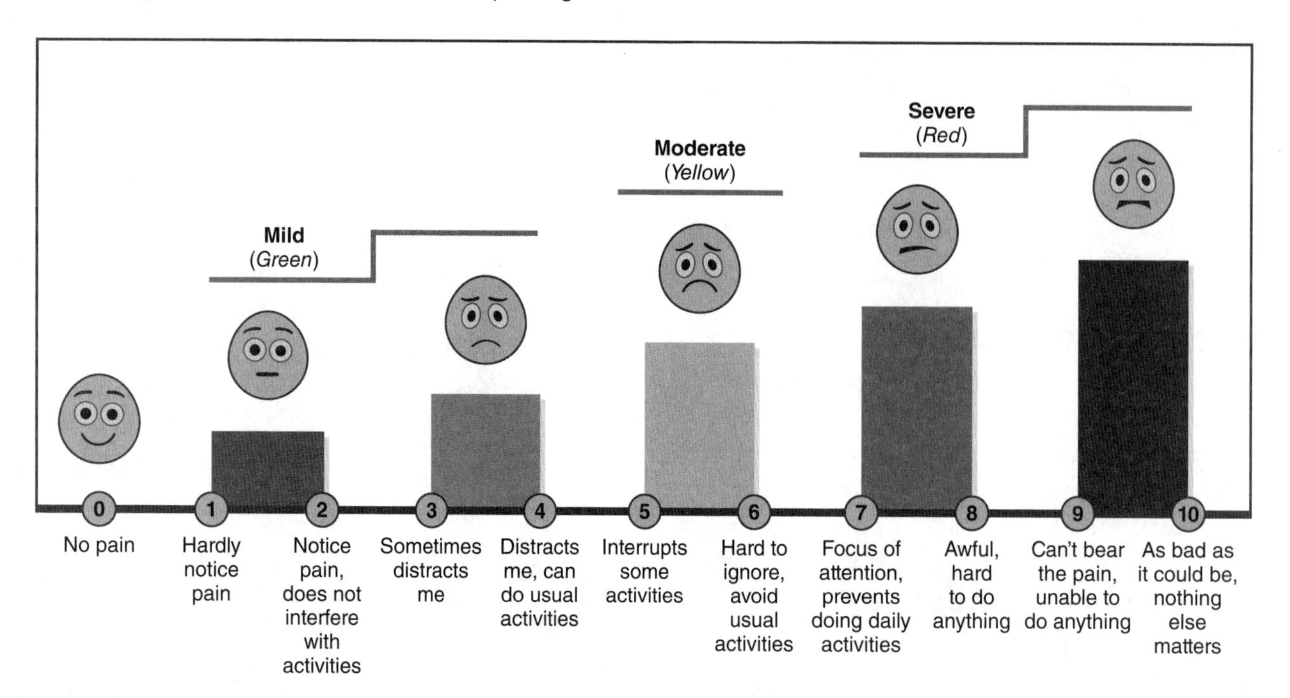

Figure 10.7 Defense and veterans pain rating scale.

Cleeland CS, Ryan KM. Pain assessment: global use of the Brief Pain Inventory. Ann Acad Med Singapore 23(2): 129-138, 1994.

so asking the respondents their age in addition to their soda drinking habits will help you with your final analysis: "Respondents in the age range of 18 to 25 years report drinking four sodas a day, whereas those over 25 years report drinking only two a day."

Hands-on Statistics 10.4: Tabulate Likert Data Using Excel

In this exercise, you will use Excel to examine Likert respondent results. You could also use R-Project or other tools to perform this task. Note that a regular Likert scale will be used here, with responses being correlated with the following numbers: Strongly disagree = 1, Disagree = 2, Neither agree nor disagree = 3, Agree = 4, and Strongly agree = 5. When the final analysis of each question is done, a higher number will suggest a higher level of agreement by the respondents.

1. Open a new, blank Excel workbook.
2. Enter the data as shown in **Figure 10.8**. Note that each possible surveyed respondent (10 people) has been assigned a number, with question numbers (five questions) noted by Q1, Q2, etc.

	A	B	C	D	E	F
1	Respondent	Q1	Q2	Q3	Q4	Q5
2	1	1	4	2	4	3
3	2	2	2	1	5	4
4	3	1	1	1	4	5
5	4	3	5	1	4	3
6	5	1	4	3	5	5
7	6	1	3	1	4	4
8	7	2	4	2	5	3
9	8	3	5	4	3	3
10	9	1	5	1	4	4
11	10	2	4	2	5	4

Figure 10.8 Using Excel to analyze Likert data: Part 1.

3. In cell B12, enter the following formula to find the mean:

=average(B2:B11)

Then click and drag the mouse pointer to the right to autofill C12 through F12 with the formula (**Figure 10.9**).

Note that for Question 1 (Q1), most of the respondents reported that they either strongly disagreed or disagreed, as indicated by the average of 1.7. This is what you might report, depending on your needs.

B12 | fx =AVERAGE(B2:B11)

	A	B	C	D
1	Respondent	Q1	Q2	Q3
2	1	1	4	2
3	2	2	2	1
4	3	1	1	1
5	4	3	5	1
6	5	1	4	3
7	6	1	3	1
8	7	2	4	2
9	8	3	5	4
10	9	1	5	1
11	10	2	4	2
12	Mean total	1.7	3.7	1.8

Figure 10.9 Using Excel to analyze Likert data: Part 2.

Types of Studies

Longitudinal Study

A **longitudinal study** is carried out over an extended period of time. Longitudinal studies can be broken down into three types: cohort studies, trend studies, and panel studies.

Cohort Study

A **cohort study** focuses on a specific population over an extended period of time. In ancient Rome, a "cohort" was 1 of 10 military divisions of a legion of soldiers, but for our purposes, a cohort is a group that has experienced the same event or events, whatever they may be. The key with a cohort study is that the same demographic group is examined several times over an extended period, not necessarily that the same individuals are examined each time. For example, a cohort study might focus on people who are baby boomers and who take a certain medication for hypertension. On initiation of the study, a group of people who fit this demographic would be examined. Then, a few years later, another group of people who belong to this demographic—not necessarily the same individuals—would be examined. Thus, cohort analysis involves an event and the variable under examination.

Trend Study

A **trend study** involves repeated surveying of a unique sample of the population, typically over a long period of time, with different groups of people in the same population and potentially with different researchers.

For example, a trend study might involve surveying nurses newly hired at a local hospital in the years 2016, 2017, and 2018 about their adjustment to their work environment. The respondents would be different for each year, but they would all be newly hired nurses.

Panel Study

A **panel study** is similar to a trend study but uses the exact same group of individuals—the panel—over the course of the study. Because the same respondents are surveyed each time, this type of study allows the researcher to find out why changes in the population are occurring. For example, a panel study might involve surveying a sample of HIT students regarding their use of an online tool several times over the course of study.

Case Study

A **case study** focuses on one individual or community rather than on a demographic group. A *prospective* case study is one in which the researcher establishes evaluation criteria first and then actively gathers data related to events as they occur. A *retrospective* case study considers data regarding events that have already happened, similar to historical research.

Documentary Study

A **documentary study** is a form of historical research that considers documents and stored records to provide data for the research. The researcher then performs content analysis on the collected data. You must take the proper precautions with your subjects and data and obtain any needed institutional review board approval for your work. (See Chapter 9, *Research Methodology and Ethics*, for more information on institutional review boards).

Data Collection Methods

Various tools, or **instruments**, are used to collect data for research purposes. Some of the most common types of survey instruments are discussed next.

Survey

As a nonexperimental and descriptive research method, the **survey** in its various forms allows collection of valuable data not otherwise readily observable. Surveys are a familiar feature in everyday life. For example, purchase receipts from grocery or discount stores often include links to online satisfaction surveys, sometimes with opportunities to win prizes or discounts for completing them. Surveys are also conducted via phone call polling regarding preferences of political candidates or consumer products. In health care, surveys are conducted to evaluate patient satisfaction with wait times in an ED, the quality of hospital food, or the timeliness of responses by nursing staff. Regardless of their content, all surveys collect summary data on a subject.

Typically, surveys are a collection instrument only. However, some groups (such as Google and Facebook) have used surveys to "educate" or bias the opinion of the survey respondent. An honest researcher would not pursue this technique.

Obtaining responses from the people surveyed is an important consideration in the design. For example, **social exchange theory** suggests that three elements must be present to maximize participation. They are as follows:

1. The time involved in responding to the survey must be minimal (e.g., tell them it is a 5-minute survey).
2. The reward for completion of the survey must be high (e.g., you are entered to win $1,000 upon completion).
3. The odds of winning the reward must be reasonable (e.g., 1 out of 5 respondents will get a new computer or other valuable prize).

Another approach to encouraging responses is to persuade people that participation in the survey is helping to address some significant social concern in the community, such as reducing illegal drug use. Thus, respondents may be motivated to participate due to a sense of altruism.

One popular type of survey is the Delphi survey, in which experts in a field are called on to help make judgments on an unknown topic or topics. These surveys might provide qualitative results as well as quantitative expert judgments. Typically used in situations in which there is a lack of knowledge about the subject, a Delphi survey initially can define the research questions and then, in successive rounds, answer these questions.

DID YOU KNOW? In 1400 BCE, the Oracle of Delphi in Greece was one of the most highly regarded shrines in the land. It was thought that the spring at Delphi was the center of the world and that knowledge flowed forth from it. People came from Greece and all over the ancient world to have their questions answered and futures predicted by the Pythia, a priestess of the god Apollo. From when to plant seedlings to questions of love and war, no questions were off limits, as long as the person asking the question had enough gold to offer the priestess. It is said that mind-altering gases emitted from the ground may have added to the cryptic messages the priestess gave, which often had ambiguous meanings.

Now you know!

Hübner-Bloder and Ammenwerth (2009) describe a great example of a Delphi survey used to determine quality indicators for hospital information systems and further provide judgmental ratings for each indicator. In this case, the Delphi survey was administered over three rounds to experts in hospital information systems. The first round was a qualitative round to determine what the performance indicators were, with the successive two rounds establishing quantitative values to attach to the specific indicators.

Interview

An **interview** is a data collection approach that involves the researcher meeting with the respondent, whether in person or remotely (e.g., via Skype, FaceTime, or other apps). Interviews are a familiar part of the job search process, in which the interviewer is in the role of the researcher and the job applicant is in the role of the respondent. Such interviews often include two or more interviewers asking 10 to 20 questions of the interviewee and rating the candidate's responses. Interviews may be conducted with a sample of people from the population. The responses from an interview are typically open-ended.

Questionnaire

A **questionnaire** is a data collection instrument that poses questions to the respondent via a paper form or website. It can be either closed or open-ended. Results from an open-ended questionnaire (as with the interview) are harder to tabulate, whereas those from a closed ("yes" or "no") instrument are easy to report numerically.

Observation

Observation is a data collection method in which the researcher watches events, behaviors, or physical characteristics and documents findings. The observation may be overt, meaning the subjects are aware they are being studied, or covert, meaning the subjects do not know they are being studied (as when the observer watches through a one-way mirror).

Appraisal

An **appraisal** is a unique data collection method in that it takes into consideration not only statistics gathered but also strengths and weaknesses of the study. As with surveys, appraisals are a common assessment method in everyday life. For instance, a home appraisal, which is an essential part of the home buying process, can determine the final price of the home. This appraisal is based on many factors, including the size, condition, and location of the home. No single piece of data should be excluded from consideration in the appraisal.

A **health care appraisal (HCA)** is similar to a home appraisal in many respects but is designed to assess an employee's health status. Factors that are evaluated include body weight, smoking and drinking habits, and work environment, all of which are considered in an overall appraisal of health. The employee's insurance co-pay then is tied to his or her health status, with lower rates associated with better health. Healthcare professionals should be aware of HCAs because they are increasing in popularity as healthcare costs to employers rise.

Survey Tools

SurveyMonkey is a popular, user-friendly online survey tool. It offers a free version with good features and an easy migration path to a paid version, which has additional features. The following Hands-on Statistics exercise provides a simple overview of how to use SurveyMonkey.

Hands-on Statistics 10.5: SurveyMonkey

This exercise provides a short introduction to the free version of SurveyMonkey. Although the paid version offers more features, the free version is still very useful.

1. Create a free account with SurveyMonkey by visiting the site's homepage (https://www.surveymonkey.com/).
2. Once logged in, click on *Create Survey* in the upper right corner.
3. Select *Start from an Expert Template*, then select *Healthcare* from the list that appears. Within the *Healthcare* menus, scroll down until you see *Diet & Exercise Template*; then click on *Use This Template*.
4. Close the Builder tip, and hover the mouse pointer over the first question. Click on the *Edit* tab, and then on *Edit* next to the question. Edit the question to read as follows: "How physically

(*continues*)

healthy do you think you are?" Ignore the warning that appears. Click *Save* at the bottom of the box.

5. For question 2, hover over it, select *Options*, check the box that requires respondents to answer the question, and save your choice as you did in step 4.
6. For question 3, hover over it and select *Logic*. Note that you can change the survey to end based on a response or skip to a certain question. For example, if you have questions that are for females only, you can adjust the settings so that the survey will skip ahead, bypassing questions only a female would need to answer, if the respondent indicates he is male. Click *Cancel* to close this question without making changes.
7. Click on *Preview & Test* in the upper right corner, and the preview window will appear. You can preview your test, which allows you to see it and take it as users would, clicking *Done* when you are finished. You can also click *Get Feedback* in the lower right and your default mail system will open, allowing you to send the preview to colleagues for review. Close the preview window.
8. Back in the *Diet & Exercise Template* window, select the tab *Collect Responses* in the upper right corner to send out your survey. We will use WebLink to email it out, but note you could use Facebook or put it on a website. Select *Next* on the lower right.
9. Type your organization name (use "Statistics Test" for this text) in the appropriate field, and click *Save*.
10. On the bottom left, click *Email*, and then type email addresses of respondents, separated by a comma. Note you are entering these manually, but you could have a predetermined list or import from your email address book.
11. Type a subject, and then edit the message as you see fit (or leave it as is for this test). Select *Next*. Verify your email, if requested, then click *Next* again.
12. Select *Send Now* to send the survey out. Note it is now scheduled to be sent out, but will not go out immediately. After it is sent, if you sent yourself a survey, check your email and complete your survey, and hopefully the other people you sent it to will also complete it.
13. After it has been completed by respondents, click the *Analyze* link on the top right. This feature will allow you to summarize the data you have collected from your respondents. Note that for multiple choice questions, a bar chart visually displays responses for each question, whereas for short answer questions, a text box displays what was entered (e.g., number of microwavable meals a day).
14. For the first question, click the *Customize* button and examine the different types of charts available to visually display the results for that question. Also note that *Export* features are available only if you upgrade to the paid version. However, you may use the Print Screen key on your computer's keyboard (PRTSC) to take a screen capture of the resulting graphic and incorporate it into a presentation or word-processing document.

Now that you have created a survey and found out how easy it can be to collect your data, in the next section, you will learn about a tool that will greatly help you in your data analysis tasks.

Pivot Tables

A **pivot table** is a tool found in spreadsheets that can sort, count, average, and summarize the data it references. Software developer Pito Salas originated the idea in the very popular spreadsheet application Lotus Improv in the mid 1980s. Generically it is called a pivot table, but Microsoft trademarked the name PivotTable to refer to their tool, and LibreOffice and OpenOffice named theirs DataPilot. All perform generally the same tasks, which allow the sorting and summarization of data in a spreadsheet. The rotation or pivoting of the data via dragging and dropping data with your mouse in the table is where the name pivot table comes from. Now that you know a bit of the history of a pivot table, use it in the next example with Excel to see its real power.

Hands-on Statistics 10.6: Exploring Data With an Excel PivotTable

In this simple example, you will learn the basics of how to use PivotTable.

1. Open a new, blank Excel workbook.
2. Enter the data shown in **Figure 10.10**.

	A	B	C	D	E	F	G
1							
2							
3	Date of Treatment	Physician	Treatment	Severity	Patients seen	Cost per visit	Total charges
4	2/1/2014	Dr. Smith	Sinus	L1	6	140	840
5	2/2/2014	Dr. Long	Check up	L3	2	200	400
6	2/3/2014	Dr. Smith	Check up	L3	5	200	1000
7	2/4/2014	Dr. Long	Back ache	L2	4	185	740
8	2/5/2014	Dr. Long	Sinus	L1	7	140	980
9	2/1/2014	Dr. Smith	Laceration	L1	2	175	350

Figure 10.10 Raw data for Excel PivotTable.

3. Highlight the data, select *Insert* from the toolbar, and then select *PivotTable*. A *Create PivotTable* window will appear. Click next to *Existing worksheet* and then on the button to the right with the red arrow icon. Move the small window out of the way and click on a cell below and to the right of the data area in the Excel worksheet. Click *OK* in the *Create PivotTable* window, and it will close.
4. Next you will design your PivotTable using the fields in the PivotTable options, which now appear on the right. Under *Field list*, to the right, select the fields shown in **Figure 10.11**.
5. All of the fields you just selected should now be in the area named *Row Labels*. Using your mouse, drag and drop certain fields to match **Figure 10.12**.

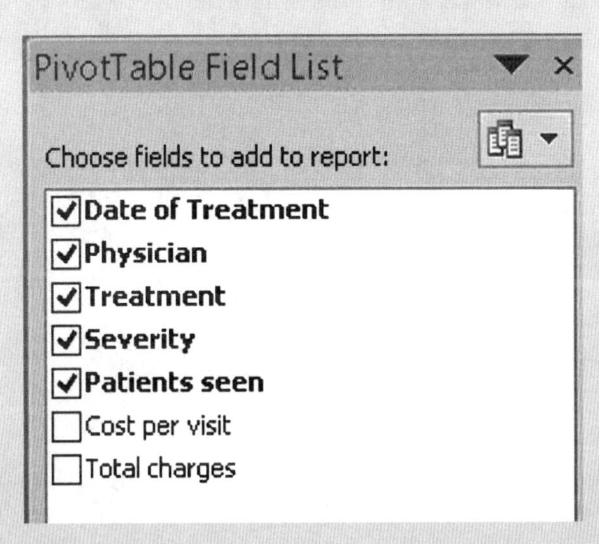

Figure 10.11 Fields to select in Excel PivotTable.

Figure 10.12 Sorting and dragging fields in Excel PivotTable.

6. The *Report Filter* will allow you to filter by all physicians or just one. Select the dropdown menu arrow in the *Physician* field in the worksheet to examine this feature (**Figure 10.13**). As for *Treatments*, you will see them across the top of the table, with *Row Labels* showing severity and count from the *Sum Values* list.

(*continues*)

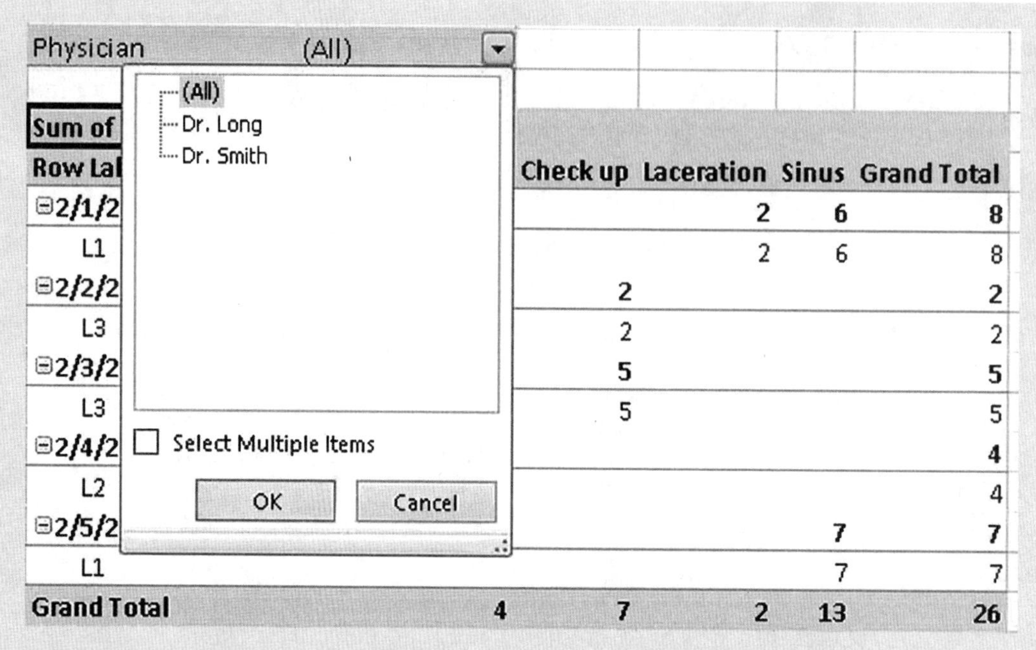

Figure 10.13 Selection of all or an individual physician in Excel PivotTable.

7. Using the dropdown menu of the *Row Labels* field, you may sort your output in various ways, as shown in **Figure 10.14**.

This exercise is but a sample of the capabilities of a pivot table. Regardless of whether you use Excel or another application, such as OpenOffice, this is a powerful tool.

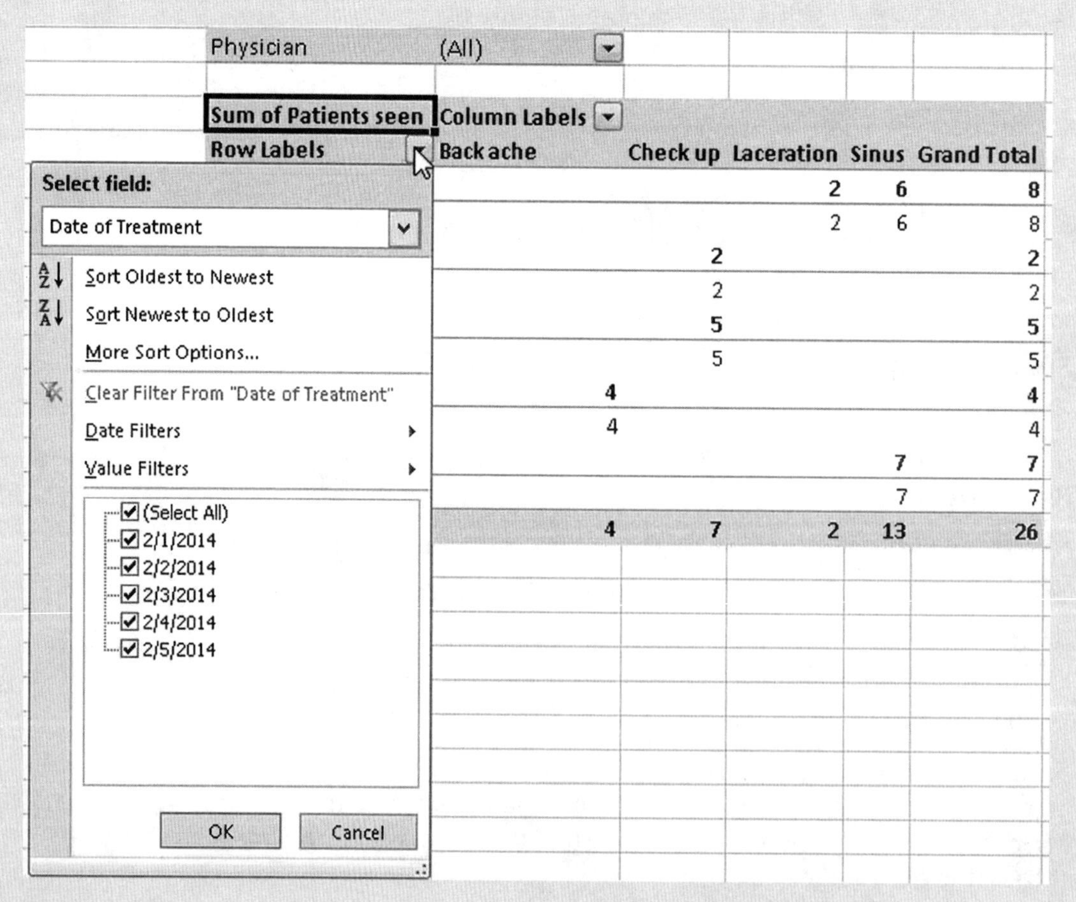

Figure 10.14 Sorting output in Excel PivotTable.

Multiple Regression Analysis

In the previous chapter, you learned about simple linear regression analysis, which involves only one independent variable and one dependent variable. Now you will examine multiple regression analysis, or multivariate regression analysis. With this statistical method, you may examine the relationship between a dependent variable and two or more independent variables. For example, a multiple regression analysis could predict the dependent variable y based on the independent variables $x1$, $x2$, $x3$, etc. Recall that an independent variable might be the amount of a drug that was given to a patient or the severity of the patient's condition, whereas a dependent variable might be the age of the patient.

An important consideration when performing multiple regression analysis is whether the data have a linear or nonlinear relationship to one another. If data can be graphed to a straight line, they are linear. If they have exponents or do not graph to a straight line, they are nonlinear. For example, suppose that in a study the variable y represents years of schooling and x represents earnings per year; if you find that y corresponds directly to x (that is, an increase in years of education always results in an increase in earnings per year), the data can be graphed to a straight line, and thus the relationship between y and x is linear. However, suppose that only the first 2 years of post-secondary education resulted in an increase in yearly income, meaning any years of education beyond that did not result in an increase in yearly income. In this case, the relationship between the variables is nonlinear, meaning it cannot be graphed as a straight line.

In the following Hands-on Statistics, we assume linearity. However, regression analysis with nonlinear data would require you to choose an option from your statistics package that allows for such data.

Various computer applications may be used to perform multiple regression analysis. One is Excel, which you will be using in the next Hands-on Statistics. Excel, however, can handle only up to 16 independent variables. If you have more, you would need a more robust application, such as R-Project, a third-party commercial add-in for Excel, or another statistical analysis application.

Hands-on Statistics 10.7: Multiple Regression Using Excel

Is there a relationship between the years of experience medical coders have, the hours they spend coding, and the revenue they generate? Follow along as we examine this question with a multiple regression analysis using Excel. Make sure you have installed the Add-in for the Data Analysis toolpack in Excel (completed in an earlier chapter of the text).

1. Open a new, blank Excel workbook.
2. Enter the data shown in **Figure 10.15**.

2	What relationship between years of coding, hours worked, and revenue?		
3	Revenue	Years coding	Hours spent coding
4	3200	8	700
5	3000	7	900
6	3500	9	800
7	4000	6	900
8	4700	8	880
9	5000	6	850
10	6000	5	1000
11	5500	6	1000
12	6500	7	950
13	7000	4	1100
14	4500	5	1200
15	7500	9	830
16	5750	3	1100
17	7250	6	1100
18	8000	8	1160

Figure 10.15 Data for Excel multiple regression.

(continues)

3. Highlight the range of data, including the column headings, and select the *Data* menu and then *Data Analysis*.
4. Select *Regression* from the dropdown list, and click *OK*. The *Regression* window will appear.
5. Click on the red button to the right of the *Input Y Range* field, and the *Regression* window will minimize.
6. Select the entire *Revenue* column in the Excel worksheet, from the heading to the last cell, and then click the red button again to maximize the *Regression* window. You will see that the range of cells you just selected now appears in the *Input Y Range* field, as shown in **Figure 10.16**.

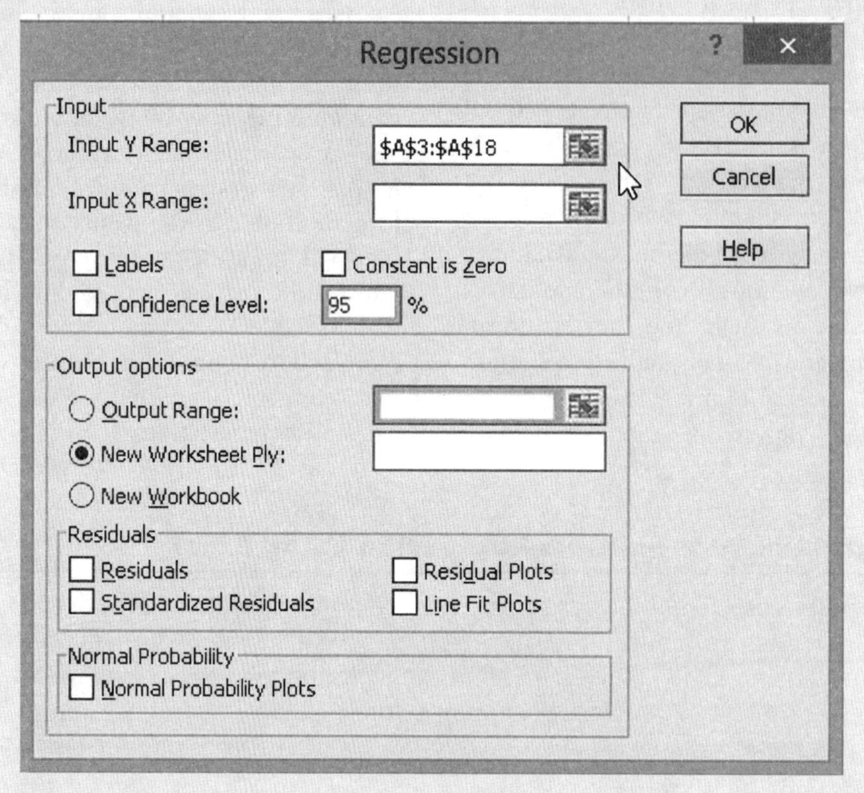

Figure 10.16 Selecting *Y* values in Excel multiple regression.

7. Next select the red button to the right of the *Input X Range* field, and select the values and column headings (cells B3 to C18) for the *X* values (**Figure 10.17**).
8. Click in the check box next to *Labels* to allow for labels in the first row (Figure 10.17).
9. Click *OK* and your results will appear on another worksheet. Resize the columns so they look similar to those in **Figure 10.18**.
10. Now you can analyze your findings. Working from the top down, the key aspects that you should be interested in are as follows:
    - The R Square value of .4309 indicates that the two *x* independent variables explain about half of the variation between revenue, years coding, and hours spent coding.
    - The *P* value for hours spent coding is .01 and for years coding is .24. Because the *P* value for both independent variables is < .05, there is 95% significance, which is good.
    - The positive coefficients of 317 for years coding and 8.17 for hours spent coding mean that for every year of coding experience, there is a 317-unit increase in revenue, and for every hour spent coding, there is an 8.17-unit increase in revenue. Therefore, the more years you code, the more revenue you will generate, and, to a lesser degree, the more hours you code, the more revenue you will generate. Thus, a coder with more years of coding experience will yield more revenue than one with fewer years of coding experience.

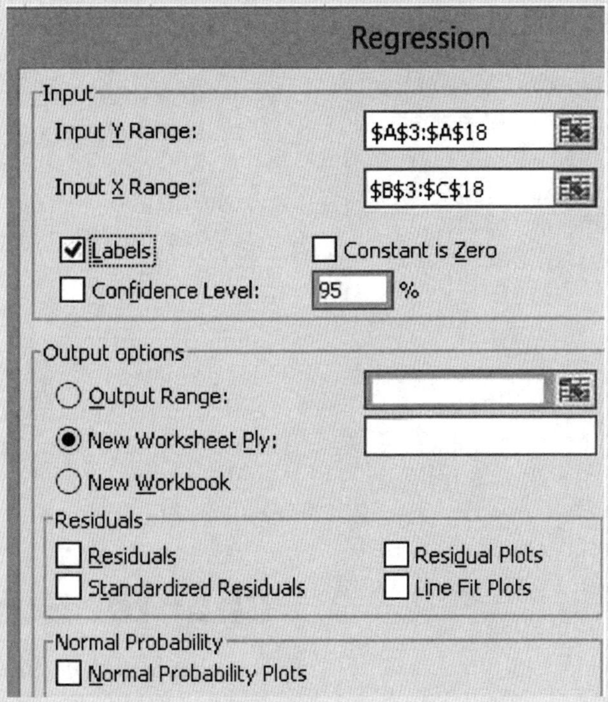

Regression

Input
Input Y Range: $A$3:$A$18
Input X Range: $B$3:$C$18
☑ Labels ☐ Constant is Zero
☐ Confidence Level: 95 %

Output options
○ Output Range:
◉ New Worksheet Ply:
○ New Workbook
Residuals
☐ Residuals ☐ Residual Plots
☐ Standardized Residuals ☐ Line Fit Plots
Normal Probability
☐ Normal Probability Plots

Figure 10.17 Selecting *X* values in Excel multiple regression.

	A	B	C	D	E	F
1	SUMMARY OUTPUT					
2						
3	*Regression Statistics*					
4	Multiple R	0.6564632				
5	R Square	0.430943932				
6	Adjusted R Square	0.336101255				
7	Standard Error	1440.085346				
8	Observations	15				
9						
10	ANOVA					
11		*df*	*SS*	*MS*	*F*	*Significance F*
12	Regression	2	18846183.7	9423092	4.543777	0.033957081
13	Residual	12	24886149.63	2073846		
14	Total	14	43732333.33			
15						
16		*Coefficients*	*Standard Error*	*t Stat*	*P-value*	*Lower 95%*
17	Intercept	-4302.816574	3786.167162	-1.13646	0.277952	-12552.16615
18	Years coding	317.0408371	258.3398983	1.227224	0.243271	-245.8334471
19	Hours spent coding	8.17530193	2.742251631	2.981237	0.011459	2.200448901

Figure 10.18 Summary output in Excel multiple regression.

Keep in mind that the data used above are hypothetical, and so the results are not necessarily applicable to the real world.

You should also consider that if the data were from a nonrandomized sample (e.g., it was all of the coders on staff, or the entire population under study), the results would not be as good a predictor of the population at large as if you had had a randomized sample of coders from a larger population. Remember this when planning your research design.

How Does Your Hospital Rate?

Now that Lori has gathered and analyzed all of the data she needs regarding the performance of her hospital, she is ready to publish it on social media. She knows that Johnson & Johnson and GlaxoSmithKline are but two of many pharmaceutical firms that extensively use social media.

Consider the following:

1. What social media approaches would you recommend that Lori use to get the word out about her hospital's performance?
2. What groups of people should she try to target in this effort?
3. What kind of ongoing maintenance will be required to ensure that her social media strategy will be effective?

Global Perspective

In a recent study, more than 40% of consumers surveyed responded that social media affects how they deal with their health care (Honigman, 2013). Users up to 25 years old are twice as likely to use social media to find health-related information as are those 45 to 54 years old. Ninety percent of respondents younger than 24 years noted that they would trust information shared on social media networks.

Yet, despite the potential influence of social media, only 31% of healthcare organizations had written guidelines in place for social media (Honigman, 2013). The Mayo Clinic is the healthcare organization with the largest social media following on Twitter, with over half a million followers. The Cleveland Clinic is has the third largest following on YouTube, with 3 million followers, and the University of Texas MD Anderson Cancer Center is 30th on Flickr (Rowe, 2013).

What do all of these statistics mean? Certainly younger users put more trust in social media, even if the sites—such as Twitter and Facebook—are not validated, as the Mayo Clinic is. Also, social media sites are a "go to" for healthcare information. There is no question healthcare facilities can leverage this marketing resource. Only the future can tell how these applications of social media will expand and change the face of health care.

Chapter Summary

This chapter has covered an array of topics related to data collection and reporting to equip you to gather, analyze, and report data. Nonprobability and probability sampling approaches were explained, and methods of generating random and pseudorandom numbers in Excel and R-Project were described. Considerations and guidelines related to designing quality research questions for data collection were covered, along with types of questions. Types of research studies and data collection methods were outlined. Tools for surveying and analyzing data were introduced, including SurveyMonkey and PivotTable, and Hands-on Statistics exercises for each tool were provided. Finally, multiple regression analysis was explained, and a related Hands-on Statistics exercise offered.

Apply Your Knowledge

1. List four examples of information collection. At least two should be related to health care. Support your choices with a reason or reasons each is an example of information collection.
2. List four examples of descriptive research. At least two should be related to health care. Support your choices with a reason or reasons each is a descriptive study.
3. For your medical facility, write a one-page narrative describing a hypothetical population, a sample frame, and a method to obtain data for a sample.
4. For one of your descriptive research examples in question 2, list considerations and instruments needed to collect your data.
5. Visit the following websites and list two items from each that would help to improve facility quality:
   - https://hcahpsonline.org
   - http://www.cdc.gov/nchs/nhcs.htm
   - http://www.leapfroggroup.org/cp
6. Visit the following web page from CMS.gov's QualityNet: https://www.qualitynet.org/#siteNavigation. At the middle bottom for *Hospital Outpatient*, list five reporting measures on

hospital quality. Use this URL for direct access: https://www.qualitynet.org/outpatient/oqr/measures.

7. Examine one of the following survey tools and prepare a short paper or presentation in which you inform the management of a hypothetical organization of the strengths and weaknesses of the tool and whether you recommend it for the organization:
   - Zoomerang
   - CrowdSignal
   - SurveyGizmo
   - SurveyPlanet
   - SoGoSurvey
   - YARP
8. Create a Likert scale for the following question: "How would you rate your treatment by the nursing staff at Orangeburg Hospital?"
9. Create a forced Likert scale for the following question: "Based on your recent medical procedure, rate on the scale how well you feel now."
10. Using any graphical tool of your choice, create a visual graphic for either of the scales you created in questions 8 or 9.
11. Using either clip art or a drawing package, create a VAS for assessment of pain for the dentist's office where you work. It should be designed for people with limited English proficiency and for children.
12. Create an Excel spreadsheet, with sample data, to find the mean score for 10 survey questions. Use a standard Likert scale.
13. Create an Excel spreadsheet, with sample data, to find the mean score for 10 survey questions. Use a forced Likert scale.
14. Create an Excel spreadsheet, with sample data, to find the mean score for 10 presurvey questions and 10 postsurvey questions. Use a standard Likert scale and conduct a *t* test to determine whether there is a significant difference between the pre- and post-test results.
15. Create a "Kwik" survey from www.kwiksurveys.com. List two pros and two cons about this website and survey tool.
16. Create a survey with Google forms by creating a free account at the following website: https://accounts.google.com/ServiceLoginAuth.
17. Create a PivotTable in Excel that includes at least five rows and three columns (including column headings) of numerical data.
18. Visit the following Medicare site on hospital satisfaction: https://www.medicare.gov/hospitalcompare/search.html. Compare two hospitals in your state. List at least two differences in satisfaction or quality.
19. Using OpenOffice or LibreOffice, recreate with DataPilot the table you created in the PivotTable exercise with Excel in question 17.
20. Create a multiple regression analysis on data of your choice and summarize the findings.
21. List two websites with information on social media use in health care.
22. Read the following article from *Becker's Hospital Review*, and summarize it in one page: http://www.beckershospitalreview.com/lists/20-hospital-and-health-system-leaders-to-follow-on-twitter.html.
23. Read the following article from *Imperial College London News*, and summarize it in one page: http://www3.imperial.ac.uk/newsandeventspggrp/imperialcollege/newssummary/news_1-5-2014-18-3-24

References

Batten, J. W. (1986). *Research in Education* (rev. ed.). Greenville, NC: Morgan Printers.

Honigman, B. (2013). Twenty-four outstanding statistics and figures on how social media has impacted the health care industry. *ReferralMD*. Retrieved May 21, 2014, from http://getreferralmd.com/2013/09/healthcare-social-media-statistics/

Hübner-Bloder, G., & Ammenwerth, E. (2009). Key performance indicators to benchmark hospital information systems: A Delphi study. *Methods of Information in Medicine, 48*(6), 508–518. doi:10.3414/ME09-01-0044

Rowe, J. (2013). Four deft ways hospitals use social media. *Healthcare IT News*. Retrieved July 18, 2013, from http://www.healthcareitnews.com/news/4-deft-ways-hospitals-use-social-media

Web Links

Library Guide to APA Style: Paper and Electronic: http://www.lonestar.edu/departments/libraries/nharrislibrary/APA_2010_final.pdf

Data Preparation: http://www.socialresearchmethods.net/kb/statprep.php

Back Pain Visual Analogue Scale: http://www.ericlinmd.com/back-vas-form.php

Welcome to the Wong-Baker FACES Foundation: http://www.wongbakerfaces.org

CHAPTER 11

The Future of Healthcare Statistics

When you meet someone who has more answers than questions, walk away, or better yet run!

—Jerome Witschger

CHAPTER OUTLINE

Introduction
The Future of Healthcare Statistics
 Radio Frequency Identification
 Automatic Medication Dispenser
Health Information Exchange
Efforts to Decrease Healthcare Costs
Time Series Analysis of Data
Forecasting Future Data
Project Management General Concepts
Analysis of Covariance
Coronavirus—The Next Pandemic?
Ebola
Global Perspective
Chapter Summary
Apply Your Knowledge
References
Web Links

LEARNING OUTCOMES

After completing this chapter, you should be able to do the following:

1. Discuss the future of health care and radio frequency identification technology.
2. Discuss costs associated with health care.
3. Demonstrate how to use time series analysis.
4. Demonstrate how to forecast data.
5. Discuss terms and concepts associated with project management.
6. Demonstrate the use of a Gantt chart.
7. Demonstrate the use of critical path method.
8. Demonstrate how to use Excel and R-Project for analysis of covariance.
9. Discuss global epidemic issues.

KEY TERMS

…dable Care Act (ACA)
… of covariance
…)
Computer-assisted coding software (CACS)
Consumer-mediated exchange
Critical path method (CPM)
Cyclical variations
Directed exchange

Forecasting
Gantt chart
Irregular fluctuation
Project management
Query-based exchange
Radio frequency identification (RFID)
Real-time locating (RTL) system
Seasonal fluctuations
Time series analysis

How Does Your Hospital Rate?

Sky Valley Hospital is looking at methods to reduce costs and minimize disruptions due to the implementation of International Classification of Diseases (ICD)-10. Currently Sky Valley uses traditional coders in the health information management (HIM) department. They have also reviewed findings from ICD-10 implementation in Canada, which indicate up to 50% losses in revenue and productivity due to the more detailed nature of ICD-10 codes and the time involved in getting current coders up-to-speed on the new methodology (Eramo, 2014). As such, Sky Valley has determined that changes need to be made in the HIM department related to coding practice and imminent ICD-10 implementation. It has been suggested that the HIM department provide an analysis of the feasibility of using **computer-assisted coding software (CACS)**, which analyzes healthcare documents and assigns diagnosis codes based on specific phrases and terms. CACS can reduce the cost of ICD-10 implementation due to loss of productivity and enhance the accuracy and speed of traditional coders.

Sarah, the director of the HIM department, has been asked to conduct this analysis. To help guide her research, she is concerned with answering the following questions:

1. How will doctors respond to a new system?
2. Will training be required for the doctors?
3. Will training be required for coders?
4. Will the role of a coder change?

Consider the following:

1. How will Sarah find answers to her questions?
2. What will she do with the information she finds?

Introduction

In the future, the healthcare world, including healthcare statistics, will be much more interconnected. You will be more aware of health care events in other countries and of the effects that diseases from other parts of the world could have on people in the United States. Data collection by the Centers for Disease Control and Prevention (CDC), National Institutes of Health, and other organizations, which is already staggering, will continue to increase due to more systems being converted to the electronic medical record and the ability to obtain data in real time. Patients will also be more connected with healthcare providers around the world. Someone in Alabama, for instance, will be able to connect remotely with a specialist in California. Additionally, expert systems such as CASC will increase in use to speed the overall medical coding and billing process and reduce patient costs.

The Future of Healthcare Statistics

In this section, we will consider two cutting-edge technological advances and changes in the overall US healthcare system that are likely to affect the future of healthcare and, thus, of healthcare statistics.

Radio Frequency Identification

Radio frequency identification (RFID) devices are devices that may be ingested or implanted in the human body and that transmit signals related to certain events occurring in the body. The first ingestible sensor, the ingestion event marker by Proteus Health, was approved by the US Food and Drug Administration in early August 2012. This device "is activated by stomach fluid, then transmits a signal through the body to the skin patch, indicating that a patient has ingested medication" (Versel, 2012).

Other purposes of implanted chips, RFID devices, and **real-time locating (RTL) systems** include patient tracking (such as with Alzheimer patients), billing, drug interaction data, and much more. Miliard, managing editor of *Healthcare IT News*, reported that these devices and the integrated systems that support them "have an even bigger role to play in the hospitals of the future" (Miliard, 2013). These devices will provide accurate and up-to-date patient information and increase the already vast amount of data we have available.

However, in the United States, some states have objected to this technology as being overly invasive and have not endorsed it. As with any technology, these devices are subject to abuse and may be used for controlling or improper purposes. When used properly, though, they can save lives and thus will likely play a significant role in the future of health.

Automatic Medication Dispenser

Another technological innovation addresses a common problem faced by older adults. As people grow older, they may find themselves on increasingly complicated medication regimens; moreover, their memories may not be as reliable as they once were. Thus, they may sometimes forget to take their medications or forget which medication they took or when they took it. Forgetting that one has already taken a medication and then inadvertently taking an extra dose can lead to a dangerous overdose and an emergency department (ED) visit.

E-pill or MedMinder automatic medication dispensers address this problem by automatically dispensing medications at prescribed times and correct dosages. The units have alarms and network connections for monitoring, and can even call a cell phone if the patient misses a dose. This emerging technology is improving how older patients receive their medications and how those monitoring them can stay informed.

Health Information Exchange

As of 2020, going to the hospital, getting access to your healthcare data online, and having the data available to your personal healthcare provider have become much easier than in previous years. Interestingly, health information exchange (HIE) has been implemented in some areas in the United States since the 1990s and as of the mid-2000s has become more common. According to the Office of the National Coordinator for Health Information Technology (ONC, 2019), "Electronic exchange of clinical information allows doctors, nurses, pharmacists, other healthcare providers, and patients to access and securely share a patient's vital medical information electronically." HIE, according to the ONC, can offer improved health care quality, more efficient health care, streamlined administrative processes, more engaged patients, and support for community health in general.

The concept of how to implement the exchange of patient information in a cost-effective manner has been daunting for healthcare providers. The main purpose of HIE is to facilitate and expand secure electronic movement of health information. As Dixon, Grannis, and Revere (2013) state, HIE "is the capacity to electronically transfer and share health information among healthcare-related stakeholders and organizations such as clinics, laboratories, payers, hospitals, pharmacies and public health." This process allows the flow of information among organizations based on nationally recognized standards.

The timely sharing of information enabled by HIE improves the overall quality of patient care and its cost effectiveness. HIE gives healthcare providers access to comprehensive health information instantly and allows patient access to their health information. HIE aids physicians in avoiding unnecessary readmissions, reducing medication errors, and improving diagnosis. In short, HIE provides a completeness of health information.

There are three forms of HIE: directed exchange, query-based exchange, and consumer-mediated exchange. The **directed exchange** is used to send information such as laboratory results, patient referrals, and discharge summaries directly to providers. The information is sent over the internet using encrypted software to ensure its security. Directed exchange is also used to send immunization data to the appropriate public health organizations for statistical purposes and to report quality measures to the Centers for Medicare and Medicaid Services.

Query-based exchange allows ED physicians to query information such as medications, X-ray images, and any problems listed. Easy access aids physicians in their decision process, reduces medication errors, and reduces duplicate testing, which allows the physician to make the best healthcare decision for the patient.

Consumer-mediated exchange allows patients to add information about themselves to improve their health care and to identify any missing or incorrect information. Patients also have the right to correct any billing mistakes and track their own health information.

One of the essential components of effective HIE is the electronic medical record (EMR). The use of EMRs reduces paperwork that can be lost in the shuffle of day-to-day operations and allows information regarding patient health to be accessed immediately in an emergency situation. It helps to coordinate the patient's care, reduces the cost of having unnecessary tests and procedures performed, and, most importantly, allows patients access to their own health information. HIE and the EMR have improved overall health care tremendously in the United States.

DID YOU KNOW? In the healthcare industry, you may hear both the terms *electronic health record (EHR)* and *electronic medical record (EMR)* used, sometimes seemingly interchangeably. There is a difference between the two, however (Garrett & Seidman, 2011).

An EMR allows the facility to do the following

1. Track data over time.
2. Identify patients who are due for preventive screenings.
3. Follow up on patients for blood pressure readings.
4. Monitor and improve overall quality of care within a practice.

An EHR serves the following purposes:

1. Ensures all members of a medical team have instant access to medical information, such as drug allergies and lab results.
2. Provides patient access for test results.

Allows clinicians to note discharge instructions and follow-up care.

Now you know!

Efforts to Decrease Healthcare Costs

In the 1950s, most families were able to pay their medical bills because the cost averaged only a few hundred dollars a year. By 2008, the average annual cost per family had reached just over $7,000 (Auerbach & Kellermann, 2011), and in 2019, the Kaiser Family Foundation reported this figure to be over $20,000. Data from the Organization for Economic Co-operation and Development (OECD) further indicate that yearly expenditures in the United States rank far and above those in the 31 partner countries in the OECD. Thus, healthcare spending has increased at an alarming rate since the 1950s and is projected to continue to increase. To address this problem, the Affordable Care Act was passed in 2010.

With the goals of expanding the number of Americans covered by health insurance, improving care, and lowering costs, the **Affordable Care Act (ACA)** has offered incentives to healthcare facilities to become more efficient, lower their readmission rates, and offer standardized rates of pay for services. Moreover, with insurance companies increasingly transferring healthcare costs to patients through higher deductibles and wellness incentive programs, patients are limiting their visits to see their physicians.

In 2013 alone there were 13 million outpatient visits, while hospital admissions increased 2.6% from 2012. Moreover, healthcare spending is expected to increase at the rate of 5.8% annually, which is only 1% higher than the growth rate of the US economy. The Obama administration, as of 2014, allocated $17 billion for health care and by the year 2024 will have spent more than $1 trillion to provide healthcare coverage for consumers (Centers for Medicare and Medicaid Services, 2019). Another factor increasing healthcare costs in the United States is the aging of the large baby boomer generation.

Time Series Analysis of Data

Time series analysis, or time-trend analysis, is a powerful way to visually display and analyze different points of data over a continuum. You might use this approach to analyze how the body weight of middle school students varies over a period of several years or how a person's blood sugar level varies over a year. In either example, all data may be charted to show whether data points decreased, increased, or stayed the same.

You may use two types of data in time series analysis: *continuous* or *discrete*. Continuous data are those that occur along a continuum, such as waveforms on a heart rate monitor, which represent every instant of time within the period being studied. Discrete data are those that correspond to specific points of time, often defined before data collection, such as every hour or at the end of the month. (For more information on continuous and discrete data, see Chapter 2, *Central Tendency, Variance, and Variability*.)

On analysis, various data patterns often become evident. **Seasonal fluctuations** in data may occur over the course of a year, such as allergy visits peaking every April and traumatic injuries occurring the least in January. Similarly, **cyclical variations** in data may occur over the course of a week or month, such as ED visits increasing exponentially on Friday nights compared with all other days of the week. An **irregular fluctuation**—known as "noise" and associated with outlier data—is a change in otherwise ordered data that does not seem to fit with the overall pattern the data displays.

In the next Hands-on Statistics exercise, you will use Excel to analyze time series data via a moving average, which will scatterplot the data and create a second plot to smooth out peaks and valleys in the data, thus representing an overall trend in the data.

Hands-on Statistics 11.1: Time Series Analysis Using Excel

The director of the HIM department wants to know the number of overtime hours worked by full-time office staff for the past year. The director's goal is to see if the number changes drastically due to new equipment upgrades last year, or if it stays the same.

For this exercise you will perform time series analysis using Excel. Imagine that you are calculating the total number of overtime hours worked by staff in the hospital over the course of a year. Note that you will need to have installed the Data Analysis pack in Excel—as explained in an earlier chapter—to complete this exercise with Excel; OpenOffice has it already installed. Follow these directions:

1. Open a new, blank Excel workbook.
2. Enter the data shown in **Figure 11.1**.
3. Select *Data Analysis* from the *Data* menu and choose *Moving Average*, then click *OK*.
4. In the *Input Range* field of the *Moving Average* window, select the data range of *B4:B15* for units used (**Figure 11.2**). Select an interval of 2 to average the previous two data points (or the *Moving Average* trend line), select an output range to the right of the data, and lastly check the *Chart output* checkbox. Click *OK*.
5. The output chart should appear as shown in **Figure 11.3**. Note that the average of the previous two data points (because you chose an interval of 2) are shown to the left. The actual data points are plotted with diamonds, with the trend average plotted over them with squares. Redoing step 4 with an interval of 4 would yield a smoother plot line.

	A	B
1	Time series Moving Average	
2		
3	Month (1=Jan.)	Units used
4	1	100
5	2	80
6	3	90
7	4	79
8	5	82
9	6	110
10	7	91
11	8	82
12	9	90
13	10	96
14	11	92
15	12	99

Figure 11.1 Initial monthly usage data.

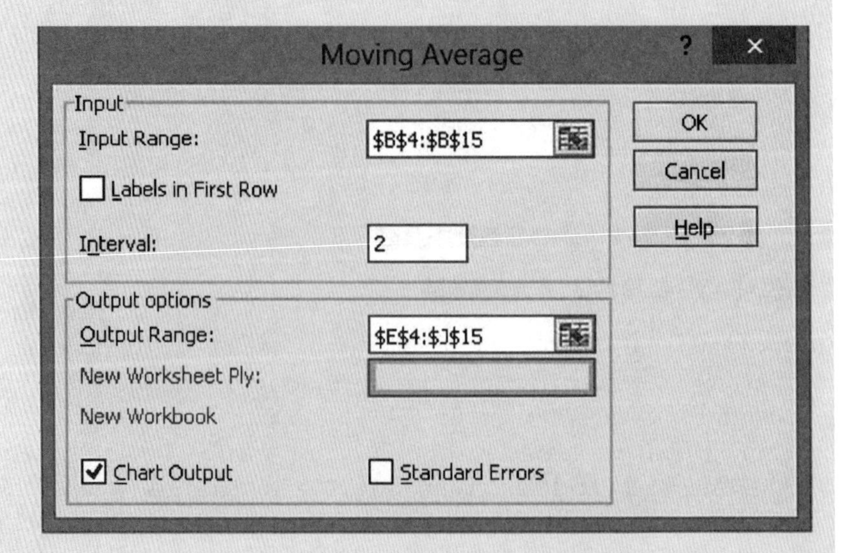

Figure 11.2 Selection of data for time series analysis.

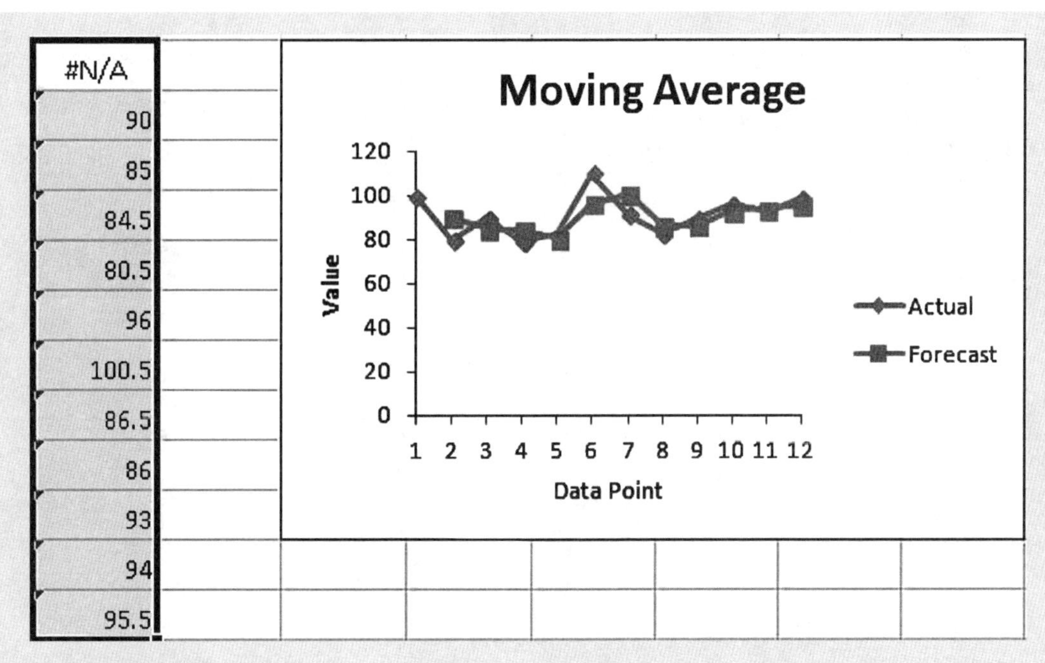

Figure 11.3 Output chart.

The graph in Figure 11.3 illustrates that a greater number of units were used from January to December, although only by a slight amount. It would be interesting to see whether unit usage went down in the next year and steadily increased again or whether it picked up from the same point in January and moved at the same continuing rate.

Note that this procedure could also be used to calculate monies collected during a week or blood pressure readings taken on a daily basis for a month; the concept is the same.

Forecasting Future Data

In addition to examining existing data for overall trend analysis, you may also predict future trends via **forecasting**. Use of forecasting tools by healthcare professionals can contribute to the streamlining of logistical processes, cost savings, minimization of supply delays, and more efficient patient care.

One example of such forecasting is the logistical chain ordering system, which eliminates the need for someone to take stock of supplies on hand before ordering. Ideally, in this system, a patient's health information is entered into the computer and associated ICD codes are entered related to any diagnoses. At the same time, consumable items related to any procedures associated with the ICD code are automatically calculated, decremented from the inventory, ordered (if the minimum threshold is met), and delivered to the department in the facility. Many large supply chain stores currently use such systems to stay competitive.

An automated supply chain might be envisioned as follows: "Through a careful integration of the hospital system's diverse data, including information captured from clinical, financial, and operational sources, the simple act of scheduling a routine medical procedure would activate an automated system" (mThink, 2004). This system would enable such acts as ordering supplies related to ICD codes and planning for contingency supplies.

Hands-on Statistics 11.2: Forecasting Using Excel

To practice forecasting using Excel, follow the directions below.

1. Open a new, blank Excel workbook.
2. Enter the data shown in **Figure 11.4**.

(continues)

	A	B
1	Forecasting Operating Costs	
2	'x' independent var	'y' dependent var
3	Month (by #)	Cost (1000's by dept)
4	1	44
5	2	45
6	3	49
7	4	49
8	5	60
9	6	62
10	7	80
11	8	76
12	9	85
13	10	88
14	11	91
15	12	79
16	13	90
17	14	93
18	15	84
19	16	
20	17	
21	18	
22	19	
23	20	
24	21	
25	22	
26	23	
27	24	

Figure 11.4 Initial data for forecast example.

B19 fx =FORECAST(A19,B4:B18,A4:A18)

	A	B	C
1	Forecasting Operating Costs		
2	'x' independent var	'y' dependent var	
3	Month (by #)	Cost (1000's by dept)	
4	1	44	
5	2	45	
6	3	49	
7	4	49	
8	5	60	
9	6	62	
10	7	80	
11	8	76	
12	9	85	
13	10	88	
14	11	91	
15	12	79	
16	13	90	
17	14	93	
18	15	84	
19	16	101.4666667	
20	17		
21	18		
22	19		
23	20		
24	21		
25	22		
26	23		
27	24		

Figure 11.5 Forecast function.

3. In cell B19, enter the following formula to predict a forecast of values for B19 through B27:

   =FORECAST(A19,B4:B18,A4:A18)

   This formula will produce a cost projection (in thousands of dollars). It relies on the independent variable *x* and the dependent variable *y*, meaning that *y* depends on *x* for its computation. The format for forecast requires, from left to right in **Figure 11.5**, the cell to base the prediction on (A19), the range for known *y* values, and the range for known *x* values.
4. Click in the lower right of cell B19 and drag the mouse pointer down to autofill the other months. Right click on the cells and select *Format Cells* to format the desired number of decimal places.

DID YOU KNOW? R-Project has many built-in datasets that allow you to practice various functions before using real data. From the R-Project prompt, type "data()" and press Enter to see all of the available datasets. To see the structure of a dataset, such as the biochemical oxygen demand (BOD), type "ls(BOD)" and press Enter. To see the data, type "BOD" and press Enter.

Now you know!

Another function in Excel, trendline, yields results similar to those from forecast, although the format used for reference data cells is different. Keep in mind that the forecast function is based on the assumption that the data points are linear, or follow along a line, and are not scattered about when plotted with an application such as Excel. In the Apply Your Knowledge section, you will have an opportunity to plot data to check for this linearity. Of course, the results are predictions, and accuracy is subject to both the linearity and amount of data available to make predictions, so use forecasting as a part, but not all, of your planning process.

Project Management General Concepts

An important area of research in healthcare statistics is project management. **Project management** is the process of directing, tracking, and problem solving. For the purposes of this chapter, we will assume that the initial steps of validation and funding for the project have already taken place and that you are charged with finding and implementing a solution to a problem. For example, the administration of your hospital has determined that your department needs new computers installed in the office, but with no interruption to services offered by the department. Assuming that initial steps of validation and funding for the project have already occurred, you now have the task of managing this upgrade and keeping the department running during the process. A discussion of the steps for managing this project follows.

Step 1 is to determine the need and feasibility of the project.

Step 2 involves the definition of and planning for the project. The time frame or scope of the project is outlined following a detailed breakdown of all key aspects of the project. Budgeting, staffing, outsourcing, scheduling, and contingency planning take place in this step.

Developed in the late 1950s, Dupont Corporation's **critical path method (CPM)** is a good way to accomplish this step and, in fact, outline the entire project (**Figure 11.6**). CPM outlines critical and noncritical tasks and helps you sequence a flow of events and resources to complete the project in a timely manner. Although you could orchestrate the entire project using CPM, here it is shown as part of a larger process. The five general steps of using the CPM for a project plan for the aforementioned upgrade are as follows:

1. List all tasks and resources that are needed to finish the project. Be as detailed as possible and sequence the tasks in order.
2. Allot the time required to finish each task, with a best case time (minimum), most likely time, and maximum time.
3. If one activity depends on another, list this.
4. List deliverables or key accomplishments at the end of each step.
5. Develop backup plans in the event tasks have completion problems such as time or cost overruns, resource allocation issues, and so forth.

Step 3 is where you start the project. Tasks are assigned and communication methods are established. This could be accomplished by having a common

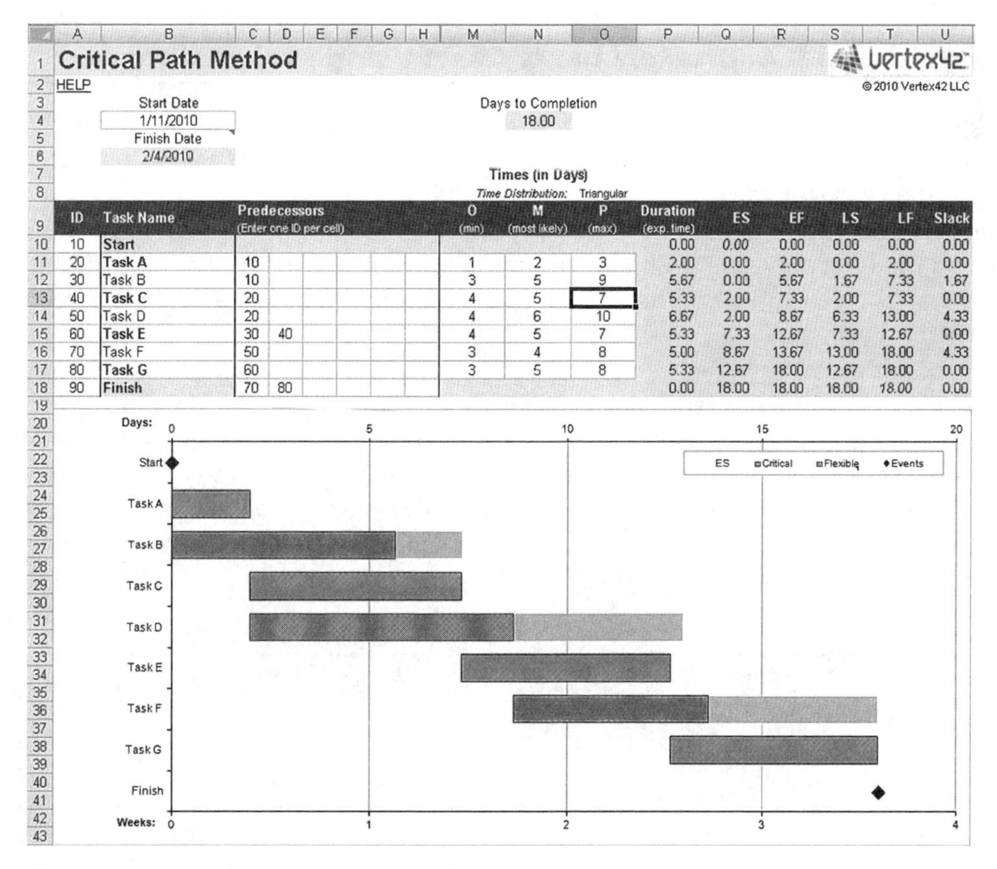

ID	Task Name	Predecessors (Enter one ID per cell)		O (min)	M (most likely)	P (max)	Duration (exp. time)	ES	EF	LS	LF	Slack
10	Start						0.00	0.00	0.00	0.00	0.00	0.00
20	Task A	10		1	2	3	2.00	0.00	2.00	0.00	2.00	0.00
30	Task B	10		3	5	9	5.67	0.00	5.67	1.67	7.33	1.67
40	Task C	20		4	5	7	5.33	2.00	7.33	2.00	7.33	0.00
50	Task D	20		4	6	10	6.67	2.00	8.67	6.33	13.00	4.33
60	Task E	30	40	4	5	7	5.33	7.33	12.67	7.33	12.67	0.00
70	Task F	50		3	4	8	5.00	8.67	13.67	13.00	18.00	4.33
80	Task G	60		3	5	8	5.33	12.67	18.00	12.67	18.00	0.00
90	Finish	70	80				0.00	18.00	18.00	18.00	18.00	0.00

Figure 11.6 Vertex 24's free CPM Excel template.

shared location where progress is updated either by workers or by a central manager who receives emails or other updates as tasks progress.

Step 4 is management of the project. The project manager tracks progress of work, comparing it to estimated time frames. Using a spreadsheet or project management software, the manager adjusts schedules, reallocates resources, and keeps the project moving forward. The project manager may also report the project status to the facility's management.

Step 5 is the project close and wrap-up. Once the last task is successfully completed, a summary evaluation of how the project progressed and processes that could make future projects run more smoothly are noted. You can find many free spreadsheet templates that allow you to use CPM for your project by searching the internet or Vertex24, as shown in Figure 11.6 and listed in the Web Links at the end of this chapter.

Next you will install ProjectLibre, an open-source and free project management application that is similar to Microsoft Project. Major companies such as IBM, Caterpillar, and Abbot use this application. In the following Hands-on Statistics exercise, you will install the software and create a Gantt chart.

Hands-on Statistics 11.3: How to Install ProjectLibre

1. Open your internet browser and enter the following into the address bar: http://www.projectlibre.org/. Click on *Download* and you will be redirected to the SourceForge distribution site. Click on the green download button. Note that there are versions for Windows, Mac, and Linux, so it is a cross-platform application. Open the default download location (or the location where you saved the installer), and locate the installer. Note, this will vary with platform, type of web browser, and browser settings.
2. Assuming you are installing to a Windows-based computer (not a Macintosh), locate in your download directory the file "projectlibre-1.x.x," in which the *x*'s represent the version you have downloaded. Double click it to start the installation process. As with any Windows installer, you may have to right click the file and select *Run As Administrator*.
3. Now that the software is installed, you will learn to use a handy feature, the **Gantt chart**.

Hands-on Statistics 11.4: Create a Gantt Chart Using ProjectLibre

1. Start ProjectLibre, accept the license agreement, and enter your email, if requested.
2. Select *Create Project* from the pop-up window that appears. Type *Upgrade Printer* in the *Project Name* field and your last name in the *Manager* field. Click *OK*.
3. In the field to the left of the main window, you will see a table with the headings *Name, Duration*, and *Start*. In the field to the right, you see a Gantt chart area. Expand the table field by clicking the center bar and dragging it to the right, until the column headings *Finish, Predecessors*, and *Resource Names* are shown.
4. On the left, under the column heading *Name*, list the following tasks, each in its own row, as shown in **Figure 11.7**: "Remove old printer," "Unbox new printer," "Install printer," "Install software," and "Test printer." Note that the *Duration, Start*, and *Finish* columns will autopopulate with default data.
5. Next you will adjust when the tasks take place and for how long. From the toolbar above, select *Task* and then *Zoom In*. The pane on the right is larger now. Click the cell *Remove old printer* to highlight it, and then select *Information* from the toolbar. A pop-up window titled *Task Information – 1* will appear. In the *Duration* field, key in "1 hour" and then click *Close*. Note the corresponding progress bar on the Gantt chart is shorter now.

	ⓘ	Name	Duration	Start	Finish	Predecessors	Resource Names	19 Oct 14 S M T W T F S
1		Remove old printer	1 day?	10/22/14 8:00 AM	10/22/14 5:00 PM			
2		Unbox new printer	1 day?	10/22/14 8:00 AM	10/22/14 5:00 PM			
3		Install printer	1 day?	10/22/14 8:00 AM	10/22/14 5:00 PM			
4		Install software	1 day?	10/22/14 8:00 AM	10/22/14 5:00 PM			
5		Test printer	1 day?	10/22/14 8:00 AM	10/22/14 5:00 PM			

Figure 11.7 Initial tasks in ProjectLibre.

6. You can also set the duration for each task by clicking and dragging on the progress bars in the Gantt chart. Assign 1 hour to the second task in this way. Note that the duration has changed to .125 days.
7. Select *Zoom In* one more time, click on the progress bar of the first task in the Gantt chart, and drag it to the progress bar of the second to create a dependency. Now you will see an arrow and the second task moved forward in time, indicating that the first task must be completed before the second starts, as shown in **Figure 11.8**.
8. Move the others to resemble **Figure 11.9**. You will click and hold on the task above (starting with task 2) and drag it to the task below, creating a dependency. Then click the right side of the progress bar and drag it to the left to shrink the amount of time required for the task.
9. Clicking on any cell within the row of a task in the table and selecting *Information* will allow you to change parameters of the task, such as duration or start or end dates or times.

You can print or use the Print Screen key (PRTSC) to capture project data for management or for use in a presentation, as needed.

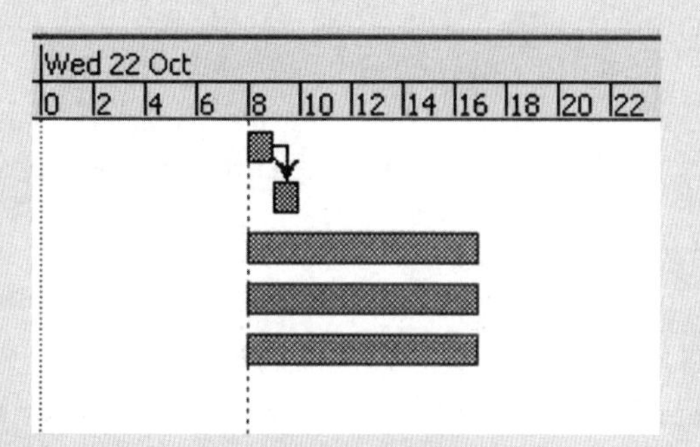

Figure 11.8 Gantt chart showing dependency between tasks and time frame.

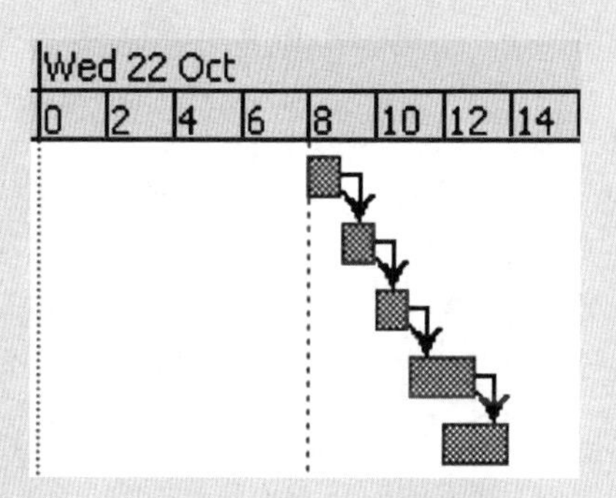

Figure 11.9 Tasks assigned to 1 day with dependencies on previous task.

DID YOU KNOW? The technological advancements of home health monitors, smart pill technology, and microchips are becoming significant measures in cutting costs and improving health care. Data from portable patient monitors (such as those measuring blood pressure, heart rate, insulin levels, etc.) can be encrypted and transmitted via wireless networks and the internet to doctors and wellness agents for monitoring.

Smart pill devices, which are still being developed, are miniature sensors that can be attached to oral medications to wirelessly transmit data regarding the type and amount of medication ingested to the patient's personal monitor and to the patient's providers via the internet. When a patient's levels are in a nonstandard range, the providers will receive an immediate alert and can notify the patient and change dosages.

Another innovation is e-pill automatic pill dispensers, which can alert patients to take their medications at the proper time during the day, and can even notify monitoring providers of missed doses. This device is expected to help reduce hospital readmission rates in the United States due to improved patient compliance with medication regimens and the subsequent decrease in medical emergencies related to overdoses and missed doses.

Microchip implants are also expected to improve care in the future. These devices will be able to reduce the need for blood samples to be taken by directly monitoring blood glucose and other levels and forwarding this information to appropriate monitoring services and healthcare providers. Additionally, the resultant data they receive will be used for trend analysis to give patients detailed information about their levels so they can incorporate that information into their exercise and dietary habits.

Now you know!

How Does Your Hospital Rate?

Sarah knows that with traditional coding, a trained and certified HIT coder reviews medical documents and, based on the findings of the doctor, assigns ICD codes appropriate for the diagnosis. With the implementation of the ACA and the more refined diagnosis codes in ICD-10, patient billing is much more accurate, but Sarah knows there will be implementation issues.

After some research, she understands that a CACS system takes the human element of analyzing the documents away, to some degree, and implements a computer system that reads the documents for both key words (such as fracture) and context (e.g., a previous fracture would be much different than a current fracture) and assigns ICD codes based on these findings. This system works similar to a grammar/spell checker in a word processing application. She also finds that there are two general types of CACS, structured input and natural language processing (NLP). With structured input systems, the doctor uses point-and-click fields to refine diagnosis results and other findings. The doctors will need to undergo retraining to use the new structured input system because the system often requires certain fields to be completed, but a more accurate EMR is obtained.

In contrast to structured input systems, NLP systems interpret what the doctor has written, and based on a rule set similar to many data mining algorithms, the system interprets the doctor's diagnosis. Sarah thinks an NLP system would create less "tension" with the doctors, though it might create some extra work for coders, who must audit and "clean up" the resulting EMRs produced with an NLP system.

Consider the following:

1. Who would initially train and test the NLP system?
2. How would doctors and coders be introduced and trained on the system?
3. What would be a good ongoing process to ensure accurate diagnosis transcription by the system?

Analysis of Covariance

Analysis of covariance (ANCOVA) is an extension of analysis of variance (ANOVA) that involves both ANOVA and regression analysis. ANCOVA is used to test for differences between group means while statistically controlling for an extraneous continuous variable. It allows for covariates, which are additional variables that ANOVA does not normally account for. Specifically, these factors are not random elements but are interval and in fact "co-vary" with the dependent variable. Age or a pre-test score is a good example of a covariate.

For example, imagine that you are trying three new methods of teaching on 60 students—one method on 20 of the students, a second on another 20, and a third on the last 20. You have final examination score data and intelligence quotient (IQ) data (interval covariate) on the students. If you were to plot regression lines for the groups, roughly parallel lines would show that the data are linear in nature. Next you could examine whether increasing IQ influences test results. With ANCOVA, you are trying to explain variation of the dependent variable (test scores) by controlling for the influence of covariates (IQ). Your goal, therefore, is to determine whether teaching style affects test scores; the influence of IQ is not the primary concern. You will practice performing ANCOVA in R-Project in the following Hands-on Statistics exercise.

Hands-on Statistics 11.5: Analysis of Covariance Using R-Project and Cholesterol Data

Imagine that you have data from a sample of people, broken into three different treatment groups. Age is the covariate with the categorical variable of the treatment and the level of cholesterol after treatment. Follow these directions to practice using ANCOVA:

1. Open a new, blank Excel workbook.
2. Enter the data shown in **Figure 11.10**.

	A	B	C
1	cholesterol	age	treatment
2	181	46	1
3	228	52	1
4	146	70	1
5	199	48	1
6	115	34	1
7	112	41	1
8	249	65	1
9	259	54	1
10	201	33	1
11	120	49	1
12	339	76	2
13	182	39	2
14	224	71	2
15	189	58	2
16	137	18	2
17	173	44	2
18	178	33	2
19	240	78	2
20	225	51	2
21	223	43	2
22	190	44	3
23	257	58	3
24	356	67	3
25	189	19	3
26	214	41	3
27	140	30	3
28	196	47	3
29	262	58	3
30	215	41	3
31	140	30	3

Figure 11.10 Raw data on cholesterol treatments.

3. Save the data to a comma-separated values (CSV) file in a location you can find and access (this is critical!). In Excel, select *File* and *Save As*, and the *Save As* window will appear. From the dropdown menu of the *Save as type* field at the bottom of this window, select *CSV (Comma delimited)*. Type "treat.csv" in the *File name* field and click *Save*.
4. Start R-Project and then type in the following to make sure you can read the data file you just created (Note: in place of *"c:/users/burton/desktop/treat.csv"*, type the file path that corresponds to where you saved your CSV file in step 3):

```
mydata <- read.csv ("c:/users/burton/desktop/treat.csv", header=TRUE, sep=",")
```

5. Type "mydata" and press Enter to verify that the data in the CSV file you created have been imported and associated with the variable *mydata* properly. If so, you should see 30 items displayed.

(continues)

6. Now you will create a scatterplot of your data. The following code extracts just test 1, 2, or 3 to a data frame and removes the number in the third column denoting the state. Type the following, pressing Enter after each line:

```
t_1_raw <- subset(mydata, treatment==1)
t_1_raw
t_2_raw <- subset(mydata, treatment==2)
t_2_raw
t_3_raw <- subset(mydata, treatment==3)
t_3_raw
t_1 <- t_1_raw[,1:2]
t_2 <- t_2_raw[,1:2]
t_3 <- t_3_raw[,1:2]
```

7. Now your data are cleaned up and in three data frames. Key in the following to create a scatterplot of your data (**Figure 11.11**). (Note that the first line plots data as solid squares; the second line allows you to add one screen and not create new ones; and the third line plots data as hollow squares.) Remember that anything after a # sign is a comment to the programmer stating what the line does; it is for the programmer's benefit and does not control the program in R in any way. The colors noted in the text are to help distinguish the data.

```
plot(t_1, pch=15, col='green') # green solid squares
```

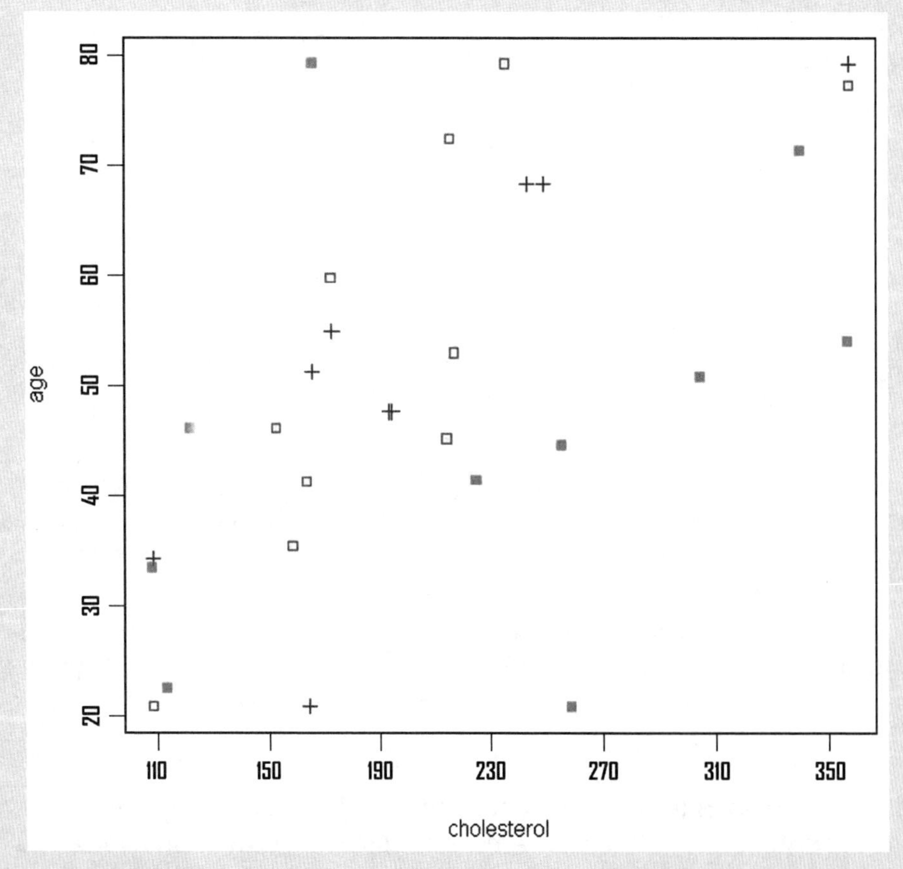

Figure 11.11 Scatterplot of three treatments.

par(new=TRUE) # allows you to add to one screen and not create new ones

plot(t_2, pch=22, col='red') # red hollow squares

par(new=TRUE)

plot(t_3, pch=3, col='blue')

8. Next you can extract the data for the *x* and *y* variables and create a regression line. Type the following, pressing Enter after each line:

t_1_chol <-t_1$cholesterol # *x* variable

t_1_age <-t_1$age $ *y* variable

t_2_chol <-t_2$cholesterol # *x* variable

t_2_age <-t_2$age $ *y* variable

t_3_chol <-t_3$cholesterol # *x* variable

t_3_age <-t_3$age $ *y* variable

9. Use the *lm* function to create a linear model for the data and *abline* to display lines on the chart. Type the following, pressing Enter after each line:

abline(lm(t_1_age ~ t_1_chol)) # line for treatment 1

par(new=TRUE) #add to chart

abline(lm(t_2_age ~ t_2_chol))

par(new=TRUE) # add to chart

abline(lm(t_3_age ~ t_3_chol))

10. Lastly, you will see whether there is a difference in the treatments. Type the following, pressing Enter (return) after each line:

output <- lm(mydata$cholesterol ~ mydata$age + mydata$treatment)

anova(output)

The function *mydata$cholesterol* specifies the data frame and column data desired.

As shown in **Figure 11.12**, the output indicates that there is significant difference at the .01 level for variation between treatments, but age does not seem to be a factor. At this point, you could conduct a Tukey honest significant difference test to determine which treatment might be most effective. Chapter 7, *Infection, Consultation, and Other Data*, provides details on this test.

```
Analysis of Variance Table

Response: mydata$cholesterol
                  Df Sum Sq Mean Sq F value    Pr(>F)
mydata$age         1  37881   37881 20.9009 9.626e-05 ***
mydata$treatment   1  11855   11855  6.5412   0.01647 *
Residuals         27  48935    1812
---
Signif. codes:  0 '***' 0.001 '**' 0.01 '*' 0.05 '.' 0.1 ' ' 1
> |
```

Figure 11.12 Output from R-Project.

How Does Your Hospital Rate?

Sarah learns from the American Health Information Management Association (AHIMA) that there are advantages in a CACS system, such as increased productivity in coding and consistent application of coding rules. Some disadvantages include the cost of the system and an increased risk of errors. AHIMA has noted that coders who used a CACS spent 22% less time coding a record than they would have using traditional coding (Fitzgerald, 2013). However, records created with a CACS that were unchecked by a coder were less accurate than a combination of the medical coding software and a manual coder, so certainly a case can be made for a hybrid system in which both computer and human are used to their best advantage. Sarah thinks a good mix of human and computer interaction would be a hybrid system in which a CACS would handle the majority of coding tasks, with certified coders working on difficult or complex situations and serving as auditors for CACS-processed documents.

Voice recognition systems are also commonly used at medical establishments for transcription, and Sarah thinks this technology should be added to the overall analysis. She recognizes that the end is fast approaching of the old approach, in which a doctor dictates notes into a recorder for manual transcription and later reviews it for approval. In fact, most computers and smartphones now offer some form of built-in voice recognition. However, they typically do not have the precision or analysis capabilities that specialized medical voice recognition software offers, so Sarah knows they will need a professional-grade solution.

By the end of her research, Sarah has arrived at the following answers to her questions to help with the final analysis she will write and give to management:

Question: How will doctors respond to a new system?

Answer: Probably not well to a structured input system, but they would respond well to something that they did not have to "relearn," such as an NLP-based system.

Question: Will training be required for the doctors?

Answer: Not really, only with respect to the detailed nature of ICD-10, which would have come about regardless of the CASC system chosen.

Question: Will training be required for coders?

Answer: Yes, because they will be conducting more audit or "cleanup" work on what the NLP system creates, but it should allow for higher throughput and will allow a more seamless transition to ICD-10. Coders will also need training on moving from ICD-9 to ICD-10.

Question: Will the role of a coder change?

Answer: Yes, they will be more of an auditor, except in cases of complex coding.

Consider the following:

1. Do you agree with her findings?
2. Which system would you have chosen?
3. If there is a local hospital near you, does its staff use CASC?
4. If so, which type do they use, structured input or NLP?
5. Did they have implementation issues?
6. Who was most affected—doctors, coders, or all HIM staff?
7. How does your hospital rate on this subject?

Coronavirus—The Next Pandemic

The coronavirus outbreak in Wuhan, China, is the latest pandemic, although the World Health Organization (WHO) lists many pandemic contenders. Similar to a scene from a zombie movie, Chinese officials closed off the city, shut down public transportation, and tried to contain the virus that has already made it around the world, with over 15,000 people sick within weeks of the first documented case. WHO (2020) declared this a Public Health Emergency of International Concern in early 2020, though it is not first on the list of pandemics. Also toward the top of the WHO list of pandemics is the Ebola virus.

Ebola

The first identified human outbreak of the Ebola virus, which was named after the Ebola River, was in Africa in 1976. Ebola, or Ebola virus disease, is highly infectious because only a small particle of fluid from someone infected can cause sickness; however, it is not an airborne virus, so it is not necessarily transmitted if someone sneezes. Additionally, the virus affects animals (it is killing off thousands of African gorillas) and exists in five subspecies.

In 2014, a doctor and nurse from North Carolina who were treating patients in Africa were infected with the virus and were quarantined at the CDC headquarters in Atlanta, Georgia, where they both made a full recovery. A short time later, a passenger who had recently arrived in the United States from Africa was found to have the disease and died in a Texas hospital. Two nurses who had cared for this patient also became infected with the virus but recovered fully.

Even after a full recovery from the disease, men can transfer the virus for up to 7 weeks via their semen. Because transmission is by bodily fluids, burial practices must allow for containment and destruction of the virus. When the first US patient with Ebola died in Texas in 2014, his body was cremated and the remains given to his family. This precaution was not at the family's request but was necessary due to the danger from disease transmission.

As of 2014, Guinea, Liberia, and Sierra Leone have experienced widespread outbreaks of Ebola, whereas other countries, including the United States, have had only a few isolated cases. To combat this threat, the United Nations, WHO, and the United States have sent teams and supplies to affected areas to train healthcare workers and help contain the disease. The United States plans to train 500 healthcare workers a week at established centers in Africa. The United States has spent hundreds of millions of dollars at home and abroad to fund this effort, so the threat is being taken seriously. Only time will tell whether Ebola becomes a pandemic.

Global Perspective

Mankind has been burdened by large outbreaks of disease since earliest recorded time. Whether it is an epidemic (local outbreak) or a pandemic (global outbreak), the result is widespread fear and death. Arguably the worst recent pandemic was the Spanish flu (also known as La Grippe) outbreak of 1918 and 1919, which is believed to have caused about 50 million deaths worldwide during this 2-year period. This number is even higher than those killed in World War I, which lasted from 1914 to 1918. By comparison, the plague of Justinian, which occurred around 450 AD, is estimated to have killed around 100 million people over a 50-year time span. Although it claimed only half as many lives, the Spanish flu pandemic spread much more rapidly and killed more people per year than the plague of Justinian, making it worse. Other critical diseases to watch around the world include SARS, Marburg, smallpox, HIV/AIDS, cholera, and Ebola, to name but a few.

From a healthcare standpoint, we now understand that preventing and containing these outbreaks is just as important as treating the effects. Until very recently, data on cases, treatment options, and containment strategies were not consistent and certainly were not shared equally by all countries. However, a new field, digital epidemiology, promises to greatly assist in the dissemination of this information.

Digital epidemiology is "a new field of increasing importance for tracking infectious disease outbreaks and epidemics by leveraging the widespread use of the internet and mobile phones" (Salathé, Freifeld, Mekaru, Tomasulo, & Brownstein, 2013). Researchers can learn about and possibly track the spread of infectious diseases from everyday people around the world with the help of news websites, blogs and microblogs (such as Twitter), and social media sites, something that was not possible only a few years ago. Sites such as HealthMap and Google Flu (as of 2019 historical data only) offer but a glimpse of what is possible with digital epidemiology.

Another promising aspect of digital epidemiology is that it can be used for other health issues and not only disease outbreaks. The technology model is really a data mining model that fits well with any topic related to health, wellness, or disease. As Voss (2015) states, "Mining these novel, big-data streams is of enormous interest. . . . Researchers and public health officials could use data-mining techniques to detect adverse drug reactions, assess mental disorders, or track health behaviors much faster than they do with traditional methods." Certainly the expansive global divide is made smaller every day by the innovative data mining of internet data.

Chapter Summary

This chapter has provided a glimpse into the future of health care based on emerging technologies. RFID, automatic medication dispensers, and the ongoing efforts to contain and reduce the costs of health care were discussed. Forecasting and time series analysis were examined so that you can leverage the power of these tools to help predict future needs or trends. Project management is important to understand in large and small facilities, so the chapter discussed key aspects of project management, in particular CPM and the Gantt chart to visually display project status. The open-source ProjectLibre software was used as a great example of a project management tool. Statistical analysis of data with multiple variants, or potentially influencing factors, with ANCOVA was covered. Lastly the Ebola virus outbreak of 2014 was reviewed and an online visualization tool provided to examine outbreak data on Ebola and other critical diseases.

Apply Your Knowledge

1. Using time series analysis in Excel, track and record your blood pressure, weight, or some other data over a month.
2. Google has an interesting tool for examining trend data. Visit the following URL and list three or more features of this tool: https://trends.google.com/trends/. (Note that you will need to create a Google account to sign in.)
3. Using the following data, predict via Excel a forecast for the missing values. In the data, 1 = Jan, 2 = Feb, and so on. Predict the following Saturday's staffing needs for the ED:

	1	5
	2	4
	3	6
	4	4
	5	4
	6	
	7	
	8	
	9	
	10	
___	11	
	12	

4. Considering forecasting, use the following data and instructions to examine whether the data are linear or nonlinear: 1, 1.9, 2.3, 2.5, 3, 4, 4.5, 5, 5.1, 5.3, 5.6, 5.7. Note that data are in CSV format.
5. Data that are "out of order" or somewhat extreme can be improved by a statistical process referred to as "smoothing." Use the Excel exponential smoothing function on the data shown in **Figure 11.13**. Add a trend line to the final result.
6. Imagine that you have been assigned the task of managing a new wellness program at the hospital where you work. This program helps workers with fitness goals such as losing weight, gaining weight, or preparing for some goal, such as getting ready for a sporting event. Be creative and use the free template from Microsoft found at the following website to track the implementation of the program: http://blogs.office.com/2011/11/30/tracking-small-projects-in-excel/.
7. Using the free Excel template (available at http://www.vertex42.com/ExcelTemplates/critical-path-method.html), create a project plan and

	A	B
1	1	#N/A
2	3	1
3	2	1
4	7	2
5	4	1.28
6	2	1.852
7	4	2.0668
8	1	2.06012
9	9	2.1287
10	3	2.81583
11	5	2.83424
12	6	3.05082
13	3	3.34574
14	1	3.31116
15	2	3.08005
16	4	2.97204
17	2	3.07484
18	9	2.96735
19	2	3.57062
20	3	3.41356
21	4	3.3722
22	1	3.43498
23	2	3.19148
24	1	3.07234
25	3	2.8651

Figure 11.13 Data for smoothing in Excel.

Gantt chart for a smoking cessation program at your workplace. Be as detailed as possible and include items such as signage, user group creation, and incentive program development.

8. Using ProjectLibre, create a complete project plan to implement the "My Fit Facility" program your administrator has assigned you to manage. Be specific and plan on a 6-month time frame for 100% implementation. Include fitness, smoking cessation, stress reduction, and team building.
9. Using the following link, download the project management template and create a project plan for the same project you completed with ProjectLibre: http://blog.worldlabel.com/2009/using-openoffice-org-calc-to-manage-schedules.html. Which application, ProjectLibre or the Excel template, was the best choice? What were the pros and cons of each?
10. Find and list the URLs for three other Excel project management templates.
11. Find one other free or open-source project management application not listed in the chapter.
12. Visit Apple's HealthKit at https://developer.apple.com/healthkit/. List three benefits of the site, and list a concern for improvement.
13. As part of an overall wellness program for the facility, your supervisor has asked you to evaluate WalkingSpree. Visit the website (http://www.walkingspree.com/) and list the pros and cons regarding adoption of this platform.
14. Create a PowerPoint or Impress presentation that outlines key details of the following application from the Mayo Clinic: http://www.mayoclinic.org/apps/mayo-clinic.
15. In one page or less, identify and describe at least two innovative ideas on health software and hardware from the article at the following website: http://www.theverge.com/2014/7/22/5923849/how-apple-and-google-plan-to-reinvent-healthcare.
16. Identify the type of information provided at the following URL: http://www.google.org/flutrends/us/#US.
17. Compare and contrast features of the following online Gantt project management tools:
    http://www.tomsplanner.com
    http://creately.com/plans
    https://www.zoho.com/projects

    Create at least one Gantt chart using one of the sites listed.
18. Use ANCOVA to solve the following problem. In response to a recent Ebola outbreak in your community, the facility where you work is trying two new methods for cleaning hospital linens. Because both methods properly clean and disinfect, the time a worker requires to complete a batch of linen is what will be measured. It is hypothesized that housekeeping workers with more years of experience will perform their cleaning tasks more quickly, regardless of the cleaning method used.

    Using ANCOVA, determine whether years of experience has an effect on the time required to clean the linen.
19. Visit the following URL and explore the MedCalc application: http://www.medcalc.org/. What are three benefits of this software compared with Excel or R-Project? Do an internet search to find at least two similar applications, and provide a short narrative describing their features.
20. Visit HealthMap (https://healthmap.org/en/) and list at least three major outbreaks around the world included on the site. You could also create a screen capture and include a graphic, showing the global hot spots and data in a presentation.

References

Auerbach, D., & Kellermann, A. (2011). A decade of health care cost growth has wiped out real income gains for an average US family. *HealthAffairs, 30*, 1630–1636, doi:10.1377/hlthaff.2011.0585

Dixon, B., Grannis, S., & Revere, D. (2013). Measuring the impact of a health information exchange intervention on provider-based notifiable disease reporting using mixed methods: A study protocol. *BMC Medical Informatics and Decision Making, 13*. Retrieved September 3, 2013, from https://bmcmedinformdecismak.biomedcentral.com/articles/10.1186/1472-6947-13-121

Eramo, L. (2014). Warning: Productivity loss ahead. *For the Record, 26*(2), 10.

Fitzgerald, J. (2013). AHIMA study examines CAC impact on coding. *HIM-HIPAA Insider*. Retrieved December 10, 2013, from http://www.hcpro.com/HIM-294608-865/AHIMA-study-examines-CAC-impact-on-coding.html

Garrett, P., & Seidman, J. (2011, January 4). EMR vs EHR—What is the difference? *HealthITBuzz*. Retrieved May 21, 2013, from https://www.healthit.gov/buzz-blog/electronic-health-and-medical-records/emr-vs-ehr-difference

Kaiser Family Foundation. (2019, September 25). 2019 Employee health benefits survey. Retrieved December 10, 2019, from https://www.kff.org/report-section/ehbs-2019-section-1-cost-of-health-insurance/

Centers for Medicare and Medicaid Services. (2019). National health expenditure data. *CMS.gov*. Retrieved October 10, 2019, from https://www.cms.gov/Research-Statistics-Data-and-Systems/Statistics-Trends-and-Reports/NationalHealthExpendData

Miliard, M. (2013). RFID and RTLS can save lives. *Healthcare IT News*. Retrieved April 3, 2014, from http://www.healthcareitnews.com/news/rfid-rtls-can-save-lives

mThink. (2004). Effective demand forecasting in the health care supply chain. Retrieved September 30, 2015, from https://mthink.com/effective-demand-forecasting-health-care-supply-chain/

Salathé, M., Freifeld, C., Mekaru, S., Tomasulo, A., & Brownstein, J. (2013). Influenza A (H7N9) and the importance of digital epidemiology. *New England Journal of Medicine*. doi:10.1056/NEJMp1307752

The Office of the National Coordinator for Health Information Technology. (2019). Why is health information exchange important? *HealthIT.gov*. Retrieved December 15, 2019, from https://www.healthit.gov/faq/why-health-information-exchange-important

Versel, N. (2012). Proteus gains de novo FDA clearance for ingestible biomedical sensor. *mobihealthnews*. Retrieved July 15, 2015, from http://mobihealthnews.com/18075/proteus-gains-de-novo-fda-clearance-for-ingestible-biomedical-sensor/

Voss, K. (2015). Tracking outbreaks and epidemics through digital epidemiology. *Penn State Science News*. Retrieved September 21, 2016, from http://epidemics.psu.edu/articles/view/tracking-outbreaks-and-epidemics-through-digital-epidemiology

World Health Organization. (2020). Statement on the second meeting of the International Health Regulations (2005) Emergency Committee regarding the outbreak of novel coronavirus (2019-nCoV). Retrieved from https://www.who.int/news-room/detail/30-01-2020-statement-on-the-second-meeting-of-the-international-health-regulations-(2005)-emergency-committee-regarding-the-outbreak-of-novel-coronavirus-(2019-ncov)

Web Links

Category Archives: Locked Automatic Pill Dispenser: https://epillreminders.wordpress.com/category/locked-automatic-pill-dispenser/

FDA Approves Digestible Microchips to be Placed in Pills: http://www.medscape.com/viewarticle/768665

Critical Path Method—CPM & PERT: http://www.vertex42.com/ExcelTemplates/critical-path-method.html

Statistics Online Computational Resource: http://socr.ucla.edu/SOCR.html

Analysis of Covariance (ANCOVA): http://www.statisticshell.com/docs/ancova.pdf

Table of Critical Values for the *F* distribution (for use with ANOVA): http://homepages.wmich.edu/~hillenbr/619/AnovaTable.pdf

Infection Prevention and Control Recommendations for Hospitalized Patients Under Investigation (PUIs) for Ebola Virus Disease (EVD) in U.S. Hospitals: http://www.cdc.gov/vhf/ebola/hcp/infection-prevention-and-control-recommendations.html

HealthMap: http://healthmap.org/en/

Analysis of Covariance (ANCOVA): http://elderlab.yorku.ca/~elder/teaching/psyc3031/lectures/Lecture%207%20Analysis%20of%20Covariance%20-%20ANCOVA%20%28GLM%202%29.pdf

The Future of Medicine: Squeezing out the Doctor: http://www.economist.com/node/21556227

The Nation's Top Choice for Healthcare Continuing Education: http://health-information.advanceweb.com/Article/Two-Types-of-Computer-assisted-Coding.aspx

Appendix A

Common Statistical Abbreviations Used in This Text

Variable Type	Abbreviation	Meaning
Population	N	Entire population
	n	Sample from a population
Hypothesis testing	H_0	Null hypothesis
	H_1 or H_a	Alternative hypothesis
	α	Alpha; significance level
	β	Beta; probability of committing a type II error (false negative or failure to reject a false null hypothesis)
Random	Z or z	Standardized score or *z* score, often found in a table
	SD	Standard deviation
	VAR	Variance
	DF	Degrees of freedom

Appendix B

Resources for Further Information

Help With Excel (Chapter 1)

- Basic Excel Skills: http://chandoo.org/wp/excel-basics/
- Open Office Calc Basic Spreadsheet Tutorial: https://www.lifewire.com/open-office-calc-basic-spreadsheet-tutorial-3123949
- Tutorial: Ten Concepts That Every Calc User Should Know: https://forum.openoffice.org/en/forum/viewtopic.php?t=39529

Historical Use of Data for Improving Health (Chapter 1)

- Aulus Cornelius Celsus Biography: http://www.faqs.org/health/bios/65/Aulus-Cornelius-Celsus.html
- John Snow Site: http://www.ph.ucla.edu/epi/snow.html

Help With R-Project (Chapter 3)

- R Programming Fundamentals: https://www.pluralsight.com/courses/r-programming-fundamentals
- Exploring US Healthcare Data: http://www.r-bloggers.com/exploring-us-healthcare-data/ (outstanding article on R and how to use it for healthcare analytics)

Microcharts and Sparklines (Chapter 3)

- Microcharts. Simple Way to Create a Nice Report: http://datamonkey.pro/blog/excel_snippets/ (includes coverage of the REPT function in Excel)

Data Mining With R and *arules* (Chapter 4)

- Microsoft R Application Network: http://www.inside-r.org/ (good general info on R and packages freely available to extend the power of R)

When to Use Certain Statistical Measures (Chapter 5)

- Statistical Tests: When to Use Which? https://towardsdatascience.com/statistical-tests-when-to-use-which-704557554740

Institutional Review Board Information From the US Department of Health and Human Services (Chapter 9)

- Institutional Review Board (IRB) Written Procedures: Guidance for Institutions and IRBs: https://www.hhs.gov/ohrp/regulations-and-policy/requests-for-comments/guidance-for-institutions-and-irbs/index.html

Use of Excel With Forecasting (Chapter 11)

- Forecasting: Basic Time Series Decomposition in Excel: https://www.capacitas.co.uk/blog/forecasting-basic-time-series-decomposition-in-excel

Appendix C

Historical Abuse of Human Research Subjects

The following are some historical accounts of the abuse of human research subjects. These examples complement the accounts included in Chapter 9 and help reveal the circumstances that led to the development of institutional review boards.

Dr. Sims

Dr. J. Marion Sims was an innovator of new techniques in gynecology and changed the field of women's reproductive health. In the 1800s, anesthesia and antiseptic procedures were minimal to nonexistent, so imagine the pain the female patients and slaves of Dr. Sims suffered undergoing forced surgery with no anesthesia, let alone any sterile procedures. Some underwent numerous surgeries, and certainly many of the slaves he operated on lost their lives to infections. In an 1857 lecture to fellow doctors, Sims stated that his operations were "not painful enough to justify the trouble" of using the newly discovered anesthesia.

World War II

World War II saw many atrocities committed on combatants and noncombatants in the name of medical science. In fact, volumes of material have been written on the subject, so the few excerpts here are in no way representative of all countries involved nor the scope of the horror. These are but a few of many experiments that occurred.

Nazi Atrocities

The Nazis practiced sterilization of "undesirables" in a methodology they acquired from academics and other medical experts in the United States from the early 19th century, especially in California. In fact, this sterilization of undesirables, fully endorsed by the famous judge Oliver Wendell Holmes, never left the United States and continued to be practiced in many states until the 1970s! North Carolina was one of the worst offenders and is paying damages to victims to this day.

Nazi medical experimentation during World War II was brought to a new and more horrific level with experiments performed on prisoners in concentration camps. Dr. Josef Mengele at Auschwitz concentration camp and others conducted experiments, under the guise of medical experimentation, that stagger the mind with how cruel and absurd humans can be. Fatal experiments were performed on twins with the goal of proving Aryan supremacy, and sterilizations were performed by the hour, without anesthesia. The Nazis also experimented on patients to try out new chemical warfare agents and weaponized diseases and tested the limits of prisoner endurance to extremes in altitude and cold so that German pilots would be better protected.

After the war, in 1947, 26 Nazi physicians were tried for crimes against humanity in Nuremberg, Germany. Others, however, were covertly relocated to the United States, the Soviet Union, and England, where they were put to work, without repercussions for wartime crimes, in secret weapons labs. Research "Operation Paperclip" to find out more. Mengele escaped with many other Nazis to South America and lived there until his death as an old man in 1979; he experienced no criminal retribution for the people he needlessly killed.

This trial by the Nazi War Crimes Tribunal resulted in the Nuremburg Code, the first internationally recognized code of research ethics. The Nuremberg Code established guidelines for "permissible medical experiments" and requirements that experiments be based on the results of animal experimentation in preliminary tests, be good for society, have results that are necessary, and be conducted by scientifically qualified persons. Additionally, human safeguards included that no experiment should put the patient at a risk for death or a disabling injury and that the medical treatment should not be more harmful than the problem being treated. Lastly, options to decline or end treatment by either the patient or the doctor were outlined. For the times, this was a landmark achievement for medical science.

Japanese Unit 731 Atrocities

Shiro Ishii, head of the terrible Japanese Unit 731, committed monstrous acts against human subjects. His unit (and others) conducted experiments on criminals, people tricked by the prospect of employment, pregnant women (often pregnant from being raped by the guards and doctors at the facility), mothers and children, and US prisoners of war.

Ishii was given command of Unit 731, known publicly under the innocuous name of Epidemic Prevention and Water Supply Unit of the Kwantung Army, in 1936. After the initial area where unit 731 was located in Manchuria was blown up by prisoners, the unit was moved to a laboratory in Pingfan. The locals were told it was a lumber mill, but in reality the "lumber" was human subjects.

Even though the unit had its main experimental camp in Manchuria, it conducted experiments on populations in many occupied areas. One experiment involved dropping deadly materials on populations of noncombatants via airplane. Up to 400,000 Chinese civilians are believed to have died due to being exposed to cholera, anthrax, and other agents. At the height of the unit's germ warfare production capabilities, 731 had the capacity to create enough bacteria to kill the entire world's population many times over. The camp at Pingfan, Manchuria, was producing 300 kg (about 660 pounds) of plague germs and bacteria per month at one point.

The details of the experiments are shocking. Ishii would use plague-infected rats to experiment with transmission of the deadly disease to subjects. Then, infected persons were put in prison cells with healthy victims to see how easily the disease spread. Some of the horrific experiments included the following:

- Men, women, children, and infants were vivisected (dissected while still alive) without anesthesia. Many of these subjects were pregnant women who had been impregnated by the doctors or guards, as daily rape of female prisoners was the norm and not the exception.
- Limbs were amputated from men, women, children, and infants to study blood loss.
- Surgery drills were conducted in which prisoners were shot so that surgeons could practice operating on them.
- Severed limbs were reattached to opposite sides of the body.
- Prisoners' limbs were frozen and amputated or frozen and then thawed to fully study the effects of gangrene.
- Prisoners had their stomachs removed while alive.
- Prisoners were hung upside down to see how long it took to die.
- Prisoners had horse urine injected into their kidneys.
- Starvation studies were conducted.
- Prisoners were subjected to extremely high temperatures to determine the relationship between temperature and human survival.
- Prisoners were exposed to lethal dosages of x-ray radiation.

Shortly after the bombing of Hiroshima in 1945, Unit 731 vanished after officials in Tokyo ordered the destruction of all incriminating materials, including the facilities in Pingfan, to avoid prosecution for war crimes. Many of these research efforts by Japan were conducted to develop biological warfare weapons, including plague, anthrax, cholera, and many other pathogens, to use against the United States.

After World War II, the United States wanted information and not retribution and granted immunity to some guilty Japanese in exchange for data. Previously unknown documents prove that General Douglas MacArthur gave these people their freedom and that the United States covered up the research that was being performed on human subjects and even gave researchers compensation instead of the trial they deserved. All of this information was suppressed from the American public, although many US servicemen were subjected to these tortures, including downed pilots from Doolittle's raid on Japan immediately following the sneak attack by the Japanese on Pearl Harbor.

Post World War II and the Cold War

Abuse of human subjects continued in the era of the Cold War—a state of hostilities, based on differing ideologies, between the United States and its allies and the Soviet Union and its allies that brought the world to the brink of nuclear annihilation. Tension and the need for ever-more diverse and powerful weapons pervaded all political and ethical thought during this period.

As has been noted, many German scientists were saved from the gallows, under the greatest secrecy, and put to work in the allied powers' research labs. One such Nazi, Dr. Hubertus Strughold, who was a pioneer in aviation medicine, obtained much of his research knowledge from human medical experiments on prisoners at Dachau concentration camp. Under US Operation Paperclip, he was brought to the United States, his past was scrubbed clean, and he was granted work as head of Aviation Medical Research for the Air Force. Later in his life, when the Air Force was in the final stages of naming a library after him, former Nazi prisoners recognized him and it was given another name. He was one of many offenders around the world who escaped criminal prosecution.

With atomic testing at a high point, many heinous human medical experiments were conducted on unknowing patients in the United States and abroad right after World War II. In the 1940s, mentally retarded and other unknowing patients were exposed to radiation in secret experiments, in which they were injected with plutonium to study its effects on the human body. It took until 1995 for the President's Advisory Committee on Human Radiation Experiments to state that *some* of the 1940s experiments were unethical.

As an example, Ebb Cade, a worker at Oak Ridge nuclear facility was involved in a bad car accident in early 1945. He was bed-bound with a broken arm and leg. Other than these injuries, the 53-year-old African American man was healthy. Once the doctors knew they had a healthy subject, they injected him—without his knowledge, of course—with plutonium to study the effects of the deadly radioactive material on him. Over the next few days, doctors collected tissue and bone samples, including pulling 15 of his teeth, to see how far the plutonium had entered his body. His night nurse checked on him, only to find that he had escaped. But he was brought back. He died of heart failure in 1953.

Billings Hospital in Chicago, the University of Rochester, and the University of California saw patients "treated" with plutonium and other radioactive isotopes. The thought was that since the patients would probably not live for more than 5 years due to their condition, it was ethical to experiment on them. Though many patients did have life-threatening illnesses, many were still living when investigations finally began in 1974. If patients or family were informed of anything during the experiments, it was only that they had received radioactive isotopes for medical treatment. After the investigations, families were informed of the truth, if possible.

Appendix D

Formulas for Healthcare Statistics

Chapter 2 Formulas

Percentage Rates

$$\frac{\text{Total number of times something actually happens} \times 100}{\text{Total number of times it could have happened}}$$

- Mean: The average of values in a dataset, determined by dividing the sum of all values in a dataset by the total number of values in the dataset
- Median: The midpoint of ordered numbers in a dataset from the lowest to the highest
- Mode: The most frequently recurring value in a dataset

Chapter 3 Formulas

Average Daily Census

$$\frac{\text{Total inpatient service days for a period (excluding newborns)}}{\text{Total number of days in the period}}$$

Average Daily Newborn Census

$$\frac{\text{Total newborn inpatient service days for a period}}{\text{Total number of days in the period}}$$

Remember the following:

- Daily inpatient census: The total number of patients that are treated during a 24-hour period
- Inpatient service day: A 24-hour period in which one inpatient receives services from the hospital
- Total inpatient service days: The sum of all inpatient service days for a period

Chapter 4 Formulas

Bed Occupancy

$$\frac{\text{Total inpatient service days in a period} \times 100}{\text{Total bed count days in the period}}$$

Bed Occupancy as a Percentage

$$\frac{\text{Total inpatient service days in a period} \times 100}{\text{Total inpatient bed count days} \times \text{Number of days in the period}}$$

Newborn Bassinet Occupancy Ratio

$$\frac{\text{Total newborn inpatient service days for a period} \times 100}{\text{Total newborn bassinet count} \times \text{Number of days in the period}}$$

Bed Turnover

Bed turnover works in conjunction with the total number of discharges in a given period.

The *direct formula* is as follows:

$$\frac{\text{Number of discharges (including deaths) for a period}}{\text{Average bed count during the period}}$$

This formula is used when there is a change in bed count during the period you are calculating.

The *indirect formula* is as follows:

$$\frac{\text{Occupancy rate} \times \text{Number of days in a period}}{\text{Average length of stay}}$$

Average Length of Stay

$$\frac{\text{Total length of stay (LOS) (discharge days)}}{\text{Total discharges (including deaths)}}$$

Remember the following:

- LOS: The number of calendar days from admission to discharge
- Total LOS: The sum of the days of any group of inpatients who were discharged in a certain time frame

Chapter 5 Formulas

Incidence Rate

$$\frac{\text{Number of new cases of a disease occurring in the population during a specified time period}}{\text{Number of persons exposed to risk of developing the disease during that period}}$$

Prevalence

$$\frac{\text{Number of cases of disease present in the population at a specified period of time}}{\text{Number of persons at risk of having the disease at that specified time}}$$

Gross Mortality

$$\frac{\text{Number of inpatient deaths (including newborns) in a period} \times 100}{\text{Number of discharges (including adult, children, and newborn deaths) in the same period}}$$

Fatality Rate

$$\frac{\text{Number of people who die of a disease in a specified period} \times 100}{\text{Number of people who have the disease}}$$

Net Mortality

$$\frac{(\text{Total number of inpatient deaths} - \text{Deaths within 48 hours of admission for a specified period}) \times 100}{\text{Total number of discharges} - \text{Deaths within 48 hours of admission for that period}}$$

Postoperative Mortality Rate

$$\frac{\text{Total number of deaths (within 10 days after surgery)} \times 100}{\text{Total number of patients who were operated on for a given period}}$$

Maternal Mortality Rate

$$\frac{\text{Number of direct maternal deaths for a period} \times 100}{\text{Number of obstetrical discharges (including deaths) for the period}}$$

Remember, when calculating maternal death rate, they are direct obstetric deaths.

Anesthesia Mortality Rate

$$\frac{\text{Total number of deaths caused by anesthetic agents} \times 100}{\text{Total number of anesthetics administered to patients}}$$

Newborn Mortality Rate

$$\frac{\text{Total number of newborn deaths for a period} \times 100}{\text{Total number of newborn discharges (including deaths) for a given period}}$$

Fetal Mortality Rate

$$\frac{\text{Total number of intermediate and/or late fetal deaths for a period} \times 100}{\text{Total number of live births} + \text{Intermediate and late fetal deaths for a given period}}$$

Cancer Mortality Rate

$$\frac{\text{Number of cancer deaths during a period} \times 1{,}000}{\text{Total number in population at risk}}$$

Hospital Standardized Mortality Ratio

$$\frac{\text{Number of observed deaths} \times 100}{\text{Number of expected deaths}}$$

Chapter 6 Formulas

Gross Autopsy Rate

$$\frac{\text{All autopsies on inpatient deaths during a particular time frame} \times 100}{\text{All inpatient deaths for the time frame}}$$

Net Autopsy Rate

$$\frac{\text{All autopsies on inpatient deaths during a particular time frame} \times 100}{\text{All inpatient deaths} - \text{Cases released to the coroner or medical examiner}}$$

Adjusted Hospital Autopsy Rate

$$\frac{\text{All hospital autopsies for a specified period} \times 100}{\text{All hospital patients who died and whose bodies are accessible for autopsy}}$$

Newborn Autopsy Rate

$$\frac{\text{Newborn autopsies performed during a particular time frame} \times 100}{\text{All newborn deaths for the same time frame}}$$

Fetal Autopsy Rate

$$\frac{\text{Autopsies conducted on intermediate and late fetal deaths for a particular time frame} \times 100}{\text{All fetal deaths (intermediate and late) for the same time frame}}$$

Chapter 7 Formulas

Nosocomial Infection Rate

$$\frac{\text{Total number of nosocomial infections for a given period} \times 100}{\text{Total number of discharges + Deaths for a given period}}$$

Infection Rate

$$\frac{\text{Total number of infections} \times 100}{\text{Total number of discharges + Deaths for a given time}}$$

Postoperative Infection Rate

$$\frac{\text{Total number of infections in clean surgical cases for a given period} \times 100}{\text{Number of surgical operations for the period}}$$

Complication Rate

$$\frac{\text{Total number of complications for a period} \times 100}{\text{Total number of discharges in the period}}$$

Cesarean Section Rate

$$\frac{\text{Total number of C-sections performed in a period} \times 100}{\text{Total number of deliveries in the period (including C-sections)}}$$

Consultation Rate

$$\frac{\text{Total number of patients receiving a consultation} \times 100}{\text{Total number of patient discharges}}$$

Chapter 8 Formulas

Case Mix Index

$$\frac{\text{Sum of the weights of MS-DRGs for patients discharged during a given period}}{\text{Total number of patients discharged}}$$

Payback Period Rate

$$\frac{\text{Total cost of project}}{\text{Annual incremental cash flow}}$$

Return on Investment (ROI)

$$\frac{\text{Average annual incremental cash flow}}{\text{Total cost of project}}$$

Other Formulas

Total Hospital Infection Rate (Morbidity)

$$\frac{\text{Total number of hospital infections} \times 100}{\text{Total number of discharges}}$$

Community-Acquired Infection Rate

$$\frac{\text{Number of community-acquired infections} \times 100}{\text{Total number of discharges}}$$

Delinquent Medical Record Rate

$$\frac{\text{Total number of delinquent records} \times 100}{\text{Average number of discharges during a completion period}}$$

Incomplete Medical Record Rate

$$\frac{\text{Total number of incomplete records} \times 100}{\text{Total number of discharges during the completion period}}$$

Percentage of Medicare Patients

$$\frac{\text{Total number of Medicare discharges} \times 100}{\text{Total number of adult and children discharges}}$$

Percentage of Medicare Discharge Days

$$\frac{\text{Total number of Medicare discharge days} \times 100}{\text{Total number of discharge days for adults and children}}$$

Readmission Rate

$$\frac{\text{Number of readmissions for a period} \times 100}{\text{Number of total admissions (including readmission)}}$$

Appendix E

CAHIIM Competency Exercises

Note: *AAS* is a 2-year associate of science degree; *BS* is a 4-year bachelor of science degree.

AAS I.6 Case Study

The Health Information Department at Sunrise Hospital runs daily reports on patient appointments for the physicians. The reports are showing errors and cannot be distributed until they are corrected.

Look at the **Table E.1** and find the errors. Hint: There are two errors.

Why is it important to have accurate information, and what are some potential consequences if inaccurate information is given?

BS I.6 Case Study

The Health Information Department at Sunrise Hospital runs daily reports on patient appointments for the physicians. The reports are showing errors and cannot be distributed until they are corrected.

Look at Table E.1 and find the errors.

Table E.1 Data Dictionary

Field Title	Field Name	Data Type	Description
Discharge Year	Discharge Year	Numeric	Year of discharge.
OSHPD_ID	OSHPD_ID	Numeric	A unique, six-digit identifier assigned to each facility by the Office of Statewide Health Planning and Development. The first two digits indicate the county in which the facility is located. The last four digits are unique within each county. Note: Hospitals consolidated under a single license may report combined data under one OSHPD ID.
Facility Name	Facility Name	Alpha-Numeric	The hospital name.
County Name	County Name	Plain Text	The county where the hospital is located.
Hospital ZIP	Hospital ZIP	Numeric	The ZIP code of the hospital.
MSDRG Code	MSDRG Code	Numeric	Three-digit Medicare Severity-Diagnosis Related Group (MSDRG) code.
MSDRG Description	MSDRG Description	Alpha	Description of the MSDRG.
Number of Valid Discharges	Discharges	Numeric	Total number of discharges where a valid charge was present. Excludes any discharges without valid charges.
Total Charges for Valid Discharges	Charges	Numeric	Total charges where a valid charge was present.

(continues)

Table E.1 Data Dictionary *(continued)*

Field Title	Field Name	Data Type	Description
Mean Charge Per Stay	Mean Charge Per Stay	Numeric	Average of the charge per stay where a valid charge was present on the record.
Mean Charge Per Day	Mean Charge Per Day	Numeric	Average of the charge per day where a valid charge was present on the record
Mean Adjusted Length of Stay	Mean Adjusted Length of Stay	Numeric	Average of the adjusted Length of Stay. Patients admitted and discharged on the same day are assigned a value of 1.
Ratio	Ratio Type	Plain Text	Top 25 MSDRG data is presented with several ratios: Mean charge per stay, Mean adjusted length of stay, and state average number of discharges. Datasets indicate which top 25 MSDRG ratios are available by type.
Rank	Rank	Numeric	Rank of MSDRG for each hospital.
Sum of Adjusted Length of Stay	Sum of Adjusted Length of Stay	Numeric	Sum of adjusted length of stay for all records. Patients admitted and discharged on the same day are assigned a value of 1.
Sum of Adjusted Length of Stay for Valid Discharges	Sum of Adjusted Length of Stay for Valid Discharges	Numeric	Sum of adjusted length of stay for records with valid charges. Patients admitted and discharged on the same day are assigned a value of 1.

Dataset: NumberofDischarges_Top25MSDRGs, MeanChargePerStay_TOP25MSDRG, and MeanLenghtofStay_TOP25MSDRG

List the six data quality standards. Evaluate why it is important to follow these standards, and list some potential consequences if these standards are not followed.

Discussion Board

AAS III.5

Name and describe three research methods used in health care. How would these methods be used in a healthcare setting?

BS III.5

Choose two research methods you would use in health care to compare and contrast. What do they have in common, and how do they differ from each other? Give three examples for how each research method is used in health care.

AAS III.1

Identify three areas of health informatics, and explain the role each would play in health information management.

BS III.1

Analyze two areas of health informatics, and explain what information can be gained from these areas to benefit a healthcare facility. Compare and contrast how each area can be used. Give examples.

AAS III.7 and BS III.7

Identify three standards used for health information exchange. Evaluate why these standards are important and what could potentially happen if they are not followed.

AAS Statistics Project

III.2, III.3, III.4, III.6

You are the Health Information Management Director for Sunshine County Hospital in Long Beach, California. Sunshine is a 150-bed general hospital. The chief financial officer has tasked you with reviewing the benchmarks from similar hospitals in your region regarding the length of stay by Medicare Severity Diagnosis-Related Group (MS-DRG).

Your responsibility as the HIM director includes gathering statistical data concerning length of stay as it relates to MS-DRGs. Gather the appropriate data from the Excel spreadsheet titled "AppendixE_Statistics_Project" in the eBook to complete the following tasks:

- Calculate the adjusted length for each MS-DRGs 3, 207 and 57.
- Calculate the mean charge per day for each MS-DRGs 3, 207 and 57.
- Calculate the mean charge per stay for each MS-DRGs 3, 207 and 57.
- Compare and contrast the MS-DRGs by the three preceding calculations.
- Build an Excel spreadsheet from your calculated data to create appropriate graphs that compare and contrast the preceding calculations.

Build a cohesive PowerPoint presentation that displays your graphs and the information you have extracted from the dataset. You will be presenting your PowerPoint to the Board of Directors of Sunshine Hospital. Here are a few helpful hints:

- Answer the questions in the order in which they are asked.
- Make multiple graphs to show data in a cohesive manner.
- Show a trend line in your graphs.
- Use the note option in PowerPoint to narrate your slides.

Presentations will be graded on the following:

- Accuracy (Both the calculations and the display construction should be accurate.)
- Display selection (You must select the appropriate display type for individual answers and, overall, use a variety of display types throughout the project.)
- Presentation length (It should be at least 10 slides.)
- Neatness and overall "look and feel" of the project (Remember, it will be presented to an executive board.)
- Creativity

BS Statistics Project

III.2, III.3, III.4, III.6

You are the Health Information Management Director for Sunshine County Hospital in Long Beach, California. Sunshine is a 150-bed general hospital. The chief financial officer has tasked you with reviewing the benchmarks from similar hospitals in your region regarding the length of stay by MS-DRG.

Your responsibility as the HIM director includes gathering statistical data concerning length of stay as it relates to MS-DRGs. Find a dataset that gives information on length of stay as it relates to different MS-DRGs. Gather the appropriate data from the dataset to complete the following tasks:

- Analyze two different hospitals on three different lengths of stay by MS-DRG.
- Compare and contrast your information with appropriate graphs with trend lines.
- Compare your data to show a benchmark. How do they differ? Why are they different?
- If Sunshine Hospital falls below your benchmark, what could be done to rise above the benchmark on length of stay by DRG?

Build a cohesive PowerPoint presentation that displays your graphs and the information you have extracted from the dataset. You must narrate and record your presentation. You will be presenting your PowerPoint to the Board of Directors of Sunshine Hospital.

Presentations will be graded on the following:

- Accuracy (Both the calculations and the display construction must be accurate.)
- Display selection (You must select the appropriate display type for individual answers and, overall, use a variety of display types throughout the project.)
- Presentation length (It should be at least 15 slides.)
- Neatness and overall "look and feel" of the project (Remember, it will be presented to an executive board.)
- Creativity

Glossary

A

a priori A Latin term that means "before the fact" and is often used to refer to knowledge that is obtained without observation or experience, typically through deductive reasoning, such as is used in solving a math equation.

Adjudicate To determine the insurer's and the member's financial responsibilities for payment, after the member's insurance benefits are applied to a medical claim, as in "insurance claims adjudication."

Adjusted hospital autopsy rate The rate of autopsies performed on inpatients, outpatients, and prior patients who have died outside the hospital and whose bodies are available for a hospital autopsy.

Admissions, discharge, and transfer (ADT) A set of calculations hospitals use to determine how many patients were admitted, discharged, or transferred each month to provide yearly totals for each patient care unit in the facility.

Affordable Care Act (ACA) A US federal law passed in 2010 with the goal of increasing the number of Americans covered by health insurance, improving health care, lowering costs, and offering incentives to healthcare facilities to become more efficient.

Algorithm A step-by-step procedure for solving a problem, often performed by a computer.

Alternative hypothesis A guess at the answer to a research question that is in the affirmative, as opposed to the null hypothesis statement, which answers it in the negative.

Analysis of covariance (ANCOVA) A statistical test that is an extension of analysis of variance (ANOVA) and that involves both ANOVA and regression analysis.

Analysis of variance (ANOVA), one-way A statistical test used to determine whether there are any significant differences between the means of three or more independent (unrelated) groups.

Applied research A type of research that is less theoretical in nature and aimed more at improving a process or practice, changing a situation, or solving a problem. Also known as action, intervention, or collaborative research.

Appraisal A unique data collection method that takes into consideration not only statistics gathered but also strengths and weaknesses of the study.

Array A group of the same type of data.

Aspect A specific category of data that is used in conjunction with your viewpoint.

Assent An agreement to participate in a research study that is given by someone other than the subject on behalf of the subject, in cases in which the subject is not able to give legal consent.

Autopsy A surgical procedure performed on patients who have died to verify the exact cause of death; also known as a postmortem examination or necropsy.

Average daily census The average number of patients in a facility on a given day.

B

Balanced design An ANOVA design in which each group has the same number of data.

Bed count The available number of hospital inpatient beds that are either vacant or occupied on any given day of the week.

Bed count day The attendance of one inpatient bed that is set up and staffed for a 24-hour time frame in the hospital.

Bed occupancy ratio A measure of the proportion of beds that are occupied at any given time at a hospital, often expressed as a percentage.

Bed turnover rate The number of times that a bed changes occupants in a given time and/or the typical number of admissions for each bed during a period of time.

Belmont report A detailed report of what should be considered when conducting research involving human subjects to ensure they are treated safely and fairly.

Benchmark The number of items that meet outcome criteria divided by the total items in the database; also known as benchmark confidence.

Beneficence The value of having the welfare of the research participants as a goal of any research studies or clinical trials.

Binary code The form of a computer program that users download, install, and run; also known as the executable or compiled code.

C

Cancer registrars Professionals who collect cancer data by working closely with physicians, administrators, and researchers.

Capital budget Money designated for the purchase of items that cost more than $500 or some other amount set by administration.

Case mix index A measure of a facility's use of health services in providing care to patients, such as Medicare, Medicaid, and other third-party payers.

Case mix report Records the amount of resources consumed by and clinical severity of specific groups of patients and compares these data against those of other groups.

Case study A study that focuses on one individual or community rather than on a demographic group.

Catheter-associated urinary tract infection (CAUTI) An infection of the urinary tract—kidney, ureters, bladder, or urethra—caused by use of a urinary catheter; the most commonly reported type of healthcare–associated infection.

Census data Data that describe patients who are currently in the hospital during a certain time frame.

Certified coding specialist (CCS) Healthcare professionals who abstract and classify medical data from patient records, typically in a hospital setting.

Certificate of need (CON) A statement issued by a government agency in many states that is required for construction or modification of a healthcare facility and that is designed to help reduce and control healthcare facility costs and prevent duplication of services.

Certified coding specialist–physician-based (CCS-P) Coding practitioners specifically trained to work in physician offices, group practices, clinics, and specialty centers.

Chi-square test A type of nonparametric test. The two types of chi-square tests include the chi-square goodness-of-fit test, a one-variable test used to determine how close a categorical variable aligns with a theoretical model, and the chi-square test of independence, a two-variable test used to determine whether there is an association between the two variables.

Chief executive officer (CEO) An administrative head of a company, responsible for overseeing the day-to-day operations.

Chief financial officer (CFO) An executive officer responsible for overseeing a company's finances, including managing financial risks, planning, and reporting.

Clinical autopsy An examination of a body after death conducted for medical purposes to better understand the conditions that caused the patient's death and to help hospitals ensure and improve quality of care at the facility.

Cluster sampling A research method in which members of the population under study are assigned to clusters, some of which are randomly selected to serve as the sample to survey.

Code (1) A special language used in computer software programming; (2) a patient diagnosis or medical procedure code (e.g., ICD-10-PCS).

Code of Federal Regulations (CFR) The codification of the general and permanent rules and regulations that are published in the *Federal Register* by executive departments and agencies of the US federal government.

Cognitive dissonance A theory of psychology that suggests that we have an inner need or drive to keep our beliefs and attitudes in harmony and avoid dissonance (disharmony); based on the work of Festinger and others.

Cohort study A type of research study that focuses on a specific population over an extended period of time.

Collaborative Institutional Training Initiative (CITI) A group of experts who offer institutional review board and other certifications.

Comma-separated values (CSV) A type of data in which data fields are separated by commas.

Commercial code A type of computer programming code that is owned by a company and that is not free to share with others.

Comorbidity condition A concomitant but unrelated pathological or disease process.

Complication rate A measure of the frequency with which complications occur.

Computer-assisted coding software (CACS) A type of computer software that analyzes healthcare documents and produces the appropriate medical codes for specific phrases and terms within the document.

Concomitant Existing or occurring together, as in "poverty and poor diet are concomitant."

Confidence A ratio that takes the support number and divides it by the number of instances where the rule may hold true.

Consent An agreement to participate in a research study that is given by someone who has reached the legal age for consent, which is 18 years old in the United States.

Consultation A second opinion regarding the diagnosis from another physician.

Consultation rate The frequency with which a physician requests consultations.

Consumer-mediated exchange A form of health information exchange that allows patients to add information about themselves to improve their health care and to identify any missing or incorrect information.

Continuous data Data that have infinite values with connected data points, which can result in an unlimited selection of data; data that occur along a continuum, such as waveforms on a heart rate monitor, which represent every instant of time within the period being studied.

Convenience sampling An approach to research study sampling based on participants who are easy to reach, such as members of an audience or passersby on the street.

Coroner A person who collects data for statistical analysis, including net death rates and other rates pertaining to death, to look for patterns in deaths from certain diseases.

Critical path method (CPM) A way of managing a project based on individual activities.

Cyclical variations Changes that occur predictably at certain points within a cycle, such as a day, a week, or a year; for example, emergency department visits increasing exponentially on Friday nights compared with all other days of the week.

D

Data Units of information, such as measurements, that can be collected and interpreted; singular, *datum*.

Data dictionary A description of a database, databases, or format for cells in a spreadsheet or database; also known as data definition matrix, content specifications, metadata repository, and other such terms.

Data-driven An approach to making decisions based on statistical or population information (e.g., "data-driven decisions").

Data element A variable that has a name, type, size, and value, such as age, integer (whole number), or bit depth; a computer programming variable.

Data mining The process of extracting information from a large set of data.

Dataset Data collected on a subject under examination.

Decision model A logic system used to determine desired actions for a business based on thresholds, conditions, or events; also known as a business rule or business logic.

Declaration of Helsinki Internationally established ethical standards for human subject protection.

Degrees of freedom A measure of the number of values that are free to vary in a statistical calculation.

Dependent variable A variable in a research study that depends on or is affected by another variable (the independent variable) and that is measured to determine the extent of this change.

Descriptive model An approach to decision making that involves identifying the values, uncertainties, and other issues relevant in a given decision, its rationality, and the resulting optimal decision.

Descriptive research A type of research concerned with describing and interpreting what is, by explaining and summarizing the data being examined.

Descriptive statistics A type of statistics concerned with mathematical quantities such as mean, median, and standard deviation, which summarize and interpret some of the properties of a set of data but do not infer necessarily the properties of the entire population from which the sample was taken.

Direct deaths A mortality rate that measures cases in which a mother dies from complications associated with a procedure that is directly related to pregnancy or birth, such as incidental damage to blood vessels during a cesarean section.

Directed exchange A form of health information exchange used to send information such as laboratory results, patient referrals, and discharge summaries directly to providers.

Discharge report A report compiled at discharge that shows the services provided to the patient by the hospital, organized by department.

Discharged not final billed (DNFB) A report listing patient accounts that are not yet billed at the time of discharge, including those that have been suspended.

Discrete data Data that are finite, or categorized such that they cannot be subdivided, such as counts of parts, a whole number range of values (such as 1 to 5), every hour, by month (1 to 12); for research purposes, these values are often defined before data collection.

Documentary study A form of historical research that considers documents and stored records to provide data for the research.

Drug trial A controlled test of new drugs or invasive medical devices in which human subjects take part; conducted under the direction of the US Food and Drug Administration before the drug or device is made available for general clinical use.

E

Early fetal death Death that occurs at less than 20 weeks of gestation with a fetal weight of 500 grams or less.

Electronic data management system (EDMS) A group of technologies that work together to manage the storage, retrieval, indexing, and disposition of records, along with information assets of the organization.

Empirical A type of knowledge gained through scientific observation and experience.

Employee compensation Benefits that an employee receives in exchange for services provided to the employer.

Enterprise report management (ERM) A type of computer application that incorporates payroll, cost accounting, managed care, and detailed patient information in one database.

Error A variable in a statistical model that is created when the model does not fully represent the actual relationship between the independent variable(s) and the dependent variable.

Ethics A field of study that involves matters of right and wrong.

Evidence-based medicine (EBM) Medical practice that is based on the best available current methods of diagnosis and treatment as revealed in research.

Expected mortality rate The predicted number of deaths in a hospital based on the various levels of illness of the patients in the hospital, with very sick or medically complicated patients having a higher expected mortality rate.

Exploratory research A type of research that focusses on learning about new subjects or those for whom there is little existing information and on developing knowledge of a topic rather than improving a process.

F

***F* test** A statistical measure developed R. A. Fisher that compares two variances or standard deviations to see whether they are equal.

***F* value** A calculation that compares two variances or standard deviations to see whether they are equal; also known as *F* calculation.

False negative A research error involving a failure to reject a false null hypothesis, associated with type II errors.

False positive A research error involving a rejection of the null hypothesis when it is true; that is, there seems to be some statistical significance in the results when there is not; associated with type I errors.

Fatality rate A measure of the number of people who die from a specific disease or accident.

Fetal autopsy rate A measure of fetal autopsies calculated as follows: the number of autopsies on intermediate and late fetal deaths (× 100) divided by total number of intermediate and late fetal deaths.

Fetal death The death of a fetus in which the fetus is expelled and does not breathe or show any evidence of life through pulsations of the umbilical cord, heartbeat, or muscle movement.

Financial reports A collection of reports about an organization's financial results, financial condition, and cash flows.

Flat-file database A database that does not contain linked tables.

Flowchart A graphical way to depict a series of actions, given a certain series of events, such as the waking up in the morning and going to work, changing a toner cartridge, or performing a patient intake.

Forced Likert scale A rating scale that does not feature a "no response" option, which forces respondents to give a positive or negative rating.

Forecasting An effort to predict future trends.

Forecasting trend analysis A method of determining future patterns based on past data. For example, a company may determine that, based on data from past years, they need extra staff on certain days of the year (e.g., holidays) or that they will spend more on electricity bills during certain months. Accurate historical information is important for reliability.

Forensic autopsy A type of autopsy conducted for legal purposes and to determine the cause and manner of death, such as when criminal activity is suspected in the person's death, as in homicide.

Frequency distribution A statistical measure that describes how often different values are found in a set.

Full-time equivalent (FTE) A ratio of the total number of paid hours during a specific time period for full-time and part-time employees and those who are contracted by the number of working hours in the time period.

G

Gantt chart A visual representation of a project schedule showing start and finish dates of the different required elements, usually as a type of bar chart.

General Public License (GPL) A type of computer software in which the original source code may be copied and distributed for free as long as the user references the original author's work.

Generalizability The ability of a research study sample to accurately represent the entire population being examined and thus to yield the same findings as if the entire population were studied.

GNU's not Unix (GNU) An open-source version of the Unix operating system.

Golden hour The principle that a victim's chances of survival are greatest if the person receives resuscitation within the first hour after a severe injury.

Graphical user interface (GUI) A computer interface that presents a visual metaphor for user interaction; all data are entered using a mouse, touch screen, or voice input, as opposed to a text-based interface, in which all commands are typed.

Gross autopsy rate A measure of all autopsies, regardless of who performed them, on inpatient deaths for a period compared with the total inpatient deaths for the same period; typically reported as a percentage.

H

Health care appraisal (HCA) An assessment of an employee's health status.

Healthcare-associated infection (HAI) An infection that patients acquire while receiving treatment for medical or surgical conditions.

Health Data Interactive A powerful online search tool for harvesting healthcare data from many areas, offered by the Centers for Disease Control and Prevention.

Health Information Exchange (HIE) As defined in 2019 by the Office of the National Coordinator for Health Information Technology: "Electronic exchange of clinical information [that] allows doctors, nurses, pharmacists, other healthcare providers, and patients to access and securely share a patient's vital medical information electronically."

Healthcare Infection Control Practices Advisory Committee (HICPAC) A committee established in 1991 by the US federal government with members being appointed by the Secretary of Health and Human Services; it provides advice and guidance related to isolation practices and serves as an advisory committee to the Centers for Disease Control and Prevention for updating guidelines and policy statements related to control of nosocomial infection.

Hospital autopsy A postmortem exam of a person who was a hospital patient, an emergency department patient, a walk-in outpatient, or a home-care patient.

Hospital standardized mortality ratio (HSMR) An adjusted hospital mortality rate, based on the mortality rate of other facilities.

Hybrid A "mixed" version of data mining that employs both *a priori* and *posteriori* techniques.

Hypothesis A statement or "educated guess" based on prior knowledge that can be proven or disproven based on data; for example, garlic keeps mosquitoes away, so eating a clove of garlic per day will keep you mosquito free.

I

Iatrogenic A disease or condition that is caused by the diagnosis, manner, or treatment of a physician.

Independent variable A variable in a research study that we can measure, manipulate, or control for to produce a change in another variable (the dependent variable); also known as an experimental or predictor variable.

Indirect obstetric death A mortality rate that measures cases in which a pregnant woman dies due to some underlying condition, such as cancer, rather than something directly related to the pregnancy.

Infection control committee A committee responsible for managing many areas related to infection control, including planning, monitoring, evaluating, educating, and dealing with reportable information. Planning involves the development of policies and procedures as well as interventions to help prevent and reduce infections.

Inferential statistics A type of statistics used to make predictions or infer about future events or trends based on past data.

Infographics A type of data visualization used by large corporations, healthcare facilities, and anyone who operates a business.

Information collection Gathering of data for nonresearch purposes, such as for records, reporting (perhaps to an advisory board), or quality improvement.

Informed consent A requirement of healthcare providers and researchers to inform patients of all risks and benefits associated with medical treatment or a research study and to request and obtain the patient's signed consent to undergo the treatment or participate in the experiment; participation must be consensual, so patients or subjects must be allowed to opt out, with no repercussions, if they do not want to participate.

Inpatient hospital autopsy A type of autopsy performed only on patients who died during hospitalization after being formally admitted.

Inpatient service days A measurement of the services that a patient receives within a 24-hour time frame.

Institutional review board (IRB) A sanctioned body that determines whether research is safe to conduct within an organization.

Instrument A tool used to collect data for research purposes.

Intensive care unit (ICU) An area of the hospital reserved for patients with severe illnesses or injuries that require constant monitoring.

Intermediate fetal death Death of a fetus that occurs at or following 20 weeks of gestation, but before 28 weeks, with a fetal weight between 501 and 1,000 grams.

International Classification of Diseases (ICD) A system of coding medical diagnoses for billing purposes.

Interquartile range A measure of the dispersion within a dataset, being the difference between the third quartile and the first quartile.

Interval data A type of data in which differences between any two options are equal.

Interview A data collection approach that involves the researcher meeting with the respondent, whether in person or remotely (e.g., via Skype, FaceTime, or other apps), and asking questions.

Irregular fluctuation A change in otherwise ordered data that does not seem to fit with the overall pattern the data display; associated with outlier data.

J

Justice The fair treatment and selection of participants.

L

Labor cost The sum of all wages paid to employees, including benefits, and payroll taxes paid by an employer.

Language A formal syntax of programming code, such as C++, BASIC, or Java.

Late fetal death The death of a fetus at birth; also known as a stillbirth.

Left-hand side (LHS) A value in an assignment statement on the left-hand side of the equation; for example, *birth_rate* = *x1* + *z3*, in which *birth_rate* is the LHS value.

Length of stay (LOS) A measure of the actual number of days that a patient is admitted in the hospital.

Lift A measure of the importance of a data mining association rule.

Lift chart A visual aid for measuring model performance.

Likert scale A rating scale that asks the respondent to select a value from a continuum of possible answers, typically presenting four to seven options.

Linear regression A type of statistical analysis in which the relationship between historical data for one independent variable and one dependent variable is used to predict future values for the dependent variable.

Longitudinal study A study that is carried out over an extended period of time.

M

Machine learning A process whereby computer programs can detect patterns in data and adjust their actions accordingly, improving their own understanding.

Mann-Whitney *U* test A type of nonparametric test that may be used in place of a t test in studies in which there are two samples, but with one population having larger values than the other. Also known as the Mann-Whitney-Wilcoxon, Wilcoxon rank-sum, or Wilcoxon-Mann-Whitney test.

Market basket analysis A method of data mining that seeks to determine customer patterns so the store can more effectively market its products to customers.

Matrix A table-type format in which data are stored.

Max The maximum or largest value in a dataset.

Mean A measure of central tendency that can be determined by mathematically calculating the average of observations (e.g., data elements) in a frequency distribution; considered the most common measure of central tendency.

Median A measure of central tendency that reflects the midpoint of a frequency distribution when observations are arranged in order from the lowest to the highest.

Microchart A graphical representation of cells with numeric data.

Min The minimum or smallest value in a dataset.

Mode A measure of central tendency that consists of the most frequent observations in a frequency distribution.

Morbidity A measure of the incidence of disease.

Mortality rate A measure of the number of deaths per unit of the overall population, with the unit typically being 1,000, 10,000, or 100,000.

Mortality ratio The observed number of actual deaths divided by the number of expected deaths in a certain population.

Multistage sampling Use of several probability sampling methods to identify a sample.

N

Naive Bayes classification (NBC) A data mining method that assumes there is independence between all the observations or variables.

National Center for Health Statistics A powerful online search tool for harvesting care data on various US health trends, including crude death rates, infant mortality rates, and other types of data.

Neonatal death The death of an infant that occurs up to 28 days following birth.

Net autopsy rate A ratio of all inpatient autopsies to all inpatient deaths minus cases that were released to the coroner or medical examiner and thus not autopsied in the hospital.

Network sampling An approach to research study sampling that uses social or other network connections to help find participants.

Newborn autopsy rate The number of newborn autopsies performed during a particular time frame × 100 divided by all newborn deaths for the same time frame; calculated separately from adults and children, depending on administration decision.

Newborn death The death of a newborn that occurs during the same hospital admission as the birth.

Nominal data A type of data that indicates categories and cannot be ordered.

Nonparametric test A type of statistical test that is based on fewer assumptions about the data in question and can be used on rank and ordinal data; compares median rather than mean because the latter would not be valid due to the non-normally distributed data.

Nonprobability sampling A sampling technique in which the samples are gathered in such a way that not all individuals in the population have an equal chance of being selected.

Normal distribution A type of data distribution that, when graphed, forms a symmetric, bell-shaped curve.

Nosocomial infection A type of infection acquired by patients during their stay in a hospital.

Nosocomial infection rate A measure of the occurrence of nosocomial infections over a certain period of time in hospitals and other healthcare facilities.

Null hypothesis In a research study, the attempt to show that no significant variation exists between variables, or that a single variable is no different than zero, meaning that one variable does not effect a change in the other.

Object linking and embedding (OLE) A method whereby an object created in one application is embedded in another but remains linked to the original application such that changes to the object in the original application will affect the image as embedded in another application.

Observation A data collection method in which the researcher watches events, behaviors, or physical characteristics and documents findings.

Observed mortality rate A measure of the number of patient deaths that occur in the hospital for any reason.

One-tailed test A statistical test that examines one direction or end of the curve.

One-way ANOVA A type of ANOVA test used when you have one categorical variable (e.g., blood type—A, B, AB, or O) and one independent variable (the variable that is manipulated by the researcher).

Open-source software A type of software for which both the source code and binary code are given away freely.

Operation A surgical intervention to save a person's life or conduct repairs.

Operational budget A type of budget that compares the amount budgeted for each department with the actual amount spent by each department.

Ordinal data A type of data that can be ordered (e.g., in terms of size), although the difference between any of the two values may not always be equal.

Outlier data Elements of a dataset that lie an abnormal distance from other values in the dataset or sample.

P

***P* value** An estimate of the probability that a result happened by accident and the datasets really are not significantly different.

Paid time off (PTO) A benefit offered by employers that provides employees a bank of hours that they may use to take paid time off from work for sickness, vacation, and personal days.

Panel study A type of research study that is similar to a trend study but uses the exact same group of individuals—the panel—over the course of the study. Because the same respondents are surveyed each time, this type of study allows the researcher to find out why changes in the population are occurring.

Parameters A statistical tool that allows researchers to make assumptions about a larger population based on a random sampling of the population.

Parametric tests A type of statistical test that assumes a normal distribution of the data under observation and that variance is the same within each dataset (homogeneity of variance).

Pathologist A physician who specializes in the study and diagnosis of diseases.

Patient care unit (PCU) The number of patients admitted that day, discharged, or transferred in or out, perhaps to another hospital or unit.

Patient encounter Any contact between the patient and a healthcare provider.

Payback period In accounting, the period of time needed to recover the cost of equipment purchased.

Percentage A type of ratio that compares a number to 100, with 100 representing a whole.

Perinatal death Death of an infant during birth, or stillbirth.

Personal time, fatigue, and delay factor (PF&D) A time allowance required by the US Department of Labor, Wage and Hour Division, in recognition that normal working fatigue prevents employees from producing at their highest rate all throughout the workday.

Pivot table A tool found in spreadsheets that can sort, count, average, and summarize the data it references.

Population A dataset that includes information on every member of the group being investigated, rather than a sample of data, which is a subset of data that statistically represents the entire population, ideally.

Post-hoc analysis A type of statistical analysis used to determine whether there is a significant difference between groups under investigation in a research study.

Posteriori After the fact.

Postmortem An examination of the body after death; an autopsy.

Postneonatal death Death of an infant that occurs from 28 days after birth to 1 year of life.

Postoperative infection rate A measure of the frequency with which infections occur in hospital patients following surgery.

Predictive analysis An area of data mining that deals with extracting information from data and using it to predict future trends and behavior patterns.

Predictive model An approach to data mining that explicitly predicts future behavior based on past trends.

Primary data Data observed or collected directly from firsthand experience.

Probability sampling A sampling technique in which the samples are gathered in such a way that all individuals in the population have an equal chance of being selected.

Productivity A measure that allows the health information technology department to categorize the hospital's labor as either a product or service.

Project management The process of directing, tracking, and problem solving.

Purposive sampling A type of nonprobability sampling; also known as judgmental, selective, or subjective sampling.

Q

Query-based exchange A form of health information exchange that allows emergency department physicians to query information such as medications, X-ray images, and any problems listed.

Questionnaire A data collection instrument that poses questions to the respondent via a paper form or website. It can be either closed- or open-ended.

R

Radio frequency identification (RFID) A type of device that may be ingested or implanted in the human body and that transmits signals related to certain events occurring in the body.

Random A type of selection in which each item of a set has an equal probability of being chosen.

Range The distance between two possible values.

Rate A type of ratio in which the two quantities being compared are of different units of measure.

Rating scale A data collection instrument that enables the respondent to select a value from a continuum of possible answers, such as "least" to "most likely" or "no pain" to "worst pain"; for example, a Likert scale.

Ratio A comparison of two quantities.

Ratio data A type of data that can be ordered and that is characterized by the distance between any two consecutive values being the same (interval data) and there being an absolute zero. This means that a meaningful negative value of interval data does not exist (in statistics).

Readmission rate report A hospital statistical report that summarizes inpatient services and tracks medical readmissions.

Real-time locating (RTL) systems A type of device that allows patient tracking (such as with Alzheimer's patients), facilitates billing, and provides drug interaction data.

Recapitulation algorithm A calculation that verifies the data either monthly or yearly, which means it provides a summary of the data collected for a given period.

Registered health information administrator (RHIA) A healthcare professional who works as a critical link between healthcare providers, payers, and patients and is an expert in managing patient health information. This certification is obtained by completing an accredited program by CAHIIM and testing through AHIMA.

Registered health information technician (RHIT) A healthcare professional who ensures the quality of medical records by verifying their completeness, accuracy, and proper entry into computer systems. This certification is obtained by completing an accredited program by CAHIIM and testing through AHIMA.

Relational database A computer database that has one or more tables that are linkable.

Release of information The process of releasing patient information by healthcare providers, which is governed by the guidelines of the Health Insurance Portability and Accountability Act (HIPAA), also known as the Privacy Rule.

Respect for persons The concept that stresses that human beings have intrinsic and unconditional moral worth and should be treated as if there is nothing of greater value than they are.

Return on investment The extent and timing of gains from investments compared against the magnitude and timing of the investment cost.

Right-hand side (RHS) A value in an assignment statement on the right-hand side of the equation; for example, *birth_rate* = $x1 + z3$, in which $x1 + z3$ is the RHS value.

Risk-adjusted mortality rate A measure of the incidence of mortality, adjusted for risks.

S

Sample A representative part or a single item from a larger whole or group.

Sample frame All of the people in a population who could be sampled.

Sampling bias A bias in which a sample is collected in such a way that some members of the intended population are less likely to be included than others.

Scalar A type of data used to differentiate a single number from a vector or matrix.

Scientific notation A format for displaying large numbers in which an exponent multiplies the preceding number by 10 to the specified power.

Script-mode interface An interface that displays code as a script, rather than line by line.

Seasonal fluctuations An attribute of data whereby they may vary in concordance with seasonal changes, such as allergy visits peaking every April and traumatic injuries occurring the least in January.

Secondary data A dataset collected by someone other than the primary researcher.

Selective sample A sample in which the researcher selects the subjects who will be investigated.

Significance A percentage measure of the odds that the result of a research study did not happen by statistical accident.

Simple random sampling A type of sampling in which each member of a subset of a statistical population has an equal probability of being chosen; meant to produce an unbiased representation of a group.

Skew To change in direction or position.

Snowball sample A type of nonprobability sampling in which you gather information on your population as best you can, but you enlist the help of the people you have contacted to refer other members of the group you are examining; also known as chain-referral.

Social exchange theory A theory that survey participation is determined by three factors: minimal time involvement, reward for completion, and reasonable odds of winning the reward.

Source code The form of a program as it was written by a programmer in a computer language that is readable by humans; also known as precompiled code.

Sparklines A type of graphical representation.

Staffing The selection and training of individuals for specific job functions, and charging them with the associated responsibilities.

Standard deviation A measure of variability that describes the deviation from the mean of a frequency distribution of data.

Statistics A series of methods to collect, analyze, and interpret masses of numerical data.

Stratified sampling A type of research study sampling involving the establishment of groups or strata of the population.

Sunset The end of the life cycle for a program.

Support The number of data points that meet a set of rules and/or assumptions.

Surgical procedure A procedure that is performed to repair an injury to a specific body part.

Survey A set of questions people are asked to gather information or find out their opinions, or the information gathered by asking many people the same questions.

Systematic random sampling A type of research study sampling having a numbered list with all of the members of a population from which numbers are randomly selected, such as every third person, or whatever value is chosen.

T

t test A test of significance that is used to compare the means of two groups to determine whether they are statistically different from each other.

Tab-delimited values Data that are separated, or delimited, by tab spaces.

Tail Either one or both extreme ends of a bell curve.

Taxonomy A classification system for levels of intellectual behavior important in learning as well as for teaching and assessing. The resulting model, known as Bloom's taxonomy, contains six levels of activity, which from the simplest to the most complex are as follows: remembering, understanding, applying, analyzing, evaluating, and creating.

Telehealth A method of communication that uses electronic information and telecommunications technologies to support long-distance clinical health care, patient and professional health-related education, public health, and health administration.

Text corpus A structured dataset of textual information that is put into a format better for data analysis.

Text mining A method to reveal hidden information in unorganized text or numeric data.

Time series analysis A powerful way to visually display and analyze different points of data over a continuum; also known as time-trend analysis.

Transcription Manual processing of voice reports dictated by physicians and other healthcare professionals into text format.

Trend study A study that involves repeated surveying of a unique sample of the population, typically over a long period of time, with different groups of people in the same population and potentially with different researchers.

Tukey honest significant difference (HSD) A pairwise comparison that is essentially the same process as that used for conducting a one-way ANOVA, except the user selects the pairwise comparison option when using R.

Two-tailed test A statistical test that examines the hypothesis versus the null hypothesis to see whether they are equal.

Two-way ANOVA A type of ANOVA test used when you have two nominal variables (also known as effects, measures, or factors) and one measurement variable.

Type I error A false positive test result, which occurs when one rejects the null hypothesis when it is true; that is, there seems to be some statistical significance in the results when there is not.

Type II error A false negative test result, which occurs when one fails to reject a false null hypothesis.

Unbalanced design An ANOVA design in which the groups have differing numbers of data.

Variable A characteristic or property of something that may take on different values.

Variance A measure of the spread of observations in a distribution of data.

Vector A row of values.

Verbal autopsy A type of autopsy in which health information is gathered about a deceased person to help determine the cause of death; commonly conducted by government agents in developing countries to collect valuable national data.

Viewpoint A specific context or approach that one intentionally adopts when examining the data that allows one to focus on relevant details in the study and ignore irrelevant data.

Visual analogue scale (VAS) A type of assessment scale in which respondents select visual representations of responses, as in a pain index, which is represented by pictures of faces showing states of happiness or sadness.

Volunteer sample A type of research study sample composed of people who choose to be a part of the sample based on their interest.

z score A measure of the deviation of a value (in standard deviations) from the mean of a dataset.

Index

A

Abbreviations, 229
Adjusted hospital autopsy rate, 114–115
Admissions, discharges, and transfer (ADT), 30
Affordable Care Act (ACA), 213
Algorithms, 70
Alternative hypothesis, 96
American College of Surgeons (ACS), 5
Analysis of Covariance (ANCOVA)
  cholesterol data, 220–221
  output, 223
  scatterplot data, 222–223
Analysis of Variance (ANOVA), 117, 119
  one-way, 119
  post-hoc analysis, 144
  two-way, 119–126
Anesthesia mortality rate, 91
Applied research, 172
Appraisal, 201
A priori, 70
Arrays, 117
Aspect, 9
Assent, 177
Autopsy
  adjusted hospital autopsy rate, 114–115
  Centers for Disease Control and prevention data, 112
  clinical autopsy, 110
  definition, 109
  fetal autopsy rate, 116
  forensic autopsy, 110
  global perspective, 126
  gross autopsy rate, 112–113
  hospital autopsy, 110, 111
  inpatient hospital autopsies, 114
  net autopsy rate, 113–114
  for newborns, 115–116
  pathologist, 110
  statistical measures
    analysis of variance, 119–126
    F test, 116–119
Average daily census, 34–35
Average length of stay, 61–66

B

Balanced design, 119
Bed count
  bed occupancy ratio, 57
  day, 57
  definition of, 56–57
  government funding and research, 57
Bed occupancy ratio, 57
Bed turnover rate, 59
Belmont Report, 173, 175–176
Benchmark, 72
Beneficence, 173
Binary code, 41
Bureau of Labor Statistics, 4
Business logic, 9–10
Business rule, 9–10

C

CAHIIM competency exercises, 237–239
Capital budget, 163–166
Case mix index, 161
Case mix report, 161
Census data
  average daily census, 34–35
  data visualization, 35–38
  financial performance, 31
  infant-to-nurse ratio, 31
  inpatient care areas, 31–32
  inpatient service days, 33–34
  microcharts, 39
  newborn services, 41
  nosocomial infection, 31
  nurse-to-patient ratio, 31
  reporting and analysis, 32
  RFID technology, 32
  sparklines, 38–40
  US Department of Veterans Affairs, 32
  VistA, 32
Centers for Medicare and Medicaid Services (CMS), 4, 52, 60, 161–162
Central tendency
  frequency distribution, 18–19
  mean, 15–16
  median, 17
  mode, 17–18
Certificate of need (CON), 5, 58–59
Cesarean section rate, 137–138
Chi-square test, 179–180
Clinical autopsy, 110
Cluster sampling, 192
Code, 41
Cohort study, 199
Cold War, 233
Comma-separated values (CSV), 45
Commercial code, 41
Complication rates, 136–137
Complications/comorbidities (CC), 60
Computer-assisted coding software (CACS), 211
CON (Certificate of Need), 5, 58–59
Confidence, 72
Consent, 177
Consultation, 138–139
Consumer-mediated exchange, 212–213
Continuous data, 18
Convenience sampling, 191
Coronavirus, 224
Coroner, 109
Critical path method (CPM), 217
Cyclical variations, 214

D

Data, 2
Data collection and methods
  appraisal, 201
  health care appraisal (HCA), 201
  information collection, 191
  interview, 201
  multiple regression analysis, 205–208
  nonprobability sampling, 191–192
  observation, 201
  pivot table, 202–204
  probability sampling, 192
  quality question design
    Bloom's taxonomy, 196–197
    case study, 200
    cohort study, 199
    documentary study, 200
    longitudinal study, 199
    panel study, 199
    trend study, 199
    types of, 197–199
  questionnaire, 201
  random sampling, 193–196
  sampling, 191
  social media networks, 208
  survey, 200–202

Data dictionary
cell properties, 177–178
definition of, 177
Data distribution, 8
Data-driven, 5
Data formats, 45–46
Data harvesting, 25–26
Data mining
algorithms, 70–71
applications, 6–7
benchmark, 72
confidence, 72
data distribution, 8
dataset, 7–8
data types, 8–9
decision model, 9–10
definition, 5
descriptive model, 9
history, 6
market basket analysis, 71–72
morbidity and mortality data, 80
naïve Bayes classification (NBC), 166–168
occupancy and utilization data, 56
predictive model, 9
primary data, 10
R-Project, association rules, 70–74
resources, 230
secondary data, 10
support, 72
text mining, 146–149
variables, 8
Dataset, 7–8
Data visualization, 35–38
Decision model, 9–10
Declaration of Helsinki, 175
Degree of freedom (df), 101
Department of Health and Human Services (DHHS), 4
Department of Veterans Affairs (VA), 32, 42
Dependent variable, 8
Descriptive model, 9
Descriptive research, 191
Descriptive statistics, 15
Directed exchange, 212
Direct obstetric death, 89
Discharge report, 162
Discrete data, 18
Documentary study, 200
Drug trials, 186

E

Early fetal deaths, 92
Ebola, 225
Electronic data management system (EDMS), 157
Empirical approach, 70
Enterprise report management (ERM), 157
E-pill, 212
Error, 97
data dictionaries, 177–178
Excel, rounding errors, 158
measures of central tendency, 16
technology and error reduction, 32
type I, 97, 119, 144
type II, 97
Ethics
Belmont Report, 173, 175–176
data collection and reporting, 191
Declaration of Helsinki, 175
defined, 173
human subjects, abuse of, 173–175
human subjects, protection for, 176–177
informed consent, 177
institutional review board (IRB), 176–177
Evidence-based medicine (EBM), 6
Expected mortality rate, 95
Experimental variable, 8
Exploratory research, 172

F

False positive/false negative, 96–97
Fatality rate, 84–86
Fetal autopsy rate, 116
Fetal deaths, 92
Fetal mortality rate, 92–94
Flat-file database, 44–45
Flowchart, 9
Forced Likert scale, 197
Forecasting, 215–216
Forensic autopsy, 110
Formulas, 234–236
Freeware software, 42–44
Frequency distribution, 18–19
F test, 116–119
Full-time equivalent (FTE), 159
Future of healthcare statistics
ACA, 213
ANCOVA
cholesterol data, 220–221
output, 223
scatterplot data, 222–223
automatic medication dispenser, 212
coronavirus, 224
CPM template, 217–218
digital epidemiology, 225
Ebola, 225
forecasting future data, 215–216
Gantt chart, 218–219
health information exchange (HIE)
consumer-mediated exchange, 212–213
directed exchange, 212
query-based exchange, 212
project management, 217–219
RFID, 211–212
time series analysis, 213–215

G

Gantt Chart, 218–219
Generalizability sample, 192
General Public License (GPL), 41
GNU's not Unix, 41
Graphical user interface (GUI), 140
Gross autopsy rate, 112–113
Gross mortality rate, 83–84

H

Health care appraisal (HCA), 201
Healthcare-associated infection (HAIs), 131–132
general occurrence rates, 139–140
nosocomial infections, 132
postoperative infection rate, 134–135
reporting and tracking of, 145–146
specific infection rate, 133–134
Surgical Care Improvement Project (SCIP), 135
Health information exchange (HIE)
consumer-mediated exchange, 212–213
directed exchange, 212
query-based exchange, 212
Health information management (HIM) department
capital budget, 163–166
financial reports, types of
discharge report, 162
readmission rate report, 161–162
functions of, 153
healthcare facility staffing, 159–161
information cost, 157–159
labor cost, 154–155
NBC, 166–168
operational budget, 162–163
productivity, 159
requirements, 153–154
transcription cost, 155–157
utilization of, 161
Health Insurance Portability and Accountability Act (HIPAA), 4
The History of the Peloponnesian War, 3
Hospital autopsy, 110
Hospital standardized mortality ratio (HSMR), 95
Hybrid (mixed) version, 70
Hypothesis, 96

I

Independent variable, 8
Indirect obstetric death, 89
Infection
general occurrence rates, 139–140
infection control committees, 131–132
nosocomial infections, 132–133
postoperative infection rate, 134–135
reporting and tracking of, 145–146
specific infection, 133–134
Surgical Care Improvement Project (SCIP), 135
Infection control committees, 131–132
Inferential statistics, 119
Infographics, 38
Information collection, 191. *See also* Data collection and methods
Informed consent, 177
Inpatient bed count
bed occupancy ratio, 57
definition of, 56–57
government funding and research, 57
Inpatient hospital autopsy, 110, 114
Inpatient service days, 33–34
Institutional review boards (IRBs)
informed consent, 177
membership, 177
review process, 176
review types, 176–177
Instrument, 200
Integrated development environment (IDE), 46–47
Intensive care unit (ICU), 34
Intermediate fetal deaths, 92
Interquartile range, 21–22
Interval data, 8–9
Interview, 201
IRBs. *See* Institutional review boards
Irregular fluctuation, 214

J

Japanese Unit 731 Atrocities, 232
Justice, 173

L

Labor cost, 154–155
Language, 41
Late fetal deaths, 92
Late mortality rate, 83
Left-hand side (LHS), 72
Length of stay (LOS)
average length of stay, 61–66
discharge days, 60–61
for hospital efficiency, 59
major complications or comorbidities (MDCs), 60
median length of stay, 66–67
MS-DRGs, 60
standard deviation, 67–68
time benefits, 59
total length of stay, 61
Lift, 72
Lift chart, 72
Likert scale, 197
Linear regression, 181–186
datasets, 184
future values, 185–186
in R-Project, 182–184
Longitudinal study, 199
LOS. *See* Length of stay

M

Machine learning, 6
Major complications or comorbidities (MCC), 60
Major diagnostic categories (MDCs), 60
Mann-Whitney *U* test, 181
Market basket analysis, 71
Maternal mortality rate, 89–90
Matrix, 48
Maximum (max), 19–20
Mean, 15–16
Median, 17
Median length of stay, 66–67
Medicare and Medicaid, Centers for (CMS), 4, 52, 161–162
Medicare severity diagnosis-related groups (MS-DRGs), 60
Microcharts, 39
Minimum (min), 19–20
Mode, 17–18
Morbidity and mortality rate
adjustment, 94–95
anesthesia, 91
for cancer, 94
definition, 81, 83
fatality rate, 84–86
fetal, 92–94, 104
global perspective, 104
gross mortality rate, 83–84
hypothesis, 96
incidence, 81
maternal mortality rate, 89–90
net mortality rate, 86–87
newborn, 92
null hypothesis, 96
postoperative mortality rate, 87–88
prevalence, 82
research design, 95–96
statistical measures
nonparametric tests, 98
normal distribution, 98
parametric tests, 98
P value, 96–97
significance, 96–97
tail, 97–98
t test, 100–103
type I errors, 97
type II errors, 97
z score, 98–100
Mortality ratio, 95
Multistage sampling, 192

N

Naive Bayes classification (NBC), 166–168
National Center for Health Statistics (NCHS), 4, 112
Nazi Atrocities, 231–232
Net autopsy rate, 113–114
Net mortality rate, 86–87
Network sampling, 192
Newborns
admissions data, 34
average daily census calculations, 34, 41, 234
average length of stay calculations, 61
birth defects, 16
gross autopsy rate calculations, 112
gross mortality calculations, 83, 235
net mortality calculations, 86
newborn autopsy rate, 112, 115–116, 235
newborn bassinet occupancy ratio, 57, 58–59, 234
newborn deaths, defined, 92
newborn mortality rate, 92, 235
newborn services, 41
Nominal data, 8
Nonparametric tests, 98
Nonprobability sampling, 191
Normal distribution, 98
Nosocomial infections, 132–133
Null hypothesis, 96

O

Observation, 201
Observed mortality rate, 95
Occupancy, 55
bed turnover rate, 59
CON, 58–59
data mining, 70–74
global perspective, 74–76
inpatient bed count
bed occupancy ratio, 57
definition of, 56–57
government funding and research, 57

Occupancy (*cont.*)
LOS
average length of stay, 61–66
discharge days, 60–61
for hospital efficiency, 59
major complications or comorbidities (MDCs), 60
median length of stay, 66–67
MS-DRGs, 60
standard deviation, 67–68
time benefits, 59
total length of stay, 61
visually represent data, Wordle
numeric data, 68
PowerPoint presentation, 69–70
Occupancy ratio, 57
Occupational Safety and Health Administration (OSHA), 4
Occurrence rates, 139–140
One-tailed test, 98, 117
One-way analysis of variance, 119
Open-source software, 41–42
Operation, 87
Operational budget, 162–163
Ordinal data, 8
Outcome variable, 8
Outlier data, 21

P

Panel study, 199
Parameters, 7
Parametric tests, 98
Patient care unit (PCU), 31
Patient data, 53
census data
average daily census, 34–35
data visualization, 35–38
financial performance, 31
infant-to-nurse ratio, 31
inpatient care areas, 31–32
inpatient service days, 33–34
microcharts, 39
newborn services, 41
nosocomial infection, 31
nurse-to-patient ratio, 31
reporting and analysis, 32
RFID technology, 32
sparklines, 38–40
US Department of Veterans Affairs, 32
VistA, 32
data formats, 45–46
data stored in R, 48–51
Dosher Memorial Hospital, 30
flat-file database, 44–45
freeware and shareware software, 42–44
global perspective, 52
open-source software, 41–42
relational database, 45
R-Project, 46–48
Patient encounter, 159
Payback period, 163, 164
Percentage, 57
Percentage of occupancy, 57
Perinatal deaths, 92
Personal Time, Fatigue, and Delay (PF&D), 156
Pivot table, 202–204
Population, 7
Posteriori, 70
Post-hoc analysis, 144
Postmortem, 114
Postneonatal deaths, 92
Predictive model, 9
Predictor variable, 8
Pretty Good Privacy (PGP), 42
Primary data, 10
Probability sampling, 191
Productivity, 159
Project management, 217–219
P value, 96–97

Q

Quality of service (QOS), 4, 5
Quality question design
Bloom's taxonomy, 196–197
case study, 200
cohort study, 199
documentary study, 200
longitudinal study, 199
panel study, 199
trend study, 199
types of
demographic questions, 197–198
multiple choice, 197
rating scales, 197
short answer questions, 197
Query-based exchange, 212
Questionnaire, 201

R

Radio frequency identification (RFID), 32, 211–212
Random number functions, 193–196
Range, 20
Rapid application development (RAD) language, 46
Rate, 57
Rating scales, 197
Ratio, 57
Ratio data, 9
R commander (Rcmdr)
GUI, 140–143
HAI reporting and tracking system, 145–146
post-hoc analysis, 144
Tukey HSD, 144–145
Readmission rate report, 161–162
Real-time locating (RTL) systems, 212
Recapitulation algorithm, 34
Relational database, 45
Release of information, 157–159
Research design, 95–96
Research methodology and ethics
applied research, 172
Belmont Report, 175–176
chi-square test, 179–180
data dictionary
cell properties, 177–178
definition of, 177
declaration of, Helsinki, 175
drug trials, 186
exploratory research, 172
gynecology, 174
Institutional Review Boards (IRBs)
informed consent, 177
membership, 177
review process, 176
review types, 176–177
linear regression
datasets, 184
future values, 185–186
in R-Project, 182–184
Mann-Whitney *U* test, 181
nonparametric methods, 178–179
process of, 172–173
smallpox vaccine, 173–174
Tuskegee study, 175
Walter Reed and yellow fever transmission, 174–175
Resources, 230
Respect for persons, 173
Return on investment, 163
Right-hand side (RHS), 72
Risk-adjusted mortality rate, 95
R-Project (R), 46–48, 146–149

S

Sample, 7. *See also* Data collection and methods
Sample frame, 191
Sampling bias, 192
Scalar, 48
Scientific notation, 184
Script-mode interface, 46
Seasonal fluctuations, 214
Secondary data, 10
Selective sample, 192
Shareware software, 42–44
Simple random sampling, 192
Skewed, 16, 17
Snowball sample, 191
Social exchange theory, 200

Source code, 41
Sparklines, 38–40
Standard deviation, 22, 67–68
Statistics
  defined, 1, 3
  history and rationale, 2–3
  producers and users, 4–5
  use of, 4
Stratified sampling, 192
Sunset, 42
Support, 72
Surgical mortality rate, 87–88
Survey method, 200
Systematic random sampling, 192

T

Tab-delimited values, 45
Tail, 97–98
Taxonomy, 196
Telehealth, 5
Text mining, 146–149
Time series analysis, 213–215
Total length of stay, 61
Transcription, 155
Trend study, 199
t test, 100–103
Tukey honest significant difference (HSD), 144–145
Two-tailed test, 117
Two-way analysis of variance, 119–126
Type I errors, 97, 119, 144
Type II errors, 97

U

Unbalanced design, 119
US Department of Veterans Affairs, 32, 42
US Department of Health and Human Services (DHHS), 4
US Federal Statistical System, 4

Variability
  data harvesting, 25–26
  global perspective, 26
  interquartile range, 21–22
  maximum (max), 19–20
  minimum (min), 19–20
  outlier data, 21
  range, 20
  standard deviation, 22
  variance, 22–25
Variables, types of, 8
Variance, 22–25
Vector, 48
Verbal autopsy, 109
Veterans Information Systems and Technology Architecture (VistA), 32
Viewpoint, 9
Visual analogue scale (VAS), 197
Visualization, 146–149
Volunteer sample, 191

Wordle, 146–149
World War II
  Cold War, 233
  Japanese Unit 731 Atrocities, 232
  Nazi Atrocities, 231–232

z score, 98–100